Svetlin G. Georgiev, Khaled Zennir
Differential and Integral Calculus

Also of Interest

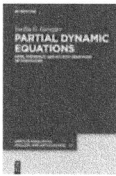

Partial Dynamic Equations. Wave, Parabolic and Elliptic Equations on Time Scales
Svetlin G. Georgiev, 2025
ISBN 978-3-11-163551-4, e-ISBN (PDF) 978-3-11-163614-6,
e-ISBN (EPUB) 978-3-11-163615-3

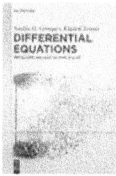

Differential Equations. Projector Analysis on Time Scales
Svetlin G. Georgiev, Khaled Zennir, 2024
ISBN 978-3-11-137509-0, e-ISBN (PDF) 978-3-11-137715-5,
e-ISBN (EPUB) 978-3-11-137771-1

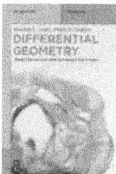

Differential Geometry. Frenet Equations and Differentiable Maps
Muhittin E. Aydin, Svetlin G. Georgiev, 2024
ISBN 978-3-11-150089-8, e-ISBN (PDF) 978-3-11-150185-7,
e-ISBN (EPUB) 978-3-11-150223-6

Functional Analysis with Applications
Svetlin G. Georgiev, Khaled Zennir, 2019
ISBN 978-3-11-065769-2, e-ISBN (PDF) 978-3-11-065772-2,
e-ISBN (EPUB) 978-3-11-065804-0

Differential Equations. Solving Ordinary and Partial Differential Equations with Mathematica®
Marian Mureşan, 2024
ISBN 978-3-11-141109-5, e-ISBN (PDF) 978-3-11-141139-2,
e-ISBN (EPUB) 978-3-11-141204-7

Svetlin G. Georgiev, Khaled Zennir

Differential and Integral Calculus

Implicit Functions, Stieltjes Integrals and Curvilinear Integrals

DE GRUYTER

Mathematics Subject Classification 2020
Primary: 26B05, 26B10, 26B12; Secondary: 26C15, 26-01

Authors

Dr. Svetlin G. Georgiev
Sorbonne University
1 Rue Victor Hugo
70005 Paris
France
svetlingeorgiev1@gmail.com

Dr. Khaled Zennir
Qassim University
Department of Mathematics
Ar-Rass
Qassim 51921
Saudi Arabia
khaledzennir4@gmail.com

ISBN 978-3-11-914462-9
e-ISBN (PDF) 978-3-11-221808-2
e-ISBN (EPUB) 978-3-11-221831-0

Library of Congress Control Number: 2025940046

Bibliographic information published by the Deutsche Nationalbibliothek
The Deutsche Nationalbibliothek lists this publication in the Deutsche Nationalbibliografie;
detailed bibliographic data are available on the Internet at http://dnb.dnb.de.

© 2025 Walter de Gruyter GmbH, Berlin/Boston, Genthiner Straße 13, 10785 Berlin
Cover image: Lidiia Moor / iStock / Getty Images Plus
Typesetting: VTeX UAB, Lithuania

www.degruyter.com
Questions about General Product Safety Regulation:
productsafety@degruyterbrill.com

Preface

This book presents an introduction to the theory of functions of several variables. The book is primarily intended for senior undergraduate students and beginning graduate students of engineering and science courses. Students in mathematical and physical sciences will find many sections of direct relevance.

This book contains six chapters, and each chapter consists of results with their proofs, numerous examples, and exercises with solutions. Each chapter concludes with a section featuring advanced practical problems with solutions followed by a section on notes and references, explaining its context within existing literature. In the book, numerous examples are included, and many problems are provided with detailed explanations or detailed solutions or answers.

In Chapter 1, the set $\mathbb{R}^n$ is defined and explored. The Cauchy–Schwarz inequality is proved. n-dimensional ball, n-dimensional sphere, and rectangular neighborhood of a point are defined. Sequences and $\mathbb{R}^n$ are defined, and some of their properties are deduced. Classifications of the points in $\mathbb{R}^n$ are represented. Open, closed, and compact sets in $\mathbb{R}^n$ are defined and investigated. The Heine–Borel theorem is formulated and proved. Multidimensional vector spaces are explored. In Chapter 2, functions of several variables are introduced. Limits of functions of several variables are defined, and some of their properties are deduced. Continuous functions of several variables are defined and investigated. Properties of continuous functions on compact sets are deduced. Uniform continuity is introduced and explored. In Chapter 3, we introduce partial derivatives of first and higher orders and differentials for functions of several variables. Some of their properties are deduced. Criteria for differentiability of functions of several variables are deduced. The gradient of a function is defined and explored. Directional derivatives are defined and investigated. In Chapter 4, higher-order partial derivatives of functions of several variables are investigated. Minimum and maximum of functions of several variables are introduced and investigated. Implicit functions are defined and explored. The method of Lagrange multipliers is introduced. In Chapters 5 and 6, many exercises are stated and solved in detail, and solutions, hints, and answers to the problems are given.

The aim of this book is to present a clear and well-organized treatment of the concepts behind the development of mathematics as well as solution techniques. The text material of this book is presented in a readable and mathematically solid format.

Paris, June 2025   Svetlin G. Georgiev and Khaled Zennir

https://doi.org/10.1515/9783112218082-202

Contents

1 The spaces $\mathbb{R}^n$

In this chapter, we define and explore the set $\mathbb{R}^n$. The Cauchy–Schwarz inequality is proved. An n-dimensional ball, an n-dimensional sphere, and a rectangular neighborhood of a point are defined. Sequences in $\mathbb{R}^n$ are defined, and some of their properties are deduced. Classifications of the points in $\mathbb{R}^n$ are represented. Open, closed, and compact sets in $\mathbb{R}^n$ are defined and investigated. The Heine–Borel theorem is formulated and proved. Multidimensional vector spaces are explored.

1.1 Definition and structures of $\mathbb{R}^n$

Euclidean space is a mathematical construct that encompasses the line, plane, and three-dimensional space as particular cases. Its elements are called vectors. Vectors can be understood in various ways: as arrows, as quantities with magnitude and direction, as displacements, or as points. However, along with a sense of what vectors are, we also need to emphasize how they interact.

Definition 1.1. For any positive integer n, the set $\mathbb{R}^n$ of all ordered n-tuples of real numbers

$$\mathbb{R}^n = \{x = (x_1, \ldots, x_n) : x_j \in \mathbb{R}, \, j \in \{1, \ldots, n\}\},$$

is called the n-dimensional Euclidean space. The elements of $\mathbb{R}^n$ are called points of $\mathbb{R}^n$ or vectors of $\mathbb{R}^n$, and the numbers $x_j, j \in \{1, \ldots, n\}$, are called the jth coordinates of these points. The number n is called the dimension of $\mathbb{R}^n$.

When $n = 1$, the parentheses and subscript in the notation (x_1) are superfluous, so we simply view the elements of $\mathbb{R}^1$ as real numbers x and write $\mathbb{R}$ for $\mathbb{R}^1$. Sometimes, elements of $\mathbb{R}^2$ and $\mathbb{R}^3$ are written (x, y) and (x, y, z) to avoid needless subscripts.

The first few Euclidean spaces $\mathbb{R}$, $\mathbb{R}^2$, and $\mathbb{R}^3$ are conveniently visualized as the line, the plane, and the space itself (see Fig. 1.1).

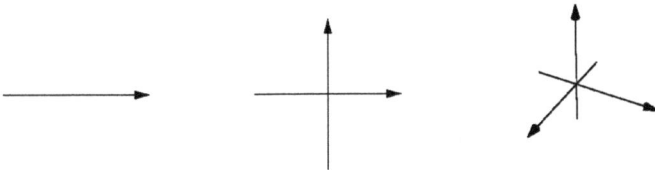

Figure 1.1: The first few Euclidean spaces.

https://doi.org/10.1515/9783112218082-001

Definition 1.2. The coordinate origin or origin of $\mathbb{R}^n$, denoted by 0, is defined as

$$0 = (0, \ldots, 0).$$

Sometimes, the origin of $\mathbb{R}^n$ will be denoted by 0_n to indicate the dimension of the space. The points $(0, \ldots, 0, x_j, 0, \ldots, 0), j \in \{1, \ldots, n\}$, are said to be the jth coordinate axes.

In the first few Euclidean spaces $\mathbb{R}, \mathbb{R}^2,$ and $\mathbb{R}^3$, a vector can be visualized as a point x or as an arrow. The arrow can have its tail at the origin and its head at the point x, or its tail at any point p and its head can be correspondingly translated to $p + x$ (see Fig. 1.2). In mathematics, the Euclidean distance between two points in the Euclidean space is the length of the line segment between them. It can be calculated from the coordinates of the points using the following definition.

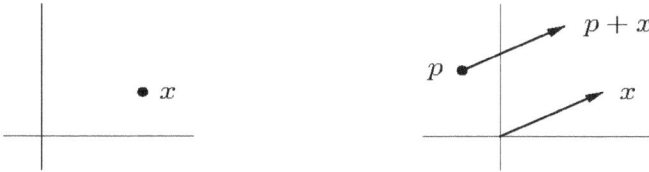

Figure 1.2: Ways to visualize a vector.

Definition 1.3. In $\mathbb{R}^n$, we define distance between its points $x = (x_1, \ldots, x_n)$ and $y = (y_1, \ldots, y_n)$ as follows:

$$d(x, y) = \sqrt{\sum_{j=1}^{n} (x_j - y_j)^2}.$$

When $y = 0$, we will write $|x| = d(x, 0)$, which is called the norm or modulus, or absolute value, or length of x.

Example 1.1. Consider $\mathbb{R}^3$ and

$$x = (-3, 0, 4), \quad y = (1, -1, 5).$$

Then the distance between the points x and y is

$$\begin{aligned}
d(x, y) &= \sqrt{(-3 - 1)^2 + (0 + 1)^2 + (4 - 5)^2} \\
&= \sqrt{16 + 1 + 1} \\
&= \sqrt{18} \\
&= 3\sqrt{2}.
\end{aligned}$$

Example 1.2. Consider $\mathbb{R}^4$ and

$$x = (-2, -1, 0, 1), \quad y = (3, -1, -1, 1).$$

Then the distance between the points x and y is computed as follows:

$$\begin{aligned}
d(x,y) &= \sqrt{(-2-3)^2 + (-1+1)^2 + (0+1)^2 + (1-1)^2} \\
&= \sqrt{25+1} \\
&= \sqrt{26}.
\end{aligned}$$

Example 1.3. Consider $\mathbb{R}^3$ and

$$x = (a, -2, -1), \quad y = (-3, 1, 1),$$

where $a \in \mathbb{R}$. We will find the parameter a such that $d(x,y) = \sqrt{13}$. By the definition of the distance between two points we have the relation

$$\begin{aligned}
\sqrt{13} &= d(x,y) \\
&= \sqrt{(a+3)^2 + (-2-1)^2 + (-1-1)^2} \\
&= \sqrt{a^2 + 6a + 9 + 9 + 4} \\
&= \sqrt{a^2 + 6a + 22}.
\end{aligned}$$

By raising to the square we obtain the quadratic equation

$$a^2 + 6a + 22 = 13$$

or

$$a^2 + 6a + 9 = 0,$$

or

$$(a+3)^2 = 0.$$

Thus $a = -3$, and for x, we get

$$x = (-3, -2, -1).$$

Exercise 1.1. Consider $\mathbb{R}^5$ and

$$x = (-3, -2, 0, 1, 1), \quad y = (-4, 1, 1, -1, 0).$$

Find $d(x,y)$.

Exercise 1.2. Consider $\mathbb{R}^3$. Find $a \in \mathbb{R}$ such that $d(x,y) = \sqrt{10}$, where

$$x = (-3, 4, a), \quad y = (-2, 1, 1).$$

Before deducting some of the properties of the distance, we need the following useful inequalities.

Lemma 1.1 (Cauchy–Schwarz[1,2] inequality). *For any $a_j, b_j \in \mathbb{R}, j \in \{1, \ldots, n\}$, we have the following inequality:*

$$\sum_{j=1}^{n} a_j b_j \leq \left(\sum_{j=1}^{n} a_j^2\right)^{\frac{1}{2}} \left(\sum_{j=1}^{n} b_j^2\right)^{\frac{1}{2}}. \tag{1.1}$$

Proof. If $a_j = 0, j \in \{1, \ldots, n\}$, then (1.1) holds. Suppose that

$$\sum_{j=1}^{n} a_j^2 > 0.$$

Define the function

$$F(t) = \sum_{j=1}^{n}(a_j t + b_j)^2, \quad t \in \mathbb{R}.$$

We have

$$F(t) = t^2 \sum_{j=1}^{n} a_j^2 + 2t \sum_{j=1}^{n} a_j b_j + \sum_{j=1}^{n} b_j^2$$

$$\geq 0, \quad t \in \mathbb{R}.$$

The last inequality is possible if the discriminant of the quadratic trinomial F is nonnegative. Thus

$$\left(\sum_{j=1}^{n} a_j b_j\right)^2 - \left(\sum_{j=1}^{n} a_j^2\right)\left(\sum_{j=1}^{n} b_j^2\right) \leq 0,$$

whereupon we get inequality (1.1). This completes the proof. $\quad\square$

---

1 Baron Augustin-Louis Cauchy (21 August 1789–23 May 1857) was a French mathematician, engineer, and physicist who made pioneering contributions to several branches of mathematics, including mathematical analysis and continuum mechanics. He was one of the first to state and rigorously prove theorems of calculus, rejecting the heuristic principle of the generality of algebra of earlier authors. He is one of the founders of complex analysis and the study of permutation groups in abstract algebra.

2 Laurent-Moïse Schwartz (5 March 1915–4 July 2002) was a French mathematician. He pioneered the theory of distributions, which gives a well-defined meaning to objects such as the Dirac delta function. He was awarded the Fields Medal in 1950 for his work on the theory of distributions. For several years, he taught at the École polytechnique.

Corollary 1.1. *For any $a_j, b_j \in \mathbb{R}, j \in \{1, \dots, n\}$, we have*

$$\left(\sum_{j=1}^{n}(a_j + b_j)^2\right)^{\frac{1}{2}} \leq \left(\sum_{j=1}^{n} a_j^2\right)^{\frac{1}{2}} + \left(\sum_{j=1}^{n} b_j^2\right)^{\frac{1}{2}}. \tag{1.2}$$

Proof. Applying (1.1), we arrive at the following chain of inequalities:

$$\sum_{j=1}^{n}(a_j + b_j)^2 = \sum_{j=1}^{n} a_j^2 + 2\sum_{j=1}^{n} a_j b_j + \sum_{j=1}^{n} b_j^2$$

$$\leq \sum_{j=1}^{n} a_j^2 + 2\left(\sum_{j=1}^{n} a_j^2\right)^{\frac{1}{2}}\left(\sum_{j=1}^{n} b_j^2\right)^{\frac{1}{2}} + \sum_{j=1}^{n} b_j^2$$

$$= \left(\left(\sum_{j=1}^{n} a_j^2\right)^{\frac{1}{2}} + \left(\sum_{j=1}^{n} b_j^2\right)^{\frac{1}{2}}\right)^2,$$

whereupon we get the desired result. This completes the proof. $\qquad\square$

Let $a \in \mathbb{R}$ and

$$x = (x_1, \dots, x_n),$$
$$y = (y_1, \dots, y_n),$$
$$z = (z_1, \dots, z_n) \in \mathbb{R}^n$$

be arbitrarily chosen. Then the defined distance in $\mathbb{R}^n$ has the following properties.

Theorem 1.1. 1. *The distance is nonnegative:*

$$d(x,y) \geq 0 \quad and \quad d(x,y) = 0 \quad if\, and\, only\, if \quad x = y.$$

2. *The distance is symmetric:*

$$d(x,y) = d(y,x).$$

3. *The triangle inequality:*

$$d(x,y) \leq d(x,z) + d(z,y).$$

The first and second properties are clearly desirable as properties of a distance function. The third property says that you cannot shorten your trip from x to z by making a stop at y.

Proof. 1. We have

$$d(x,y) = \sqrt{\sum_{i=1}^{n}(x_i - y_i)^2}$$

$$\geq 0.$$

Next, $d(x,y) = 0$ if and only if

$$\sqrt{\sum_{i=1}^{n}(x_i - y_i)^2} = 0,$$

whereupon $d(x,y) = 0$ if and only if

$$\sum_{i=1}^{n}(x_i - y_i)^2 = 0.$$

Hence $d(x,y) = 0$ if and only if $x_i = y_i$, $i \in \{1,\ldots,n\}$, i.e., $d(x,y) = 0$ if and only if $x = y$.

2. Let

$$a_i = x_i - z_i,$$
$$b_i = z_i - y_i, \quad i \in \{1,\ldots,n\}.$$

Then

$$a_i + b_i = x_i - z_i + z_i - y_i$$
$$= x_i - y_i, \quad i \in \{1,\ldots,n\}.$$

Now, applying (1.2), we get

$$d(x,y) = \sqrt{\sum_{i=1}^{n}(x_i - y_i)^2}$$

$$= \sqrt{\sum_{i=1}^{n}(a_i + b_i)^2}$$

$$\leq \sqrt{\sum_{i=1}^{n}a_i^2} + \sqrt{\sum_{i=1}^{n}b_i^2}$$

$$= \sqrt{\sum_{i=1}^{n}(x_i - z_i)^2} + \sqrt{\sum_{i=1}^{n}(z_i - y_i)^2}$$

$$= d(x,z) + d(z,y).$$

This completes the proof. □

The modulus has the following properties.

Theorem 1.2. 1. *The modulus is nonnegative:*

$$|x| \geq 0 \quad and \quad |x| = 0 \quad if \ and \ only \ if \quad x = 0.$$

2. *The triangle inequality:*

$$|x + y| \leq |x| + |y|.$$

Proof. 1. This property follows directly from the nonnegativity of the distance between two points.

2. We have

$$|x + y| = d(x + y, 0)$$

$$= \sqrt{\sum_{j=1}^{n}(x_j + y_j)^2}$$

$$\leq \sqrt{\sum_{j=1}^{n}x_j^2} + \sqrt{\sum_{j=1}^{n}y_j^2}$$

$$= |x| + |y|.$$

This completes the proof. $\qquad\qquad\qquad\qquad\qquad\qquad\qquad\qquad\qquad\square$

Like other symbols, the absolute value signs are now overloaded. Their meaning can be inferred from context, as in the second property. When n is 1, 2, or 3, the modulus gives the distance from 0 to the point x or the length of x viewed as an arrow (see Fig. 1.3). The triangle inequality name is explained by its geometric interpretation in $\mathbb{R}^2$. View x as an arrow at the origin, y as an arrow with tail at the head of x, and $x + hy$ as an arrow at the origin. These three arrows form a triangle, and the statement is that the lengths of two sides sum to at least the length of the third. Let $k, n \in \mathbb{N}$ and $k \leq n$. Define the set

$$M = \{x \in \mathbb{R}^m : x = (x_1, \ldots, x_k, 0, \ldots, 0)\}.$$

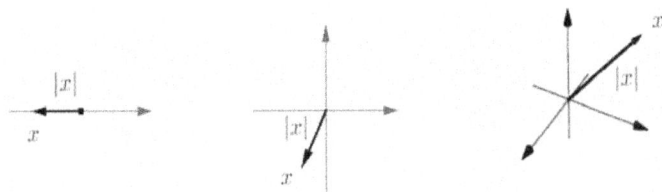

Figure 1.3: Modulus as length.

For $x = (x_1, \ldots, x_k, 0, \ldots, 0) \in M$, we denote $x' = (x_1, \ldots, x_k)$. Define the map $\phi : M \to \mathbb{R}^k$ as follows:

$$\phi(x) = x', \quad x \in M.$$

We will prove that this map is a bijection. For this aim, we will prove that it is an injection and surjection. To prove that it is an injection, take $x, y \in M$ such that $x \neq y$. Then, there is $j \in \{1, \ldots, k\}$ such that $x_j \neq y_j$. Hence $x' \neq y'$. Consequently, $\phi(x) \neq \phi(y)$, and $\phi : M \to \mathbb{R}^k$ is an injection. Next, we will prove that ϕ is a surjection. Let $z = (z_1, \ldots, z_k) \in \mathbb{R}^k$ be arbitrarily chosen. Denote

$$v = (z_1, \ldots, z_k, 0, \ldots, 0).$$

Then $\phi(v) = z$, and $\phi : M \to \mathbb{R}^k$ is a surjection. Therefore $\phi : M \to \mathbb{R}^k$ is a bijection. Now we will prove that the map ϕ preserves the distance. To this aim, take arbitrary $x, y \in M$. Then, we have

$$d(x, y) = \sqrt{\sum_{i=1}^{n} (x_i - y_i)^2}$$

$$= \sqrt{\sum_{i=1}^{k} (x_i - y_i)^2}$$

$$= d(x', y')$$

$$= d(\phi(x), \phi(y)).$$

Therefore the map $\phi : M \to \mathbb{R}^k$ preserves the distance. Thus the set M also denotes $\mathbb{R}^k$, and therefore, under this agreement, $\mathbb{R}^k \subseteq \mathbb{R}^n$, $k \leq n$.

Assume that we have two rectangular coordinate systems J_1 and J_2 such that a point A has coordinates (x, y) with respect to J_1 and coordinates (ξ, η) with respect to J_2, i.e.,

$$A = (x, y) = (\xi, \eta).$$

If we map (x, y) to (ξ, η), then we get a bijection between the set of all ordered pairs (x, y) and the set of all ordered pairs (ξ, η). If

$$A' = (x', y') = (\xi', \eta')$$

and

$$A'' = (x'', y'') = (\xi'', \eta''),$$

then

$$d(A', A'') = \sqrt{(x'' - x')^2 + (y'' - y')^2}$$
$$= \sqrt{(\xi'' - \xi')^2 + (\eta'' - \eta')^2}.$$

This motivates the following definition.

Definition 1.4. Let any point $x = (x_1, \ldots, x_n) \in \mathbb{R}^n$ be mapped to an n-tuple $\xi = (\xi_1, \ldots, \xi_n)$ so that for any two points $x' = (x_1', \ldots, x_n')$ and $x'' = (x_1'', \ldots, x_n'')$ and their corresponding n-tuples

$$\xi' = (\xi_1', \ldots, \xi_n') \quad \text{and} \quad \xi'' = (\xi_1'', \ldots, \xi_n''),$$

respectively, we have the equality

$$\sum_{i=1}^{n} (x_i'' - x_i')^2 = \sum_{i=1}^{n} (\xi_i'' - \xi_i')^2.$$

Then the n-tuple $(\xi_1, \ldots, \xi_n)$ is also called the coordinates of the point x.

By the above definition it follows that the distance between two points does not depend on the given rectangular coordinate system.

1.2 Neighborhoods of a point

In mathematics, a neighborhood is one of the basic concepts in the Euclidean space. It is closely related to the concepts of an open set and interior. Intuitively speaking, a neighborhood of a point is a set of points containing that point where we can move some amount in any direction away from that point without leaving the set. There are several kinds of neighborhoods of a given point. We start with the first one.

Definition 1.5. Let $x \in \mathbb{R}^n$ and $\epsilon > 0$. The set of all points $y \in \mathbb{R}^n$ such that

$$d(x, y) < \epsilon$$

is called the n-dimensional ball with center x and radius ϵ, or the ϵ-neighborhood of x. It is denoted by $U(x, \epsilon)$. In other words,

$$U(x, \epsilon) = \{y \in \mathbb{R}^n : d(x, y) < \epsilon\}.$$

Example 1.4. Let $n = 1$. Then

$$U(x, \epsilon) = \{y \in \mathbb{R} : |x - y| < \epsilon\}$$

is the interval of length 2ϵ with center x (see Fig. 1.4).

Figure 1.4: *n*-dimensional neighborhood of *x*.

Example 1.5. Let $n = 2$. Then

$$U(x, \epsilon) = \{y = (y_1, y_2) \in \mathbb{R}^2 : (x_1 - y_1)^2 + (x_2 - y_2)^2 < \epsilon^2\}$$

is the circle with center *x* and radius *e* (see Fig. 1.5).

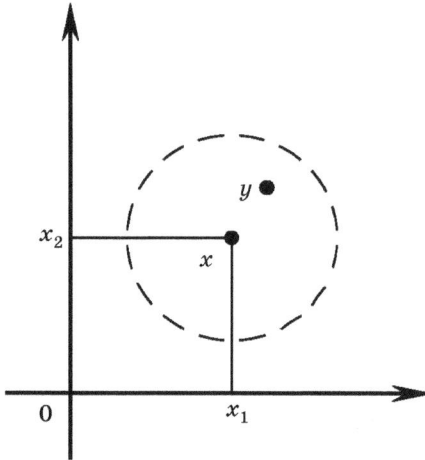

Figure 1.5: A circle with center *x* and radius ϵ.

Definition 1.6. Let $x = (x_1, \ldots, x_n) \in \mathbb{R}^n$ and $\delta_j > 0, j \in \{1, \ldots, n\}$. The set

$$P(x, \delta_1, \ldots, \delta_n) = \{y = (y_1, \ldots, y_n) \in \mathbb{R}^n : |x_j - y_j| < \delta_j\}$$

is called the *n*-dimensional parallelogram with a center at the point *x*. If

$$\delta_1 = \delta_2 = \cdots = \delta_n = \delta,$$

then $P(x, \delta, \delta, \ldots, \delta)$ is called the *n*-dimensional cube with center *x*. It is denoted by $P(x, \delta)$. Any *n*-dimensional parallelogram $P(x, \delta_1, \delta_2, \ldots, \delta_n)$ is said to be a rectangular neighborhood of *x*.

The first important property of rectangular neighborhoods is the so-called "size bounds". It reads as follows.

Theorem 1.3 (Size bounds). *Let $x \in \mathbb{R}^n$ and $\delta_j > 0, j \in \{1, \ldots, n\}$. Denote*

$$\delta_0 = \min_{j \in \{1, \ldots, n\}} \delta_j, \quad \delta = \max_{j \in \{1, \ldots, n\}} \delta_j.$$

Then

$$P(x, \delta_0) \subset P(x, \delta_1, \ldots, \delta_n) \subset P(x, \delta).$$

Proof. Note that

$$\delta_0 \leq \delta_j \leq \delta, \quad j \in \{1, \ldots, n\}.$$

Firstly, we will prove the first inclusion. For this aim, take arbitrary $y = (y_1, \ldots, y_n) \in P(x, \delta_0)$. Then

$$|x_j - y_j| < \delta_0$$
$$\leq \delta_j, \quad j \in \{1, \ldots, n\}.$$

Thus $y \in P(x, \delta_1, \ldots, \delta_n)$. Because $y \in P(x, \delta_0)$ was arbitrarily chosen and it is an element of $P(x, \delta_1, \ldots, \delta_n)$, we get the inclusion

$$P(x, \delta_0) \subset P(x, \delta_1, \ldots, \delta_n).$$

Now we will prove the second inclusion. Let now $z = (z_1, \ldots, z_n) \in P(x, \delta_1, \ldots, z_n)$. Then

$$|x_j - z_j| < \delta_j$$
$$\leq \delta, \quad j \in \{1, \ldots, n\}.$$

So $z \in P(x, \delta)$. Because $z \in P(x, \delta_1, \ldots, \delta_n)$ was arbitrarily chosen and it is an element of $P(x, \delta)$, we arrive at the inclusion

$$P(x, \delta_1, \ldots, \delta_n) \subset P(x, \delta).$$

This completes the proof. □

In the next statement, we give relations between ϵ-neighborhoods and rectangular neighborhoods.

Theorem 1.4. *Any ϵ-neighborhood of a point of $\mathbb{R}^n$ contains a rectangular neighborhood of this point and is contained in a rectangular neighborhood of this point. Any rectangular neighborhood of a point of $\mathbb{R}^n$ contains an ϵ-neighborhood of this point and is contained in an ϵ-neighborhood of this point.*

Proof. Let $x = (x_1, \ldots, x_n) \in \mathbb{R}^n$ and $\epsilon > 0$. For $y = (y_1, \ldots, y_n) \in \mathbb{R}^n$, using the definition of a distance, we have

$$|x_j - y_j| \leq d(x, y)$$
$$= \sqrt{(x_1 - y_1)^2 + \cdots + (x_n - y_n)^2} \qquad (1.3)$$
$$\leq |x_1 - y_1| + \cdots + |x_n - y_n|, \quad j \in \{1, \ldots, n\}.$$

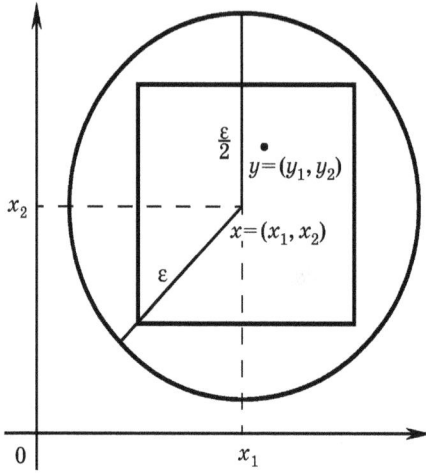

Figure 1.6: A rectangular neighborhood.

Let $y = (y_1, \ldots, y_n) \in P(x, \frac{\epsilon}{n})$. Then, using the definition of a rectangular neighborhood (see Fig. 1.6), we get

$$|x_j - y_j| < \frac{\epsilon}{n}, \quad j \in \{1, \ldots, n\}.$$

Hence, applying the right-hand side of inequality (1.3), we find

$$d(x, y) \le |x_1 - y_1| + \cdots + |x_n - y_n|$$
$$< \frac{\epsilon}{n} + \cdots + \frac{\epsilon}{n}$$
$$= \epsilon,$$

whereupon $y \in U(x, \epsilon)$. Since $y \in P(x, \frac{\epsilon}{n})$ was arbitrarily chosen and it is an element of $U(x, \epsilon)$, we obtain the inclusion

$$P\left(x, \frac{\epsilon}{n}\right) \subset U(x, \epsilon). \tag{1.4}$$

Let now $z = (z_1, \ldots, z_n) \in U(x, \epsilon)$. Then

$$d(x, z) < \epsilon.$$

Applying the left-hand side of inequality (1.3), we find

$$|x_j - z_j| \le d(x, z)$$
$$< \epsilon, \quad j \in \{1, \ldots, n\}.$$

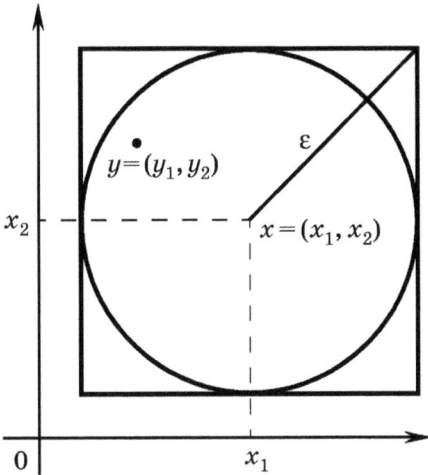

Figure 1.7: A rectangular neighborhood.

Therefore $z \in P(x, \epsilon)$. Because $z \in U(x, \epsilon)$ was arbitrarily chosen and it is an element of $P(x, \epsilon)$, we arrive at the inclusion

$$U(x, \epsilon) \subset P(x, \epsilon). \tag{1.5}$$

Take arbitrary $w = (w_1, \ldots, w_n) \in P(x, \epsilon)$. Then

$$|x_j - w_j| < \epsilon, \quad j \in \{1, \ldots, n\}.$$

Hence (see Fig. 1.7)

$$\begin{aligned}
d(x, w) &= \sqrt{(x_1 - w_1)^2 + \cdots + (x_n - w_n)^2} \\
&< \sqrt{\epsilon^2 + \cdots + \epsilon^2} \\
&= \sqrt{n\epsilon^2} \\
&\leq \sqrt{n^2\epsilon^2} \\
&= n\epsilon,
\end{aligned} \tag{1.6}$$

that is, $w \in U(x, n\epsilon)$. Since $w \in P(x, \epsilon)$ was arbitrarily chosen and it is an element of $U(x, n\epsilon)$, we arrive at the inclusion

$$P(x, \epsilon) \subset U(x, n\epsilon).$$

By the last inclusion and inclusions (1.4) and (1.5) we get the chain

$$P\left(x, \frac{\epsilon}{n}\right) \subset U(x, \epsilon) \subset P(x, \epsilon) \subset U(x, n\epsilon). \tag{1.7}$$

This completes the proof. □

Remark 1.1. Let $n = 2$. Then by Theorem 1.4 we conclude that any circle can be inscribed in a rectangle and any rectangle can be inscribed in a circle so that the center of the rectangle coincides with the center of the circle (see Fig. 1.8).

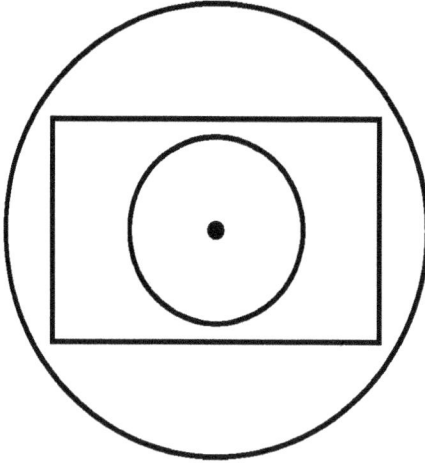

Figure 1.8: A rectangular neighborhood.

Next relations between rectangular neighborhoods and ϵ-neighborhoods are as follows.

Theorem 1.5. *For any $\epsilon > 0$ and $x \in \mathbb{R}^n$, we have the chain*

$$P\left(x, \frac{\epsilon}{\sqrt{n}}\right) \subset U(x, \epsilon) \subset P(x, \epsilon) \subset U(x, \epsilon\sqrt{n}). \tag{1.8}$$

Proof. We will prove the following two relations:

$$P\left(x, \frac{\epsilon}{\sqrt{n}}\right) \subset U(x, \epsilon) \tag{1.9}$$

and

$$P(x, \epsilon) \subset U(x, \epsilon\sqrt{n}). \tag{1.10}$$

The other parts of the chain (1.8) follow by the chain (1.7). Let $z = (z_1, \ldots, z_n) \in P(x, \frac{\epsilon}{\sqrt{n}})$. Then

$$|x_j - z_j| < \frac{\epsilon}{\sqrt{n}}, \quad j \in \{1, \ldots, n\}.$$

Hence

$$d(x, z) = \sqrt{(x_1 - z_1)^2 + \cdots + (x_n - z_n)^2}$$

$$< \sqrt{\left(\frac{\epsilon}{\sqrt{n}}\right)^2 + \cdots + \left(\frac{\epsilon}{\sqrt{n}}\right)^2}$$

$$= \sqrt{\frac{\epsilon^2}{n} + \cdots + \frac{\epsilon^2}{n}}$$

$$= \sqrt{\epsilon^2}$$

$$= \epsilon,$$

that is, $z \in U(x, \epsilon)$. Because $z \in P(x, \frac{\epsilon}{\sqrt{n}})$ was arbitrarily chosen and it is an element of $U(x, \epsilon)$, we arrive at the inclusion (1.9). Let now $w = (w_1, \ldots, w_n) \in P(x, \epsilon)$. Then

$$|x_j - w_j| < \epsilon, \quad j \in \{1, \ldots, n\}.$$

From (1.6) we find

$$d(x, w) < \epsilon \sqrt{n},$$

that is, $w \in U(x, \epsilon \sqrt{n})$. Since $w \in U(x, \epsilon)$ was arbitrarily chosen and it is an element of $U(x, \epsilon \sqrt{n})$, we get the inclusion (1.10). This completes the proof. □

Remark 1.2. Let $x \in \mathbb{R}^n$ and $\epsilon > 0$.
1. For $n = 1$, we have $P(x, \epsilon) = U(x, \epsilon)$.
2. Let $n \in \mathbb{N}$, $n \geq 2$. Take $y = (y_1, \ldots, y_n) \in P(x, \epsilon)$ such that

$$|x_j - y_j| = \frac{\epsilon}{\sqrt{n-1}}, \quad j \in \{1, \ldots, n\}.$$

Because $n \geq 2$, we have

$$\frac{\epsilon}{\sqrt{n-1}} < \epsilon.$$

Thus $y \in P(x, \epsilon)$. On the other hand, we have

$$d(x, y) = \sqrt{(x_1 - y_1)^2 + \cdots + (x_n - y_n)^2}$$

$$= \sqrt{\frac{\epsilon^2}{n-1} + \cdots + \frac{\epsilon^2}{n-1}}$$

$$= \sqrt{\epsilon^2 \frac{n}{n-1}}$$

$$= \epsilon \sqrt{\frac{n}{n-1}}$$

$$> \epsilon,$$

whereupon $y \notin U(x, \epsilon)$. Therefore in the general case the inclusion $P(x, \epsilon) \subset U(x, \epsilon)$ does not hold.

1.3 Sequences in $\mathbb{R}^n$

Sequences and series are very important concepts in mathematics. They have powerful applications in finance, statistics, and physics. In mathematics, sequences are used for studying function spaces and structures using the convergence of sequences. Some of the applications in finance include variations of financial contracts such as annuity and perpetuity contracts and bonds. Other applications include the calculation of mortgages. Sequences are not only a concept used in the field of mathematics. In fact, everyone uses sequences, irrespective of whether they use mathematics on a daily basis.

Definition 1.7. Let A be a subset of $\mathbb{R}^n$. A sequence in A is an infinite list of vectors $\{x^1, \ldots, x^m, \ldots\}$ in A, often written $\{x^m\}_{m \in \mathbb{N}}$. Since a vector in $\mathbb{R}^n$ has n entries, each vector x^m in the sequence takes the form

$$(x_1^m, x_2^m, \ldots, x_n^m).$$

Example 1.6. The Fibonacci[3] sequence is a sequence in $\mathbb{R}$ starting at zero, the second element is 1, and every the next element is determined by the sum of two preceding elements:

$$\{0, 1, 1, 2, 3, 5, 8, 13, 21, 34, \ldots\}.$$

The Fibonacci sequence is used in number theory, algebra, and geometry. It has applications in the analysis of financial markets and computer algorithms. The Fibonacci sequence appears in biological settings such as the branching of trees, the arrangement of leaves on a stem, the flowering of artichokes, and the spiral arrangement of seeds in sunflowers.

Example 1.7. Let A denote the amount of annuity payment received at the end of each year, r denote the spot-rate, and let n payments be made by the contract. Then the sequence of predicted values of all the annuity payments is a sequence in $\mathbb{R}$ given by

$$A, \frac{A}{1+r}, \frac{A}{(1+r)^2}, \frac{A}{(1+r)^3}, \ldots, \frac{A}{(1+r)^{n+1}}, \ldots.$$

---

[3] Fibonacci (born around 1170) was an Italian mathematician from the Republic of Pisa, considered to be "the most talented Western mathematician of the Middle Ages".

Example 1.8. The set

$$\left\{\left(\frac{1}{m}, \frac{m+1}{m+2}, m\right)\right\}_{m\in\mathbb{N}}$$

is a sequence in $\mathbb{R}^3$.

Example 1.9. The set

$$\left\{\frac{1}{m}, e^m, \cos m, \sin m, \frac{1}{m+1}\right\}_{m\in\mathbb{N}}$$

is a sequence in $\mathbb{R}^5$.

Definition 1.8. The sequence $\{x^{m_k}\}_{k\in\mathbb{N}}$ formed by the elements of the sequence $\{x^m\}_{m\in\mathbb{N}}$ keeping in mind the order in which they are followed, is said to be a subsequence of the sequence $\{x^m\}_{m\in\mathbb{N}}$. In this way, if $\{x^{m_k}\}_{k\in\mathbb{N}}$ is a subsequence of the sequence $\{x^m\}_{m\in\mathbb{N}}$, then the inequality $k_1 < k_2$ implies the inequality $m_{k_1} < m_{k_2}$ and vice versa.

Example 1.10. Consider the sequence in $\mathbb{R}^3$

$$\left\{\left(\frac{1}{m}, \frac{3}{m+1}, \frac{4}{m+2}\right)\right\}_{m\in\mathbb{N}}.$$

Then the set

$$\left\{\left(\frac{1}{2m}, \frac{3}{2m+1}, \frac{2}{m+1}\right)\right\}_{m\in\mathbb{N}}$$

is its subsequence.

Example 1.11. Consider the sequence in $\mathbb{R}^5$

$$\left\{\left(3m, m^2+1, \frac{2}{m}, \frac{4}{m+5}\right)\right\}_{m\in\mathbb{N}}.$$

Then the set

$$\left\{\left(9m, 9m^2+1, \frac{2}{3m}, \frac{4}{3m+5}\right)\right\}_{m\in\mathbb{N}}$$

is its subsequence.

Definition 1.9. The point $x \in \mathbb{R}^n$ is said to be the limit of a sequence $\{x^m\}_{m\in\mathbb{N}}$ of $\mathbb{R}^n$ and we write

$$x = \lim_{m\to\infty} x^m \tag{1.11}$$

if

$$\lim_{m \to \infty} d(x^m, x) = 0. \tag{1.12}$$

If (1.11) holds, then we say that the sequence $\{x^m\}_{m \in \mathbb{N}}$ converges to x. A sequence of $\mathbb{R}^n$ that converges to a point of $\mathbb{R}^n$ is said to be a convergent sequence.

Remark 1.3. Suppose that $\{x^m\}_{m \in \mathbb{N}}$ is a sequence of $\mathbb{R}^n$ and $x \in \mathbb{R}^n$. By (1.12) it follows that the sequence $\{x^m\}_{m \in \mathbb{N}}$ converges to x if and only if for any $\epsilon > 0$, there is $\delta = \delta(\epsilon)$ such that $m > \delta$ implies $x^m \in U(x, \epsilon)$. By the chains (1.7) and (1.8) it follows that the sequence $\{x^m\}_{m \in \mathbb{N}}$ converges to x if and only if for any $\delta_j > 0, j \in \{1, \ldots, n\}$, there is $m_0 \in \mathbb{N}$ such that $x^m \in P(x, \delta_1, \ldots, \delta_n)$ for all $m > m_0$.

The term limit of a sequence of $\mathbb{R}^n$ can be reduced to the term limit of a sequence of $\mathbb{R}$. We will see this in the next result.

Theorem 1.6 (Componentwise nature of convergence). *Let* $\{x^m = (x_1^m, \ldots, x_n^m)\}_{m \in \mathbb{N}}$ *be a sequence of* $\mathbb{R}^n$, *and let* $x = (x_1, \ldots, x_n) \in \mathbb{R}^n$. *Then the sequence* $\{x^m\}_{m \in \mathbb{N}}$ *converges to* x *if and only if each sequence* $\{x_j^m\}_{m \in \mathbb{N}}, j \in \{1, \ldots, n\}$, *converges to* $x_j, j \in \{1, \ldots, n\}$.

Proof. Let $\epsilon > 0$ be arbitrarily chosen. Then $\{x^m\}_{m \in \mathbb{N}}$ converges to x if and only if there is $\delta = \delta(\epsilon) > 0$ such that $m > \delta$ implies the inequality

$$d(x^m, x) < \epsilon.$$

Since

$$|x_j^m - x_j| \le d(x^m, x), \quad j \in \{1, \ldots, n\}, \quad m \in \mathbb{N},$$

the sequence $\{x^m\}_{m \in \mathbb{N}}$ converges to x if and only if there is $\delta = \delta(\epsilon) > 0$ such that $m > \delta$ implies the inequality

$$|x_j^m - x_j| < \epsilon.$$

This completes the proof. $\qquad \square$

Example 1.12. Consider the sequence in $\mathbb{R}^3$

$$\{x^m\}_{m \in \mathbb{N}} = \left\{ \left(\frac{2^m + 3^m}{4^m}, \sqrt{m+1} - \sqrt{m}, \left(\frac{m+2}{m} \right)^m \right) \right\}_{m \in \mathbb{N}}.$$

We will find $\lim_{m \to \infty} x^m$. Here

$$x_1^m = \frac{2^m + 3^m}{4^m},$$
$$x_2^m = \sqrt{m+1} - \sqrt{m},$$
$$x_3^m = \left(\frac{m+2}{m} \right)^m, \quad m \in \mathbb{N}.$$

Then

$$\lim_{m\to\infty} x_1^m = \lim_{m\to\infty} \frac{2^m + 3^m}{4^m}$$

$$= \lim_{m\to\infty} \left(\frac{2^m}{4^m} + \frac{3^m}{4^m} \right)$$

$$= \lim_{m\to\infty} \frac{1}{2^m} + \lim_{m\to\infty} \frac{3^m}{4^m}$$

$$= 0 + 0$$

$$= 0,$$

$$\lim_{m\to\infty} x_2^m = \lim_{m\to\infty} (\sqrt{m+1} - \sqrt{m})$$

$$= \lim_{m\to\infty} \frac{(\sqrt{m+1} - \sqrt{m})(\sqrt{m+1} + \sqrt{m})}{\sqrt{m+1} + \sqrt{m}}$$

$$= \lim_{m\to\infty} \frac{m+1-m}{\sqrt{m+1} + \sqrt{m}}$$

$$= \lim_{m\to\infty} \frac{1}{\sqrt{m+1} + \sqrt{m}}$$

$$= 0,$$

and

$$\lim_{m\to\infty} x_3^m = \lim_{m\to\infty} \left(\frac{m+2}{m} \right)^m$$

$$= \lim_{m\to\infty} \left(1 + \frac{2}{m} \right)^m$$

$$= \lim_{m\to\infty} \left(1 + \frac{1}{\frac{m}{2}} \right)^{\frac{m}{2}\cdot 2}$$

$$= \left(\lim_{m\to\infty} \left(1 + \frac{1}{\frac{m}{2}} \right)^{\frac{m}{2}} \right)^2$$

$$= e^2.$$

Consequently,

$$\lim_{m\to\infty} x^m = (0, 0, e^2).$$

Example 1.13. Consider the sequence in $\mathbb{R}^4$

$$\{x^m\}_{m\in\mathbb{N}} = \left\{ \left(\sum_{j=1}^{m} \frac{1}{j(j+1)}, \prod_{j=2}^{m} \left(1 - \frac{1}{j^2} \right), m(\sqrt{m^2+1} - m), \left(\frac{m^2-1}{m^2-m-6} \right)^m \right) \right\}_{m\in\mathbb{N}}.$$

We will find $\lim_{m\to\infty} x^m$. Here

$$x_1^m = \sum_{j=1}^{m} \frac{1}{j(j+1)},$$

$$x_2^m = \prod_{j=2}^{m} \left(1 - \frac{1}{j^2}\right),$$

$$x_3^m = m(\sqrt{m^2+1} - m),$$

$$x_4^m = \left(\frac{m^2-1}{m^2-m-6}\right)^m, \quad m \in \mathbb{N}.$$

Firstly, we will simplify x_1^m and x_2^m, $m \in \mathbb{N}$. We have

$$x_1^m = \sum_{j=1}^{m} \frac{1}{j(j+1)}$$

$$= \sum_{j=1}^{m} \left(\frac{1}{j} - \frac{1}{j+1}\right)$$

$$= \left(1 - \frac{1}{2}\right) + \left(\frac{1}{2} - \frac{1}{3}\right) + \cdots + \left(\frac{1}{m} - \frac{1}{m+1}\right)$$

$$= 1 - \frac{1}{m+1}$$

$$= \frac{m+1-1}{m+1}$$

$$= \frac{m}{m+1}, \quad m \in \mathbb{N},$$

and

$$x_2^m = \prod_{j=2}^{m} \left(1 - \frac{1}{j^2}\right)$$

$$= \prod_{j=2}^{m} \frac{j^2-1}{j^2}$$

$$= \prod_{j=2}^{m} \frac{(j-1)(j+1)}{j^2}$$

$$= \prod_{j=2}^{m} \left(\frac{j-1}{j} \cdot \frac{j+1}{j}\right)$$

$$= \left(\frac{1}{2} \cdot \frac{3}{2}\right) \cdot \left(\frac{2}{3} \cdot \frac{4}{3}\right) \cdots \left(\frac{m-1}{m} \cdot \frac{m+1}{m}\right)$$

$$= \frac{1}{2} \cdot \frac{m+1}{m}$$

$$= \frac{m+1}{2m}, \quad m \in \mathbb{N}.$$

Hence

$$\lim_{m \to \infty} x_1^m = \lim_{m \to \infty} \frac{m}{m+1}$$
$$= 1,$$

$$\lim_{m \to \infty} x_2^m = \lim_{m \to \infty} \frac{m+1}{2m}$$
$$= \frac{1}{2},$$

$$\lim_{m \to \infty} x_3^m = \lim_{m \to \infty} \left(m(\sqrt{m^2+1} - m) \right)$$
$$= \lim_{m \to \infty} \frac{m(\sqrt{m^2+1} - m)(\sqrt{m^2+1} + m)}{\sqrt{m^2+1} + m}$$
$$= \lim_{m \to \infty} \frac{m(m^2+1-m^2)}{\sqrt{m^2+1} + m}$$
$$= \lim_{m \to \infty} \frac{m}{m\left(\sqrt{1 + \frac{1}{m^2}} + 1\right)}$$
$$= \lim_{m \to \infty} \frac{1}{\sqrt{1 + \frac{1}{m^2}} + 1}$$
$$= \frac{1}{1+1}$$
$$= \frac{1}{2},$$

and

$$\lim_{m \to \infty} x_4^m = \lim_{m \to \infty} \left(\frac{m^2-1}{m^2-m-6} \right)^m$$
$$= \lim_{m \to \infty} \left(\frac{m^2-m-6+m+5}{m^2-m-6} \right)^m$$
$$= \lim_{m \to \infty} \left(1 + \frac{m+5}{m^2-m-6} \right)^m$$
$$= \lim_{m \to \infty} \left(1 + \frac{1}{\frac{m^2-m-6}{m+5}} \right)^{\frac{m^2-m-6}{m+5} \cdot \frac{m(m+5)}{m^2-m-6}}$$
$$= \left(\lim_{m \to \infty} \left(1 + \frac{1}{\frac{m^2-m-6}{m+5}} \right)^{\frac{m^2-m-6}{m+5}} \right)^{\lim_{m \to \infty} \frac{m(m+5)}{m^2-m-6}}$$
$$= e.$$

Consequently,

$$\lim_{m \to \infty} x^m = \left(1, \frac{1}{2}, \frac{1}{2}, e \right).$$

Example 1.14. Consider the sequence in $\mathbb{R}^3$

$$\{x^m\}_{m\in\mathbb{N}} = \left\{\left(\frac{\sqrt[3]{1+a_m}-1}{a_m}, \left(1-\frac{1}{m}\right)^m, \frac{\sqrt{m^2+1}}{m}\right)\right\}_{m\in\mathbb{N}},$$

where $\lim_{m\to\infty} a_m = 0$, $a_m \neq 0$, $m \in \mathbb{N}$. We will find $\lim_{m\to\infty} x^m$. Here

$$x_1^m = \frac{\sqrt[3]{1+a_m}-1}{a_m},$$

$$x_2^m = \left(1-\frac{1}{m}\right)^m,$$

$$x_3^m = \frac{\sqrt{m^2+1}}{m}, \quad m \in \mathbb{N}.$$

We have

$$\lim_{m\to\infty} x_1^m = \lim_{m\to\infty} \frac{\sqrt[3]{1+a_m}-1}{a_m}$$

$$= \lim_{m\to\infty} \frac{(\sqrt[3]{1+a_m}-1)(\sqrt[3]{(1+a_m)^2}+\sqrt[3]{1+a_m}+1)}{a_m(\sqrt[3]{(1+a_m)^2}+\sqrt[3]{1+a_m}+1)}$$

$$= \lim_{m\to\infty} \frac{1+a_m-1}{a_m(\sqrt[3]{(1+a_m)^2}+\sqrt[3]{1+a_m}+1)}$$

$$= \lim_{m\to\infty} \frac{1}{\sqrt[3]{(1+a_m)^2}+\sqrt[3]{1+a_m}+1}$$

$$= \frac{1}{1+1+1}$$

$$= \frac{1}{3},$$

$$\lim_{m\to\infty} x_2^m = \lim_{m\to\infty} \left(1-\frac{1}{m}\right)^m$$

$$= e^{-1},$$

and

$$\lim_{m\to\infty} x_3^m = \lim_{m\to\infty} \frac{\sqrt{m^2+1}}{m}$$

$$= \lim_{m\to\infty} \frac{m\sqrt{1+\frac{1}{m^2}}}{m}$$

$$= \lim_{m\to\infty} \sqrt{1+\frac{1}{m^2}}$$

$$= 1.$$

Consequently,

$$\lim_{m\to\infty} x^m = \left(\frac{1}{3}, e^{-1}, 1\right).$$

Exercise 1.3. Find $\lim_{m\to\infty} x^m$, where

1.

$$\{x^m\}_{m\in\mathbb{N}} = \left\{\left(\sum_{j=1}^m \frac{1}{j(j+1)(j+2)}, \frac{2^{m+1}+3^{m+1}}{2^m+3^m}, \sqrt{m^2+1}-m\right)\right\}_{m\in\mathbb{N}}.$$

2.

$$\{x^m\}_{m\in\mathbb{N}} = \left\{\left(\sqrt{m^2+1}-\sqrt{m^2-1}, \frac{a_m^2-5a_m+6}{a_m^2-7a_m+10}, \left(\frac{m+3}{m}\right)^m, \left(\frac{m^2+m+1}{m^2+3m+1}\right)^m\right)\right\}_{m\in\mathbb{N}},$$

where $\lim_{m\to\infty} a_m = 2, a_m \neq 2, 5, m \in \mathbb{N}$.

3.

$$\{x^m\}_{m\in\mathbb{N}} = \left\{\left(\left(\frac{m^3+m^2+3m+1}{m^3+m^2+2m+1}\right)^m, \sqrt{m^2+m+1}-\sqrt{m^2-m+1}\right)\right\}_{m\in\mathbb{N}}.$$

4.

$$\{x^m\}_{m\in\mathbb{N}} = \{(\sqrt[3]{1-m^3}+m, m^{\frac{3}{2}}(\sqrt{m+1}+\sqrt{m-1}-2\sqrt{m}), 2, 3, 4)\}_{m\in\mathbb{N}}.$$

5.

$$\{x^m\}_{m\in\mathbb{N}} = \left\{\left(0, m^3(\sqrt{m^2+\sqrt{m^4+1}}-m\sqrt{2}), \frac{m}{m+2}, 3\right)\right\}_{m\in\mathbb{N}}.$$

In the next result, we show that any convergent sequence has a unique limit.

Theorem 1.7. *Let $\{x^m\}_{m\in\mathbb{N}}$ be a convergent sequence in $\mathbb{R}^n$. Then there is a unique $x \in \mathbb{R}^n$ such that*

$$\lim_{m\to\infty} x^m = x. \tag{1.13}$$

Proof. We will present two ways for the proof of this statement.

First way. Since $\{x^m\}_{m\in\mathbb{N}}$ is a convergent sequence in $\mathbb{R}^n$, by Theorem 1.6 it follows that $\{x_j^m\}_{m\in\mathbb{N}}$ is a convergent sequence in $\mathbb{R}$ for each $j \in \{1,\dots,n\}$. Then there is a unique $x_j \in \mathbb{R}, j \in \{1,\dots,n\}$, such that

$$\lim_{m\to\infty} x_j^m = x_j, \quad j \in \{1,\dots,n\}.$$

Set $x = (x_1, \dots, x_n)$, and the proof completes.
Second way. Assume that there are $x, y \in \mathbb{R}^n$ such that (1.13) holds and

$$\lim_{m\to\infty} x^m = y.$$

Then

$$\lim_{m\to\infty} d(x,x^m) = \lim_{m\to\infty} d(x^m,y)$$
$$= 0.$$

By the properties of the distance d in $\mathbb{R}^n$ we have the following inequality:

$$d(x,y) \le d(x^m,x) + d(x^m,y)$$

for all $m \in \mathbb{N}$. Hence

$$d(x,y) \le \lim_{m\to\infty} d(x^m,x) + \lim_{m\to\infty} d(x^m,y)$$
$$= 0 + 0$$
$$= 0.$$

Therefore $d(x,y) = 0$, and $x = y$. This completes the proof. □

Now we will introduce a class of sequences that has a wide range of applications in topology and ordinary and partial differential equations.

Definition 1.10. We say that a sequence $\{x^m\}_{m\in\mathbb{N}}$ in $\mathbb{R}^n$ is a Cauchy sequence if for any $\epsilon > 0$, there is $\delta = \delta(\epsilon) > 0$ such that

$$d(x^m,x^p) < \epsilon$$

for all $m, p > \delta, m, p \in \mathbb{N}$.

In the next result, we give a necessary and sufficient condition for the convergence of a sequence.

Theorem 1.8 (The Cauchy criterion). *A sequence $\{x^m\}_{m\in\mathbb{N}}$ in $\mathbb{R}^n$ is a convergent sequence in $\mathbb{R}^n$ if and only if it is a Cauchy sequence in $\mathbb{R}^n$.*

Proof. 1. Let $\{x^m\}_{m\in\mathbb{N}}$ be a convergent sequence in $\mathbb{R}^n$. By Theorem 1.6 it follows that any sequence $\{x_j^m\}_{m\in\mathbb{N}}, j \in \{1,\ldots,n\}$, is a convergent sequence in $\mathbb{R}$. By the Cauchy criterion for sequences in $\mathbb{R}$ we get that every sequence $\{x_j^m\}_{m\in\mathbb{N}}, j \in \{1,\ldots,n\}$, is a Cauchy sequence in $\mathbb{R}$. Take $\epsilon > 0$ arbitrarily. Then for each $j \in \{1,\ldots,n\}$, there is $\delta_j = \delta_j(\epsilon) > 0$ such that

$$|x_j^m - x_j^p| < \frac{\epsilon}{\sqrt{n}}$$

for all $m, p > \delta_j, m, p \in \mathbb{N}$. Let

$$\delta = \max_{j\in\{1,\ldots,n\}} \delta_j.$$

Then

$$d(x^m, x^p) = \sqrt{(x_1^m - x_1^p)^2 + \cdots + (x_n^m - x_n^p)^2}$$

$$< \sqrt{\frac{\epsilon^2}{n} + \cdots + \frac{\epsilon^2}{n}}$$

$$= \epsilon$$

for all $m, p > \delta$, $m, p \in \mathbb{N}$. Thus $\{x^m\}_{m\in\mathbb{N}}$ is a Cauchy sequence in $\mathbb{R}^n$.

2. Let $\{x^m\}_{m\in\mathbb{N}}$ be a Cauchy sequence in $\mathbb{R}^n$. Take $\epsilon > 0$ arbitrarily. Then there is $\delta = \delta(\epsilon) > 0$ such that

$$d(x^m, x^p) < \epsilon$$

for all $m, p > \delta$, $m, p \in \mathbb{N}$. Hence, since

$$|x_j^m - x_j^p| < \epsilon$$

for all $m, p > \delta$, $m, p \in \mathbb{N}$, and $j \in \{1, \ldots, n\}$, we have that $\{x_j^m\}_{m\in\mathbb{N}}$ is a Cauchy sequence in $\mathbb{R}$ for all $j \in \{1, \ldots, n\}$. By the Cauchy criterion for sequences in $\mathbb{R}$ we conclude that each sequence $\{x_j^m\}_{m\in\mathbb{N}}$, $j \in \{1, \ldots, n\}$, is a convergent sequence in $\mathbb{R}$. Now, applying Theorem 1.6, we conclude that $\{x^m\}_{m\in\mathbb{N}}$ is a convergent sequence in $\mathbb{R}^n$. This completes the proof. □

Another class of sequences is the class of bounded sequences.

Definition 1.11. A sequence $\{x^m\}_{m\in\mathbb{N}}$ in $\mathbb{R}^n$ is said to be bounded if there is a constant $M \geq 0$ such that

$$|x^m| \leq M, \quad m \in \mathbb{N}.$$

One of the properties of bounded sequences reads as follows.

Theorem 1.9. *Let $\{x^m = (x_1^m, \ldots, x_n^m)\}_{m\in\mathbb{N}}$ be a bounded sequence in $\mathbb{R}^n$. Then there is a subsequence of the sequence $\{x^m\}_{m\in\mathbb{N}}$ that is convergent.*

Proof. Since the sequence $\{x^m\}_{m\in\mathbb{N}}$ is bounded, each sequence $\{x_j^m\}_{m\in\mathbb{N}}$, $j \in \{1, \ldots, n\}$, is bounded. Now, applying the Bolzano–Weierstrass theorem, we get that there is a subsequence $\{x_1^{m_{k_1}}\}_{k_1\in\mathbb{N}}$ of the sequence $\{x_1^m\}_{m\in\mathbb{N}}$ that is convergent. Note that the sequence $\{x_2^{m_{k_1}}\}_{k_1\in\mathbb{N}}$ is a subsequence of the sequence $\{x_2^m\}_{m\in\mathbb{N}}$. Then it is bounded, and applying the Bolzano–Weierstrass theorem, we get a subsequence $\{x_2^{m_{k_1,k_2}}\}_{k_2\in\mathbb{N}}$ of the sequence $\{x_2^{m_{k_1}}\}_{k_1\in\mathbb{N}}$ that is convergent, etc. Consider the sequence $\{x_n^{m_{k_1,\ldots,k_{n-1}}}\}_{k_{n-1}\in\mathbb{N}}$ that is a subsequence of the sequence $\{x_n^m\}_{m\in\mathbb{N}}$. Then it is bounded, and applying the Bolzano–Weierstrass theorem, we obtain a subsequence $\{x_n^{m_{k_1,\ldots,k_{n-1},k_n}}\}_{k_n\in\mathbb{N}}$ of the sequence $\{x_n^{m_{k_1,\ldots,k_{n-1}}}\}_{k_{n-1}\in\mathbb{N}}$ that is convergent. Consequently, the sequence

$$\{x_1^{m_{k_1,\dots,k_n}}, \dots, x_n^{m_{k_1,\dots,k_n}}\}_{k_n \in \mathbb{N}}$$

is a subsequence of the sequence $\{x^m\}_{m \in \mathbb{N}}$ that is convergent. This completes the proof.

□

So far, we have considered sequences converging to a finite limit. Now we will give a definition of convergence of a sequence to infinity.

Definition 1.12. We say that a sequence $\{x^m\}_{m \in \mathbb{N}}$ of $\mathbb{R}^n$ converges to ∞ and write

$$\lim_{m \to \infty} x^m = \infty \tag{1.14}$$

if

$$\lim_{m \to \infty} d(x^m, 0) = \infty, \tag{1.15}$$

where 0 is the origin of the coordinate system of $\mathbb{R}^n$.

Suppose that (1.14) holds. Take $x \in \mathbb{R}^n$ arbitrarily. Then

$$d(x^m, 0) \le d(x^m, x) + d(x, 0), \quad m \in \mathbb{N},$$

whereupon

$$d(x^m, x) \ge d(x^m, 0) - d(x, 0), \quad m \in \mathbb{N},$$

and thus

$$\lim_{m \to \infty} d(x^m, x) = \infty.$$

Example 1.15. Consider the sequence in $\mathbb{R}^3$

$$\{x^m\} = \{(m, 1, e^{-m})\}_{m \in \mathbb{N}}.$$

Here

$$x_1^m = m,$$
$$x_2^m = 1,$$
$$x_3^m = e^{-m}, \quad m \in \mathbb{N}.$$

Then

$$d(x^m, 0) = \sqrt{(x_1^m)^2 + (x_2^m)^2 + (x_3^m)^2}$$
$$= \sqrt{m^2 + 1 + e^{-2m}}, \quad m \in \mathbb{N}.$$

Hence

$$\lim_{m\to\infty} d(x^m, 0) = \lim_{m\to\infty} \sqrt{m^2 + 1 + e^{-2m}}$$
$$= \infty,$$

and (1.14) holds.

Now we will give a necessary and sufficient condition for the convergence of a sequence to infinity.

Theorem 1.10. *A sequence $\{x^m = (x_1^m, \ldots, x_n^m)\}_{m\in\mathbb{N}}$ of $\mathbb{R}^n$ converges to ∞ if and only if there is $j \in \{1, \ldots, n\}$ such that*

$$\lim_{m\to\infty} x_j^m = \pm\infty. \tag{1.16}$$

Proof. 1.   Let (1.14) hold. Then

$$\lim_{m\to\infty} \sqrt{(x_1^m)^2 + \cdots + (x_n^m)^2} = \infty.$$

Hence, there is $j \in \{1, \ldots, n\}$ such that

$$\lim_{m\to\infty} (x_j^m)^2 = \infty,$$

whereupon we get (1.16).

2.   Let (1.16) hold. Then (1.15) holds, and hence (1.14) holds. This completes the proof.
□

Example 1.16. Consider the sequence in $\mathbb{R}^3$

$$\{x^m\} = \left\{\left(\frac{m^2 + 1}{m}, m, \frac{1}{m}\right)\right\}_{m\in\mathbb{N}}.$$

Here

$$x_1^m = \frac{m^2 + 1}{m},$$
$$x_2^m = m,$$
$$x_3^m = \frac{1}{m}, \quad m \in \mathbb{N}.$$

Since

$$\lim_{m\to\infty} x_2^m = \lim_{m\to\infty} m$$
$$= \infty,$$

applying Theorem 1.10, we obtain that $\lim_{m\to\infty} x^m = \infty$.

Exercise 1.4. Prove that $\lim_{m\to\infty} x^m = \infty$, where

1.
$$\{x^m\} = \{(1, e^m, 2)\}_{m\in\mathbb{N}}.$$

2.
$$\{x^m\} = \left\{\left(\frac{m}{m^2+1}e^m, e^{-m}\right)\right\}_{m\in\mathbb{N}}.$$

3.
$$\{x^m\} = \left\{\left(e^{-m^2}, m^2+1, \frac{1}{m^2+2}\right)\right\}_{m\in\mathbb{N}}.$$

4.
$$\{x^m\} = \left\{\left(e^{m^2+1}, \frac{m+2}{m^2+3}, m, \frac{1}{m}, m^3\right)\right\}_{m\in\mathbb{N}}.$$

5.
$$\{x^m\} = \left\{\left(\sin m, \cos m, \frac{m^2+1}{2+\cos m}\right)\right\}_{m\in\mathbb{N}}.$$

1.4 Sets in $\mathbb{R}^n$

With ϵ-neighborhoods, it is easy to describe the points that are approached by a set $A \subset \mathbb{R}^n$.

Definition 1.13. A point $x \in A$ is said to be an interior point of A if there is its ϵ-neighborhood contained in A, that is, if there is $\epsilon > 0$ such that $U(x, \epsilon) \subset A$. The set of all interior points of A is said to be the interior of A and is denoted by A_{int}.

Example 1.17. Let A be the square region in $\mathbb{R}^2$ containing all points $(x, y) \in \mathbb{R}^2$ such that

$$0 < x < 1, \quad 0 < y < 1$$

(see Fig. 1.9). We will prove that any point of A is an interior point. Take $(x, y) \in A$ arbitrarily. Let z be the smallest of

$$x, \quad y, \quad 1-x, \quad 1-y.$$

Then the open disc with center (x, y) and radius z is contained in Q.

Example 1.18. Let A be the unit disc

$$x^2 + y^2 \leq 1$$

centered at the origin in x, y plane (see Fig. 1.10). We will prove that any point P whose distance from the origin is less than 1 is an interior point for the set A. Let

$$\epsilon = 1 - |P|,$$

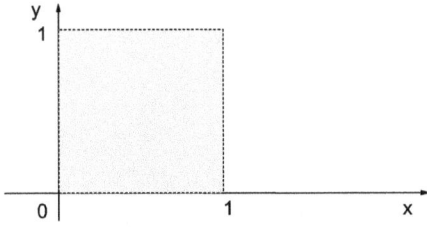

Figure 1.9: The set A in Example 1.17.

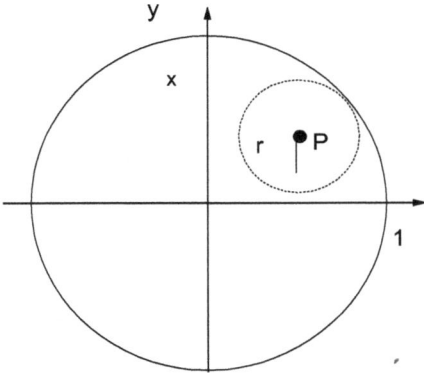

Figure 1.10: The set A in Example 1.18.

and let Q be a point in the open disc of radius ϵ centered at P. Now using the triangle inequality, we arrive at

$$|Q| = |Q - P| + |P|$$
$$< 1 - |P| + |P|$$
$$= 1.$$

Therefore all points that are within ϵ of P are all in A. Thus P is an interior point.

Definition 1.14. A set whose all points are interior points is said to be an open set. By definition we consider the empty set as an open set.

By the above definition it follows that A is an open set of $\mathbb{R}^n$ if and only if $A = A_{\text{int}}$.

Theorem 1.11. *For any $\epsilon > 0$, the ϵ-neighborhood of any point of $\mathbb{R}^n$ is an open set of $\mathbb{R}^n$.*

Proof. Let $\epsilon > 0$ and $x \in \mathbb{R}^n$. Consider $U(x, \epsilon)$. Take arbitrary $y \in U(x, \epsilon)$. Then

$$d(x, y) < \epsilon.$$

Set

$$\delta = \epsilon - d(x,y) > 0.$$

Consider $U(y, \delta)$. Let $z \in U(y, \delta)$. Then

$$d(z, y) < \delta.$$

Using the properties of the distance in $\mathbb{R}^n$, we arrive at the following chain of inequalities:

$$d(x, z) \le d(x, y) + d(y, z)$$
$$< \epsilon - d(x, y) + d(x, y)$$
$$= \epsilon.$$

Therefore $z \in U(x, \epsilon)$. Because $z \in U(y, \delta)$ was arbitrarily chosen and it is an element of $U(x, \epsilon)$, we obtain the inclusion (see Fig. 1.11)

$$U(y, \delta) \subset U(x, \epsilon).$$

Consequently, y is an interior point of $U(x, \epsilon)$. So any point of $U(x, \epsilon)$ is an interior point. Thus $U(x, \epsilon)$ is an open set. This completes the proof. $\square$

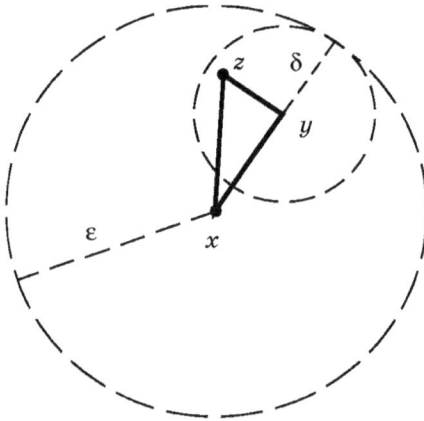

Figure 1.11: The neighborhood U.

Theorem 1.12. *Any rectangular neighborhood of any point of $\mathbb{R}^n$ is an open set.*

Proof. Let $\delta_j > 0, j \in \{1, \ldots, n\}$, and let $x \in \mathbb{R}^n$. Consider $P(x, \delta_1, \ldots, \delta_n)$. Take arbitrary $y \in P(x, \delta_1, \ldots, \delta_n)$. Then

$$|y_j - x_j| < \delta_j, \quad j \in \{1, \ldots, n\}.$$

Set

$$e_j = \delta_j - |y_j - x_j|, \quad j \in \{1, \ldots, n\},$$
$$e = \min_{j \in \{1, \ldots, n\}} e_j.$$

Consider $U(y, e)$. Let $z \in U(y, e)$. Then

$$|y_j - z_j| \le d(y, z)$$
$$< e.$$

Hence

$$|z_j - x_j| \le |z_j - y_j| + |y_j - x_j|$$
$$< e + |y_j - x_j|$$
$$\le e_j + |y_j - x_j|$$
$$= \delta_j - |y_j - x_j| + |y_j - x_j|$$
$$= \delta_j, \quad j \in \{1, \ldots, n\}.$$

Therefore $z \in P(x, \delta_1, \ldots, \delta_n)$. Because $z \in U(y, e)$ was arbitrarily chosen and it is an element of $P(x, \delta_1, \ldots, \delta_n)$, we obtain the inclusion

$$U(y, e) \subset P(x, \delta_1, \ldots, \delta_n).$$

Therefore y is an interior point of $P(x, \delta_1, \ldots, \delta_n)$. Thus any point of $P(x, \delta_1, \ldots, \delta_n)$ is an interior point, and so $P(x, \delta_1, \ldots, \delta_n)$ is an open set. This completes the proof. ☐

Corollary 1.2. *Any n-dimensional cube is an open set of $\mathbb{R}^n$.*

Exercise 1.5. Let $e > 0$ and $x^0 \in \mathbb{R}^n$. Prove that the set

$$B = \{x \in \mathbb{R}^n : d(x, x^0) > e\}$$

is an open set in $\mathbb{R}^n$.

Exercise 1.6. Let $n \ge 3$ and $e > 0$. Prove that

$$B = \{x \in \mathbb{R}^n : x_1^2 + x_2^2 < e^2, \ x_j = 0, \ j \in \{3, \ldots, n\}\}$$

is an open set in $\mathbb{R}^n$.

Exercise 1.7. Let $n = 1, f \in \mathcal{C}(\mathbb{R})$, and $x^0 \in \mathbb{R}$. Prove that the set

$$B = \{x \in \mathbb{R} : f(x) > x^0\}$$

is an open set in $\mathbb{R}$.

Theorem 1.13. *The union of a family of open sets is an open set.*

Proof. Let $\{A_\alpha : \alpha \in I\}$ be a family of open sets, and let

$$B = \bigcup_{\alpha \in I} A_\alpha.$$

Take arbitrary $x \in B$. Then there is $\beta \in I$ such that $x \in A_\beta$. Since A_β is an open set, there is $\epsilon > 0$ such that $U(x, \epsilon) \subset A_\beta$. Hence

$$U(x, \epsilon) \subset \bigcup_{\alpha \in I} A_\alpha = B.$$

Because $x \in B$ was arbitrarily chosen, we conclude that B is an open set. This completes the proof. □

Theorem 1.14. *The intersection of a finite number of open sets is an open set.*

Proof. Let $A_1, \ldots, A_k$ be open sets. Set

$$B = \bigcap_{j=1}^{k} A_j.$$

Take arbitrary $x \in B$. Then $x \in A_j, j \in \{1, \ldots, k\}$. Since $A_j, j \in \{1, \ldots, k\}$, are open sets, there are $\epsilon_j > 0, j \in \{1, \ldots, k\}$, such that $U(x, \epsilon_j) \subset A_j, j \in \{1, \ldots, k\}$. Set

$$\epsilon = \min_{j \in \{1, \ldots, k\}} \epsilon_j.$$

Then $U(x, \epsilon) \subset A_j, j \in \{1, \ldots, k\}$. Therefore

$$U(x, \epsilon) \subset \bigcap_{j=1}^{k} A_j = B.$$

Because $x \in B$ was arbitrarily chosen, we conclude that B is an open set. This completes the proof. □

Example 1.19. Let $n = 1$ and

$$A_k = \left(-\frac{1}{2^k}, 1 + \frac{1}{2^k}\right), \quad k \in \mathbb{N}.$$

We have that $A_k, k \in \mathbb{N}$, are open sets in $\mathbb{R}$ and

$$\bigcap_{k \in \mathbb{N}} A_k = [0, 1],$$

which is not an open set in $\mathbb{R}$.

Definition 1.15. Let $x \in \mathbb{R}^n$. Any open set containing x is said to be a neighborhood of x and will be denoted by $U(x)$.

Using the above definition, we can reformulate the definition of a limit of a sequence as follows.

Definition 1.16. We say that a sequence $\{x^m\}_{m \in \mathbb{N}}$ in $\mathbb{R}^n$ converges to $x \in \mathbb{R}^n$ if for any neighborhood $U(x)$ of x, there is $m_0 \in \mathbb{N}$ such that $x^m \in U(x)$ for all $m > m_0$, $m \in \mathbb{N}$.

Definition 1.17. A point $x \in \mathbb{R}^n$ is said to be an adherent point of the set A if every its neighborhood contains at least one element of the set A.

Example 1.20. Consider the open disc $S \subset \mathbb{R}^2$ of radius 1 centered at the origin $(0, 0)$. We claim that any point on the unit circle is an adherent point of S. Geometrically, this is intuitively obvious as all discs centered at a point $x \in \mathbb{R}^2$ on the unit circle will intersect the disc S or contain all of S. Hence there exists $s \in S$ such that s belongs to all balls of radius r centered at x for any $r > 0$ (see Fig. 1.12).

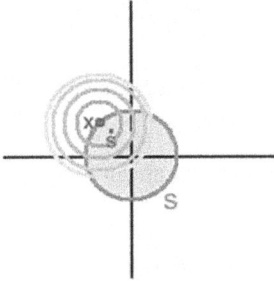

x is an adherent point of S.

Figure 1.12: The point x is an adherent point for S.

Example 1.21. Each point $x \in A$ is an adherent point of A because any neighborhood of x contains x.

Example 1.22. Let $n = 1$ and $A = (a, b)$. Then $a, b \notin A$, and a and b are adherent points of A.

Now we will give a necessary and sufficient condition for adherence of a point.

Theorem 1.15. *A point $x \in \mathbb{R}^n$ is an adherent point of A if and only if there is a sequence $\{x^m\}_{m \in \mathbb{N}} \subset A$ such that*

$$\lim_{m \to \infty} x^m = x. \tag{1.17}$$

Proof. 1. Suppose that there is a sequence $\{x^m\}_{m \in \mathbb{N}} \subset A$ such that (1.17) holds. Then for each $\epsilon > 0$, there is $m_0 \in \mathbb{N}$ such that $x^{m_0} \in U(x, \epsilon)$. Thus x is an adherent point of A.

2. Let x be an adherent point of A. Take a sequence $\{\epsilon_k\}_{k\in\mathbb{N}} \subset \mathbb{R}$ such that $\epsilon_k > 0$, $k \in \mathbb{N}$, and

$$\lim_{k\to\infty} \epsilon_k = 0.$$

Since x is an adherent point of A, there is $x^k \in A$ such that $x^k \in U(x, \epsilon_k)$ for all $k \in \mathbb{N}$. Thus we get a sequence $\{x^k\}_{k\in\mathbb{N}} \subset A$ such that

$$d(x, x^k) < \epsilon_k, \quad k \in \mathbb{N}.$$

Hence

$$0 \le \lim_{k\to\infty} d(x, x^k)$$
$$\le \lim_{k\to\infty} \epsilon_k$$
$$= 0,$$

that is, (1.17) holds. This completes the proof. $\qquad\square$

Definition 1.18. A point $x \in A$ for which there is a neighborhood that does not contain any other point of A different from x is said to be an isolated point of A.

Note that any isolated point of A is an adherent point of A because any of its neighborhood contains itself.

Example 1.23. For the set $A = \{0\} \cup [1, 2]$, 0 is an isolated point (see Fig. 1.13).

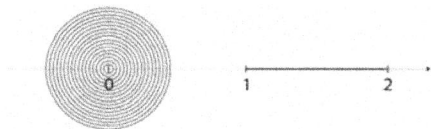

Figure 1.13: 0 is an isolated point for the set $\{0\} \cup [1, 2]$.

Example 1.24. Let $n = 2$ and

$$A = \{x = (x_1, x_2) \in \mathbb{R}^2 : x_1^2 + x_2^2 < 1\} \bigcup \{(3, 4)\}.$$

Then $U((3, 4), \frac{1}{10^{10}})$ does not contain any point of the unit circle in $\mathbb{R}^2$. Therefore $(3, 4)$ is an isolated (and adherent) point of A.

Another kind of points are the limit points.

Definition 1.19. A point $x \in \mathbb{R}^n$ is said to be a limit point, accumulation point, or cluster point of A if any of its neighborhoods contains at least one point of A different from x.

By the above definition it follows that any limit point of A is an adherent point of A.

Example 1.25. Consider the set shown in Fig. 1.14. Then the points 2 and 3 are limit points. The points 4 and 5 are not limit points.

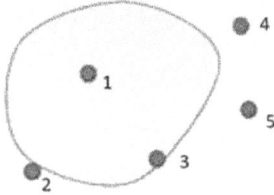

Figure 1.14: The points 2 and 3 are limit points.

Definition 1.20. Let $x \in \mathbb{R}^n$. A deleted neighborhood of x is any set obtained by a neighborhood of x minus itself. If $U(x)$ is a neighborhood of x, then its deleted neighborhood is denoted by $\overset{\circ}{U}(x)$.

Using the above definition, we can give another definition of a limit point.

Definition 1.21. A point $x \in \mathbb{R}^n$ is said to be a limit point of a set A if each its deleted neighborhood contains at least one point of A.

Remark 1.4. For any adherent point of the set A, we have that there is a neighborhood containing only one point of A, that is, the point x, or any neighborhood of x contains at least one point of A different from x. Therefore any adherent point of A is either an interior point of A, or an isolated point of A, or a limit point of A.

Now we give a necessary and sufficient condition for a point to be a limit point.

Theorem 1.16. *A point $x \in \mathbb{R}^n$ is a limit point of a set A if and only if there is a sequence $\{x^m\}_{m \in \mathbb{N}}$ of elements of A such that*

$$\lim_{m \to \infty} x^m = x, \quad x^m \in A, \quad m \in \mathbb{N}, \quad x^{m_1} \neq x^{m_2}, \quad m_1, m_2 \in \mathbb{N}, \quad m_1 \neq m_2. \qquad (1.18)$$

Proof. 1. Let (1.18). Then for each $\epsilon > 0$, there is $l \in \mathbb{N}$ such that $x^l \in U(x, \epsilon)$ and $x^l \neq x$. Thus x is a limit point of A.

2. Suppose that x is a limit point of A. Then there is $x^1 \in A$ such that $x^1 \in U(x, 1)$, $x^1 \neq x$. We have

$$d(x, x^1) < 1.$$

Set

$$\epsilon_1 = \min\left\{\frac{1}{2}, d(x, x^1)\right\}.$$

If $y \in U(x, \epsilon_1)$, then we have

$$d(x,y) < \epsilon_1$$
$$\leq d(x,x^1),$$

and then $y \neq x^1$. Because x is a limit point of A, there is $x^2 \in U(x, \epsilon_1)$ such that $x^2 \neq x$ and $x^2 \in A$. We have

$$d(x,x^2) < \frac{1}{2}$$

and $x^2 \neq x^1$. Let

$$\epsilon_2 = \min\left\{\frac{1}{2^2}, d(x,x^2)\right\}.$$

If $y \in U(x, \epsilon_2)$, then we have

$$d(x,y) < \epsilon_2$$
$$\leq d(x,x^2)$$
$$< d(x,x^1),$$

and therefore $y \neq x^1, y \neq x^2$. Since x is a limit point of A, there is $x^3 \in U(x, \epsilon_2)$ such that $x^3 \neq x$ and $x^3 \in A$. Moreover, we have that $x^3 \neq x^1$ and $x^3 \neq x^2$. Continuing in this way, we get a sequence $\{x^m\}_{m\in\mathbb{N}}$ such that

$$d(x,x^m) < \frac{1}{2^{m-1}}, \quad m \in \mathbb{N}, \tag{1.19}$$

and $x^{m_1} \neq x^{m_2}, m_1, m_2 \in \mathbb{N}, m_1 \neq m_2$. By inequality (1.19) we obtain

$$0 \leq \lim_{m\to\infty} d(x,x^m)$$
$$\leq \lim_{m\to\infty} \frac{1}{2^m}$$
$$= 0,$$

and then

$$\lim_{m\to\infty} x^m = x.$$

Thus (1.18) holds. This completes the proof. □

Remark 1.5. By Theorem 1.16 it follows that any neighborhood of any limit point of A contains an infinite number of elements of A.

Definition 1.22. The set of all adherent points of A is said called the closure of A. It is denoted by $\bar{A}$.

Evidently, we have $A \subset \bar{A}$.

Definition 1.23. The set A is said to be closed if $A = \bar{A}$.

Example 1.26. The rectangle

$$0 \le x \le 1, \quad 0 \le y \le 1$$

(see Fig. 1.15) is a closed set in $\mathbb{R}^2$.

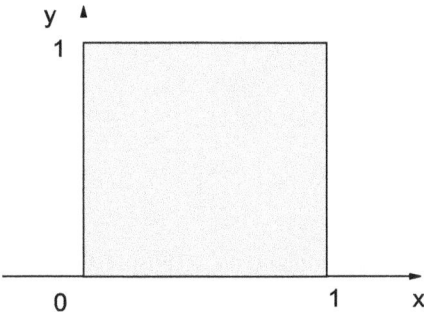

Figure 1.15: A closed set in $\mathbb{R}^2$.

Since an adherent point of the set A is an isolated point of A or a limit point of A, and since any isolated point of A is its element, the condition that all adherent points of A belong to A is equivalent to the condition that any limit points of A belong to A. Thus the set A is a closed set if and only if it contains all its limit points.

Theorem 1.17. *Let A be closed, and let $\{x^m\}_{m \in \mathbb{N}} \subset A$ converge to $x_0 \in \mathbb{R}^n$. Then $x^0 \in A$.*

Proof. Since $\{x^m\}_{m \in \mathbb{N}}$ converges to x^0, any neighborhood of x^0 contains elements of $\{x^m\}_{m \in \mathbb{N}}$. Now since $\{x^m\}_{m \in \mathbb{N}} \subset A$, we conclude that any neighborhood of x^0 contains elements of A. Therefore x^0 is an adherent point of A. Because A is closed, we get $x^0 \in A$. This completes the proof. $\square$

Theorem 1.18. *The closure $\bar{A}$ of a set A is a closed set.*

Proof. We have

$$\bar{A} \subset \bar{\bar{A}}. \tag{1.20}$$

Let x be an adherent point of $\bar{A}$. Then $x \in \bar{\bar{A}}$. Hence any neighborhood $U(x)$ of x contains at least one element y of $\bar{A}$, that is, $y \in U(x) \cap \bar{A}$. Since $U(x)$ is a neighborhood of x, we have that it is an open set. Thus $U(x)$ is an open set that contains y. From this we conclude that $U(x)$ is a neighborhood of y. Because $y \in \overline{A}$, it is an adherent point of A. Then there is an element $z \in U(x) \cap A$. So $U(x)$ is a neighborhood of x that contains at least one element of A. Therefore x is an adherent point of A, that is, $x \in \overline{A}$. Because $x \in \overline{\overline{A}}$ was arbitrarily chosen and it is an element of $\overline{A}$, we obtain the inclusion

$$\overline{\overline{A}} \subset \overline{A}.$$

By the last inclusion and (1.20) we have

$$\overline{A} = \overline{\overline{A}}.$$

This completes the proof. $\qquad\qquad\qquad\qquad\qquad\qquad\qquad\qquad\qquad\qquad\qquad\qquad\quad$ $\square$

Corollary 1.3. *Any adherent point of A is an adherent point of $\overline{A}$ and vice versa.*

Example 1.27. Any n-dimensional ball

$$Q^n(a, r) = \left\{ x = (x_1, \ldots, x_n) \in \mathbb{R}^n : \sum_{j=1}^n (x_j - a_j) < r^2 \right\}$$

with center $a = (a_1, \ldots, a_n)$ and radius $r > 0$ is an open set. It is said to be an n-dimensional open ball. Its closure

$$\overline{Q}^n(a, r) = \left\{ x = (x_1, \ldots, x_n) \in \mathbb{R}^n : \sum_{j=1}^n (x_j - a_j) \leq r^2 \right\}$$

is a closed set, and it is said to be an n-dimensional closed ball. The set

$$S^{n-1}(a, r) = \overline{Q}^n \backslash Q^n$$

$$= \left\{ x = (x_1, \ldots, x_n) \in \mathbb{R}^n : \sum_{j=1}^n (x_j - a_j) = r^2 \right\}$$

is said to be the $(n-1)$-dimensional sphere with center a and radius r.

Example 1.28. Any n-dimensional parallelogram

$$P^n(a, \delta_1, \ldots, \delta_n) = \{ x = (x_1, \ldots, x_n) \in \mathbb{R}^n : |x_j - a_j| < \delta_j, \, j \in \{1, \ldots, n\} \}$$

is an open set. Its closure

$$\overline{P}^n(a, \delta_1, \ldots, \delta_n) = \{ x = (x_1, \ldots, x_n) \in \mathbb{R}^n : |x_j - a_j| \leq \delta_j, \, j \in \{1, \ldots, n\} \}$$

is a closed set, and it is said to be an n-dimensional closed parallelogram.

Definition 1.24. The set $A^c = \mathbb{R}^n \backslash A$ is said to be the complement of A with respect to $\mathbb{R}^n$.

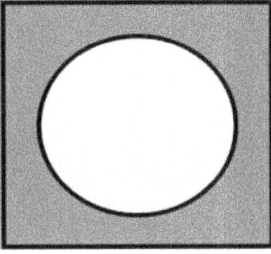

Figure 1.16: The complement of a set.

Example 1.29. The complement of the white disc is the red region (see Fig. 1.16).

Theorem 1.19. *The set A is an open set of $\mathbb{R}^n$ if and only if its complement A^c is a closed set of $\mathbb{R}^n$.*

Proof. 1. Let A be an open set. Assume that there is an element $x \in A$ that is an adherent point of A^c. Since A is an open set, we have that A is a neighborhood of x. By the definition of an adherent point it follows that A contains at least one point of A^c. This is a contradiction. Hence all adherent points of A^c are contained in A^c. Therefore A^c is a closed set.

2. Let A^c be a closed set of $\mathbb{R}^n$. Take arbitrary $x \in A$. Since A^c is a closed set, we have that $x \notin A^c$. Hence there is a neighborhood $U(x)$ of x such that $U(x) \cap A^c = \emptyset$. Therefore $U(x) \subset A$, and x is an interior point of A. Because $x \in A$ was arbitrarily chosen, we conclude that all points of A are its interior points. Consequently, A is an open set. This completes the proof. □

Corollary 1.4. *The set A is a closed set if and only if its complement A^c is an open set.*

Proof. Note that

$$A = \mathbb{R}^n \backslash A^c,$$

that is, A is the complement of A^c. Hence A is a closed set if and only if A^c is an open set. This completes the proof. □

Corollary 1.5. *The union of a finite number of closed sets is a closed set.*

Proof. Let $A_j, j \in \{1, \ldots, k\}$, be closed sets. Then by Corollary 1.4, $A_j^c, j \in \{1, \ldots, k\}$, are open sets. Hence by Theorem 1.14 we have that $\bigcap_{j=1}^{k} A_j^c$ is an open set. Now, using that

$$\left(\bigcup_{j=1}^{k} A_j \right)^c = \bigcap_{j=1}^{k} A_j^c,$$

we conclude that $(\bigcup_{j=1}^{k} A_j)^c$ is an open set. Applying Corollary 1.4, we obtain that $\bigcup_{j=1}^{k} A_j$ is a closed set. This completes the proof. □

Corollary 1.6. *The intersection of a family of closed sets is a closed set.*

Proof. Let $\{A_\alpha : \alpha \in I\}$ be a family of closed sets. Then A_α^c, $\alpha \in I$, are open sets. By Theorem 1.13 it follows that $\bigcup_{\alpha \in I} A_\alpha^c$ is an open set. Now, using that

$$\left(\bigcap_{\alpha \in I} A_\alpha\right)^c = \bigcup_{\alpha \in I} A_\alpha^c,$$

we have that $(\bigcap_{\alpha \in I} A_\alpha)^c$ is an open set. Applying Corollary 1.4, we get that $\bigcap_{\alpha \in I} A_\alpha$ is a closed set. This completes the proof. ☐

Exercise 1.8. Let B be an open set, and let C be a closed set. Prove that $B \backslash C$ is an open set.

Theorem 1.20. *Let A and B be closed nonempty sets such that $A \cap B = \emptyset$ and A is bounded. Then there is $\epsilon > 0$ such that*

$$d(x,y) > \epsilon$$

for all $x \in A$ and $y \in B$.

Proof. Suppose that such $\epsilon > 0$ does not exist. Then, for all $m \in \mathbb{N}$, there are $x^m \in A$ and $y^m \in B$ for which

$$d(x^m, y^m) < \frac{1}{m}.$$

Thus we get the sequences $\{x^m\}_{m \in \mathbb{N}} \subset A$ and $\{y^m\}_{m \in \mathbb{N}} \subset B$. Since A is a bounded set, we obtain that $\{x^m\}_{m \in \mathbb{N}}$ is a bounded sequence. Hence, applying Theorem 1.9, we conclude that there is a subsequence $\{x^{m_k}\}_{k \in \mathbb{N}}$ of the sequence $\{x^m\}_{m \in \mathbb{N}}$ that converges to $x^0 \in \mathbb{R}^n$. Because A is a closed set of $\mathbb{R}^n$, we obtain that $x^0 \in A$. We have the following chain of inequalities:

$$d(x^0, y^{m_k}) \le d(x^0, x^{m_k}) + d(x^{m_k}, y^{m_k})$$
$$\le d(x^0, x^{m_k}) + \frac{1}{m_k}, \quad k \in \mathbb{N}.$$

Hence

$$0 \le \lim_{k \to \infty} d(x^0, y^{m_k})$$
$$\le \lim_{k \to \infty} d(x^0, x^{m_k}) + \lim_{k \to \infty} \frac{1}{m_k}$$
$$= 0,$$

that is,

$$\lim_{k\to\infty} d(x^0, y^{m_k}) = 0.$$

Therefore x^0 is an adherent point of B. Because B is a closed set of $\mathbb{R}^n$, we have that $x^0 \in B$. Therefore $x^0 \in A \cap B$, a contradiction. This completes the proof. □

Definition 1.25. Let $B, C \subset \mathbb{R}^n$. The number

$$d(B, C) = \inf_{x\in B, y\in C} d(x, y)$$

is called the distance between the sets B and C.

By Theorem 1.20 it follows that if $B, C \subset \mathbb{R}^n$ are nonempty closed sets such that $B \cap C = \emptyset$, and B or C is bounded, then the distance between B and C is a positive number.

Definition 1.26. Any open set B that contains the set A is said to be a neighborhood of A. It is denoted by $U(A)$.

Definition 1.27. The union of the ϵ-neighborhoods of all points of the set A is called the ϵ-neighborhood of A. It is denoted by $U(A, \epsilon)$. In other words,

$$U(A, \epsilon) = \bigcup_{x\in A} U(x, \epsilon).$$

Theorem 1.21. *Let A be a bounded set, and let $x \in \mathbb{R}^n$ be such that*

$$d(x, A) = \epsilon \tag{1.21}$$

for some positive number ϵ. Then there is $y \in A$ such that

$$d(x, y) = \epsilon. \tag{1.22}$$

Proof. Note that

$$\begin{aligned} \epsilon &= d(x, A) \\ &= \inf_{z\in A} d(x, z) \tag{1.23} \\ &\le d(x, z) \end{aligned}$$

for all $z \in A$. Hence it follows that there is a sequence $\{y^m\}_{m\in\mathbb{N}} \subset A$ such that

$$d(x, y^m) < \epsilon + \frac{1}{m} \tag{1.24}$$

for all $m \in \mathbb{N}$. Thus $y^m \in U(x, \epsilon+1)$ for all $m \in \mathbb{N}$. Then $\{y^m\}_{m\in\mathbb{N}} \subset U(x, \epsilon+1)$ and $\{y^m\}_{m\in\mathbb{N}}$ are bounded sequences. Applying Theorem 1.9, we conclude that there is a subsequence $\{y^{m_k}\}_{k\in\mathbb{N}}$ of the sequence $\{y^m\}_{m\in\mathbb{N}}$ that converges to some $y \in \mathbb{R}^n$. Because A is a closed set of $\mathbb{R}^n$, we get that $y \in A$. By inequality (1.24) we have

$$d(x, y^{m_k}) < \epsilon + \frac{1}{m_k}, \quad k \in \mathbb{N}.$$

From this it follows that

$$d(x, y) \le d(x, y^{m_k}) + d(y^{m_k}, y)$$
$$< \epsilon + \frac{1}{m_k} + d(y^{m_k}, y)$$

and

$$d(x, y) \le \epsilon + \lim_{k \to \infty} \frac{1}{m_k} + \lim_{k \to \infty} d(y^{m_k}, y)$$
$$= \epsilon.$$

By the last inequality and inequality (1.24) we obtain (1.22). This completes the proof. $\square$

Example 1.30. Let $n = 3$ and

$$A = \{x = (x_1, x_2, x_3) \in \mathbb{R}^3 : x_1 = 3 + t, \ x_2 = 1 - t, \ x_3 = 2 + 2t, \ t \in \mathbb{R}\},$$
$$B = \{x = (x_1, x_2, x_3) \in \mathbb{R}^3 : x_1 = -t, \ x_2 = 2 + 3t, \ x_3 = 3t, \ t \in \mathbb{R}\}.$$

We will find the distance between the sets A and B. Let

$$x = (x_1, x_2, x_3) = (3 + t, \ 1 - t, \ 2 + 2t) \in A,$$
$$y = (y_1, y_2, y_3) = (-\tau, 2 + 3\tau, 3\tau) \in B, \quad t, \tau \in \mathbb{R},$$

be arbitrarily chosen. Then

$$
\begin{aligned}
(d(x, y))^2 &= (x_1 - y_1)^2 + (x_2 - y_2)^2 + (x_3 - y_3)^2 \\
&= (3 + t + \tau)^2 + (1 - t - 2 - 3\tau)^2 + (2 + 2t - 3\tau)^2 \\
&= (3 + t + \tau)^2 + (-1 - t - 3\tau)^2 + (2 + 2t - 3\tau)^2 \\
&= (3 + t + \tau)^2 + (1 + t + 3\tau)^2 + (2 + 2t - 3\tau)^2 \\
&= 9 + t^2 + \tau^2 + 6t + 6\tau + 2t\tau + 1 + t^2 + 9\tau^2 + 2t + 6\tau + 6t\tau \\
&\quad + 4 + 4t^2 + 9\tau^2 + 8t - 12\tau - 12t\tau \\
&= 19\tau^2 + 6t^2 - 4t\tau + 16t + 14 \\
&= (\sqrt{19}\tau)^2 - 2(\sqrt{19}\tau)\left(\frac{2}{\sqrt{19}}t\right) + \frac{4}{19}t^2 - \frac{4}{19}t^2 + 6t^2 + 16t + 14 \\
&= \left(\sqrt{19}\tau - \frac{2}{\sqrt{19}}t\right)^2 + \frac{110}{19}t^2 + 16t + 14 \\
&= \left(\sqrt{19}\tau - \frac{2}{\sqrt{19}}t\right)^2 + \left(\sqrt{\frac{110}{19}}t\right)^2 + 2\left(\sqrt{\frac{110}{19}}t\right)\left(8\sqrt{\frac{19}{110}}\right) + 64 \cdot \frac{19}{110} \\
&\quad - 64 \cdot \frac{19}{110} + 14
\end{aligned}
$$

$$= \left(\sqrt{19}\tau - \frac{2}{\sqrt{19}}t \right)^2 + \left(\sqrt{\frac{110}{19}}t + 8\sqrt{\frac{19}{110}} \right)^2$$
$$+ \frac{14 \cdot 110 - 64 \cdot 19}{110}$$

$$= \left(\sqrt{19}\tau - \frac{2}{\sqrt{19}}t \right)^2 + \left(\sqrt{\frac{110}{19}}t + 8\sqrt{\frac{19}{110}} \right)^2 + \frac{162}{55}.$$

Hence we conclude that $d(x,y)$ achieves its infimum when

$$\sqrt{19}\tau - \frac{2}{\sqrt{19}}t = 0,$$

$$\sqrt{\frac{110}{19}}t + 8\sqrt{\frac{19}{110}} = 0,$$

or

$$\tau = \frac{2}{19}t,$$

$$t = -\frac{76}{55},$$

or

$$\tau = -\frac{8}{55},$$

$$t = -\frac{76}{55}.$$

Consequently, $d(A,B) = \sqrt{\frac{162}{55}}$. It is achieved for

$$x = \left(3 - \frac{76}{55}, 1 + \frac{76}{55}, 2 - 2 \cdot \frac{76}{55} \right)$$
$$= \left(\frac{89}{55}, \frac{131}{55}, -\frac{42}{55} \right)$$

and

$$y = \left(\frac{8}{55}, 2 - \frac{24}{55}, -\frac{24}{55} \right)$$
$$= \left(\frac{8}{55}, \frac{86}{55}, -\frac{24}{55} \right).$$

Example 1.31. Let $n = 3$ and

$$A = \{ x = (x_1, x_2, x_3) : x_1 = x_2 = x_3 = t \in \mathbb{R} \},$$
$$B = \{ x = (x_1, x_2, x_3) : x_1 + x_2 = 1, \ x_3 = 0 \}.$$

We will find the distance between the sets A and B. Let

$$x = (x_1, x_2, x_3) = (t, t, t) \in A,$$
$$y = (y_1, y_2, y_3) = (\tau, 1 - \tau, 0) \in B, \quad t, \tau \in \mathbb{R}.$$

We have

$$
\begin{aligned}
(d(x,y))^2 &= (x_1 - y_1)^2 + (x_2 - y_2)^2 + (x_3 - y_3)^2 \\
&= (t - \tau)^2 + (t + \tau - 1)^2 + t^2 \\
&= t^2 - 2t\tau + \tau^2 + t^2 + \tau^2 + 1 + 2t\tau - 2t - 2\tau + t^2 \\
&= 2\tau^2 - 2\tau + 3t^2 - 2t + 1 \\
&= (\sqrt{2}\tau)^2 - 2(\sqrt{2}\tau)\left(\frac{1}{\sqrt{2}}\right) + \frac{1}{2} - \frac{1}{2} \\
&\quad + (\sqrt{3}t)^2 - 2(\sqrt{3}t)\left(\frac{1}{\sqrt{3}}\right) + \frac{1}{3} - \frac{1}{3} + 1 \\
&= \left(\sqrt{2}\tau - \frac{1}{\sqrt{2}}\right)^2 + \left(\sqrt{3}t - \frac{1}{\sqrt{3}}\right)^2 - \frac{5}{6} + 1 \\
&= \left(\sqrt{2}\tau - \frac{1}{\sqrt{2}}\right)^2 + \left(\sqrt{3}t - \frac{1}{\sqrt{3}}\right)^2 + \frac{1}{6}.
\end{aligned}
$$

Then $(d(x,y))^2$ achieves its infimum when

$$\sqrt{2}\tau - \frac{1}{\sqrt{2}} = 0,$$
$$\sqrt{3}t - \frac{1}{\sqrt{3}} = 0,$$

that is,

$$\tau = \frac{1}{2},$$
$$t = \frac{1}{3}.$$

Therefore $d(A, B) = \frac{1}{\sqrt{6}}$, and it is achieved for

$$x = \left(\frac{1}{3}, \frac{1}{3}, \frac{1}{3}\right),$$
$$y = \left(\frac{1}{2}, \frac{1}{2}, 0\right).$$

Example 1.32. Let $n > 3$ and

$$A = \{x = (x_1, \ldots, x_n) : x_j = t \in \mathbb{R}, \, j \in \{1, \ldots, n\}\},$$

$$B = \{x = (x_1, \ldots, x_n) : x_1 = t, \ x_2 = 1 - t, \ x_j = 0, \ j \in \{3, \ldots, n\}\}.$$

We will find the distance between A and B. Let

$$x = (x_1, \ldots, x_n) = (t, \ldots, t) \in A,$$
$$y = (y_1, \ldots, y_n) = (\tau, 1 - \tau, 0, \ldots, 0) \in B, \quad t, \tau \in \mathbb{R}.$$

We have

$$
\begin{aligned}
\left(d(x,y)\right)^2 &= (x_1 - y_1)^2 + (x_2 - y_2)^2 + \cdots + (x_n - y_n)^2 \\
&= (t - \tau)^2 + (t - 1 + \tau)^2 + t^2 + \cdots + t^2 \\
&= t^2 - 2t\tau + \tau^2 + t^2 + 1 + \tau^2 - 2t + 2t\tau - 2\tau + (n - 2)t^2 \\
&= 2\tau^2 - 2\tau + nt^2 - 2t + 1 \\
&= 2\left(\tau^2 - 2 \cdot \frac{1}{2} \cdot \tau + \frac{1}{4} - \frac{1}{4}\right) + (\sqrt{n}t)^2 - 2(\sqrt{n}t)\left(\frac{1}{\sqrt{n}}\right) + \frac{1}{n} - \frac{1}{n} + 1 \\
&= 2\left(\tau - \frac{1}{2}\right)^2 - \frac{1}{2} + \left(\sqrt{n}t - \frac{1}{\sqrt{n}}\right)^2 - \frac{1}{n} + 1 \\
&= 2\left(\tau - \frac{1}{2}\right)^2 + \left(\sqrt{n}t - \frac{1}{\sqrt{n}}\right)^2 + \frac{1}{2} - \frac{1}{n} \\
&= 2\left(\tau - \frac{1}{2}\right)^2 + \left(\sqrt{n}t - \frac{1}{\sqrt{n}}\right)^2 + \frac{n - 2}{2n}.
\end{aligned}
$$

Then $D(A, B)$ achieves its infimum when

$$\tau - \frac{1}{2} = 0,$$
$$\sqrt{n}t - \frac{1}{\sqrt{n}} = 0,$$

that is,

$$\tau = \frac{1}{2}$$
$$t = \frac{1}{n}.$$

Therefore $d(A, B) = \sqrt{\frac{n-2}{n}}$, and it is achieved for

$$x = \left(\frac{1}{\sqrt{n}}, \ldots, \frac{1}{\sqrt{n}}\right),$$
$$y = \left(\frac{1}{2}, \frac{1}{2}, 0, \ldots, 0\right).$$

Exercise 1.9. Let $n = 4$ and

$$A = \{x = (x_1, x_2, x_3, x_4) : x_1 = 1 + 2t, \; x_2 = -2t,$$
$$x_3 = 2 + 2t, \; x_4 = 2t, \; t \in \mathbb{R}\},$$
$$B = \{x = (x_1, x_2, x_3, x_4) : x_1 = 1, \; x_2 = t, \; x_3 = 1 + 2t,$$
$$x_4 = t, \; t \in \mathbb{R}\}.$$

Find $d(A, B)$.

Definition 1.28. A point $x \in \mathbb{R}^n$ is said to be a contour or boundary point of a set A if any neighborhood of A contains points of the sets A and A^c (see Fig. 1.17). The set of all contour points of the set A is called the contour or boundary of A and is denoted by ∂A.

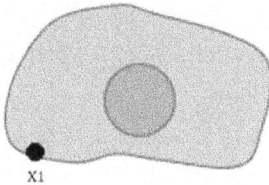

X1

Figure 1.17: Boundary point x_1 of a set.

A counterpoint then is neither an interior point (which is definitely in a set) nor an exterior point (which is definitely outside the set). Note that any contour point of A is an adherent point of A, but it is not an interior point of A. Therefore

$$\partial A \subset \overline{A} \quad \text{and} \quad \partial A \bigcap A_{\text{int}} = \emptyset.$$

Any point of the set A is an interior or contour point of A. Thus

$$A \subset \partial A \bigcup A_{\text{int}}.$$

Next, any point of the set $\overline{A}$ is an interior or contour point of A. Therefore

$$\overline{A} = A_{\text{int}} \bigcup \partial A.$$

Moreover, we have

$$\overline{A} = A \bigcup \partial A.$$

If A is a closed set, then $A \bigcap \partial A \neq \emptyset$. If A is an open set, then $A \bigcap \partial A = \emptyset$.

Theorem 1.22. *The set ∂A is a closed set.*

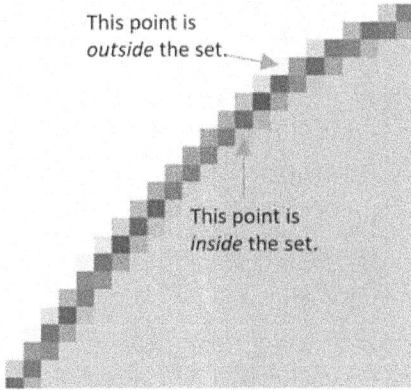

Proof. Let $x \in \mathbb{R}^n$ be an adherent point of ∂A. Then any neighborhood $U(x)$ of x contains at least one element of ∂A. We have the following two cases.

1. $\partial A \cap U(x) = \{x\}$. Then $x \in \partial A$.
2. There is an element $z \in \partial A \cap U(x)$ such that $z \neq x$. Since $z \in U(x)$ and $U(x)$ is an open set containing z, we have that $U(x)$ is a neighborhood of z. Because $z \in \partial A$, $U(x)$ contains points of A and A^c. Therefore x is a contour point of A, that is, $z \in \partial A$.

Because $x \in \mathbb{R}^n$ was an arbitrarily chosen adherent point of ∂A, we conclude that all points of ∂A are adherent, and thus ∂A is a closed set. This completes the proof. □

Example 1.33. Consider $U(0, 1)$. Then $\partial U(0, 1) = S^{n-1}(0, 1)$.

Definition 1.29. Let $f_j \in \mathcal{C}([a, b]), j \in \{1, \dots, n\}$. The map $f : [a, b] \to \mathbb{R}^n$ defined by

$$f(t) = (f_1(t), \dots, f_n(t)), \quad t \in [a, b],$$

is said to be a path in $\mathbb{R}^n$.

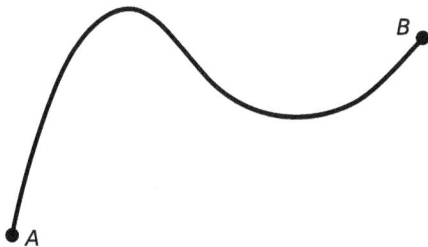

Figure 1.18: A Path in $\mathbb{R}^2$.

Definition 1.30. Two paths $f : [a, b] \to \mathbb{R}^n$ and $g : [\alpha, \beta] \to \mathbb{R}^n$ are said to be equivalent if there is a continuous monotonic function $\phi : [\alpha, \beta] \to [a, b]$ such that

$$f(\phi(\tau)) = g(\tau), \quad \tau \in [\alpha, \beta].$$

Any class of equivalent paths in $\mathbb{R}^n$ is said to be a curve in $\mathbb{R}^n$.

Definition 1.31. Let $x^0 = (x_1^0, \ldots, x_n^0) \in \mathbb{R}^n$ and $a_j \in \mathbb{R}, j \in \{1, \ldots, n\}$, be such that $\sum_{j=1}^n a_j^2 > 0$. The set of the points $x = (x_1, \ldots, x_n) \in \mathbb{R}^n$ for which

$$x_j = x_j^0 + a_j t, \quad t \in \mathbb{R}, \tag{1.25}$$

is said to be a straight line in $\mathbb{R}^n$ through the point x^0. The part of the straight line (1.25) lying in $[t_1, t_2]$ is called the segment in $\mathbb{R}^n$ with end points

$$x_j^1 = x_j^0 + a_j t_1,$$
$$x_j^2 = x_j^0 + a_j t_2, \quad j \in \{1, \ldots, n\}.$$

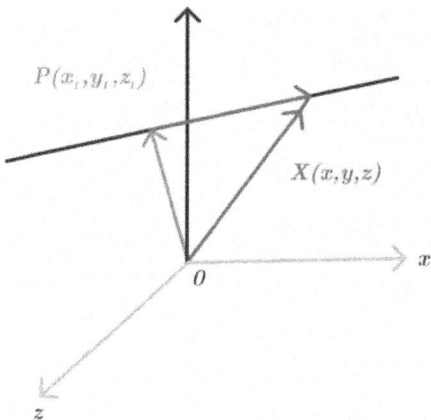

Figure 1.19: A straight line in $\mathbb{R}^3$.

Definition 1.32. The set A is said to be linearly connected if any two of its points can be connected with a curve lying in A.

Example 1.34. The set $[a, b]$ is a linearly connected set.

Example 1.35. The set $\{(0, -1), (-1, 3)\}$ is not a linearly connected set.

Theorem 1.23. *Let A be a linearly connected set in $\mathbb{R}^n$, and let $B \subset \mathbb{R}^n$ be such that*

$$A \bigcap B \neq \emptyset, \quad A \bigcap B^c \neq \emptyset.$$

Then

$$A \bigcap \partial B \neq \emptyset.$$

Proof. Let $x^1 \in A \bigcap B$ and $x^2 \in A \bigcap B^c$. Then $x^1, x^2 \in A$. Since A is a linearly connected set in $\mathbb{R}^n$, there is a curve $f : [a,b] \rightarrow \mathbb{R}^n$ connecting x^1 and x^2 and lying in A. We have

$$f(a) = x^1,$$
$$f(b) = x^2.$$

Let t_1 be the largest element of $[a,b]$ such that $f(t_1) \in B$. Then any neighborhood of $f(t_1)$ contains elements of B and B^c. Therefore $f(t_1) \in \partial B$. Hence $f(t_1) \in A \bigcap \partial B$. This completes the proof. $\qquad\square$

Definition 1.33. Any open linear connected set in $\mathbb{R}^n$ is said to be a domain in $\mathbb{R}^n$.

Example 1.36. Any open interval (a,b) is a domain in $\mathbb{R}$.

Example 1.37. Let (a,b) and (a_1, b_1) be such that $(a,b) \bigcap (a_1, b_1) = \emptyset$. Then the set

$$(a,b) \bigcup (a_1, b_1)$$

is not a domain in $\mathbb{R}$.

Example 1.38. For any $n \in \mathbb{N}$, $x \in \mathbb{R}^n$, and $r > 0$, the n-dimensional ball $U(x,r)$ is a domain in $\mathbb{R}^n$.

Definition 1.34. A set A is said to be convex if any two of its points can be connected with a segment lying in A.

Example 1.39. Any n-dimensional ball is convex.

Example 1.40. The set $(0,1) \bigcup (2,3)$ is not convex in $\mathbb{R}$.

Definition 1.35. Any set in $\mathbb{R}^n$ that is the closure of a domain in $\mathbb{R}^n$ is said to be a closed domain in $\mathbb{R}^n$.

Definition 1.36. Two sets B and C of $\mathbb{R}^n$ are said to be separable if no adherent points of B are adherent points of C.

Example 1.41. The intervals $(0,1)$ and $(2,3)$ are separable sets in $\mathbb{R}$.

Definition 1.37. The set A is said to be connected in $\mathbb{R}^n$ if it cannot be represented as a union of two separable sets.

Exercise 1.10. Determine if the set $A \subset \mathbb{R}^2$ is connected, linearly connected, open, and a domain, where
1. $A = \{x = (x_1, x_2) \in \mathbb{R}^2 : x_1^2 + x_2^2 > 1\}$.

2. $A = \{x = (x_1, x_2) \in \mathbb{R}^2 : x_1^2 + x_2^2 = 1\}$.
3. $A = \{x = (x_1, x_2) \in \mathbb{R}^2 : x_1^2 + x_2^2 \neq 1\}$.

1.5 Compact sets

A compact set is a fundamental concept in mathematics, particularly within the field of topology, denoting a set that is closed and bounded, ensuring that every open cover has a finite sub-cover. This characteristic enables various important results in analysis and topology, such as the Heine–Borel theorem. Understanding the properties of compact sets provides a solid foundation for deeper exploration into mathematical analysis and its applications.

Definition 1.38. A set $A \subset \mathbb{R}^n$ is said to be compact if any sequence of its elements contains a subsequence that converges to an element of A.

Example 1.42. The interval $[0, 1]$ is a compact set in $\mathbb{R}$.

Example 1.43. The rectangle

$$\{(x, y) \in \mathbb{R}^2 : -1 \leq x \leq 3, \ 0 \leq y \leq 5\}$$

is a compact set in $\mathbb{R}^2$.

Below we give a sufficient and necessary condition for compactness of a set.

Theorem 1.24. *A set $A \subset \mathbb{R}^n$ is a compact set if and only if it is closed and bounded.*

Proof. 1. Let A be a compact set in $\mathbb{R}^n$. Suppose that A is not bounded. Then, for any $m \in \mathbb{N}$, there is $x^m \in A$ such that

$$d(x^m, 0) > m.$$

Thus we get a sequence $\{x^m\}_{m \in \mathbb{N}}$ such that

$$\lim_{m \to \infty} x^m = \infty,$$

which is a contradiction because A is compact and from any sequence we can choose a convergent subsequence. Thus A is bounded. Now assume that A is not closed. Then there is a point $x^0 \in \mathbb{R}^n$ that is an adherent point of A and $x^0 \notin A$. By Theorem 1.15 it follows that there is a sequence $\{x^m\}_{m \in \mathbb{N}} \subset A$ such that

$$\lim_{m \to \infty} x^m = x^0.$$

Because A is compact, we get $x^0 \in A$. This is a contradiction. Therefore A is closed.

2. Let A be bounded and closed. Take an arbitrary sequence $\{x^m\}_{m\in\mathbb{N}} \subset A$. Since A is bounded, the sequence $\{x^m\}_{m\in\mathbb{N}}$ is bounded. Then there is a subsequence $\{x^{m_k}\}_{k\in\mathbb{N}}$ of the sequence $\{x^m\}_{m\in\mathbb{N}}$ that converges to a point $x^0 \in \mathbb{R}^n$. Because A is closed, we get that $x^0 \in A$. Therefore A is a compact set. This completes the proof. □

Definition 1.39. Let $x^k \in \mathbb{R}^n$ and $\epsilon_k > 0$, $k \in \mathbb{N}$. Set

$$Q_k = \overline{P(x^k, \delta_k)}, \quad k \in \mathbb{N}.$$

The sequence $\{Q_k\}_{k\in\mathbb{N}}$ is said to be a sequence of embedded closed cubes if

$$Q_1 \subset Q_2 \subset \cdots \subset Q_n \subset \cdots.$$

Theorem 1.25. *Let $\{Q_k\}_{k\in\mathbb{N}}$ be a sequence of embedded closed cubes such that*

$$\lim_{k\to\infty} \delta_k = 0.$$

Then there is a unique point $y \in \mathbb{R}^n$ such that $y \in Q_k$ for all $k \in \mathbb{N}$.

Proof. For any $k \in \mathbb{N}$ and $y \in Q_k$, we have

$$|y_j - x_j^k| \le \delta_k, \quad j \in \{1, \ldots, n\},$$

or

$$x_j^k - \delta_k \le y_j \le x_j^k + \delta_k, \quad j \in \{1, \ldots, n\}.$$

Thus the sequence of intervals $\{[x_j^k - \delta_k, x_j^k + \delta_k]\}_{k\in\mathbb{N}}$ is a sequence of embedded closed intervals for all $j \in \{1, \ldots, n\}$. Hence, for all $j \in \{1, \ldots, n\}$, there is a unique $y_j \in [x_j^k - \delta_k, x_j^k + \delta_k]$ for all $k \in \mathbb{N}$. Set

$$y = (y_1, \ldots, y_n).$$

Then $y \in Q_k$ for all $k \in \mathbb{N}$, and this point is unique. This completes the proof. □

Definition 1.40. The family

$$\Omega = \{A_\alpha : \alpha \in I\}$$

of sets $A_\alpha \subset \mathbb{R}^n$, where I is an index set, is said to be a cover of the set A if

$$A \subset \bigcup_{\alpha \in I} A_\alpha.$$

The cover Ω is said to be finite if it contains a finite number of elements A_α. If all sets A_α, $\alpha \in I$, are open, then the cover Ω is said to be an open cover of A.

The Heine–Borel theorem is significant in understanding the structure of bounded sets with applications to topology, measure theory, and functional analysis. The theorem is important in finding the best approximations for a given bounded set. This can be done by first defining an open subset of the Euclidean space and then applying the Heine–Borel theorem to decide that the said open subset is 'smaller' than the given closed set. The Heine–Borel theorem can be used in many real-life situations. It is used in problem solving, statistical analysis, approximation, and interpolation methods. The theorem is helpful in understanding the structure of various bounded sets and also gives a general method of proving the existence and uniqueness of a least upper bound and a greatest element in any arbitrary set of finite measure (length).

Theorem 1.26 (The Heine–Borel[4] theorem). *The set A is a compact set if and only if any of its open covers contains a finite open sub-cover.*

Proof. 1. Let A be a compact set, and let the family

$$\Omega = \{A_\alpha : \alpha \in I\}$$

be its open cover. Assume the contrary that Ω does not contain any finite open sub-cover of A. Because A is a compact set, by Theorem 1.24 it follows that A is closed and bounded. Hence there is a closed cube

$$Q = \{x = (x_1, \ldots, x_n) \in \mathbb{R}^n : a_j \le x_j \le a_j + b, \, j \in \{1, \ldots, n\}\}$$

that covers A for some $a_j \in \mathbb{R}, j \in \{1, \ldots, n\}$, and $b > 0$. We divide Q into 2^n equal closed cubes $Q_j, j \in \{1, \ldots, 2^n\}$ (in Fig. 1.20 the case $n = 2$ is shown). Then

$$Q = \bigcup_{j=1}^{2^n} Q_j.$$

Note that Ω is an open cover of any set $A \cap Q_j, j \in \{1, \ldots, 2^n\}$. By our assumption it follows that there is $j_1 \in \{1, \ldots, 2^n\}$ such that $A \cap Q_{j_1} \ne \emptyset$ and there is no finite sub-cover of Ω that covers $A \cap Q_{j_1}$. We divide Q_{j_1} into 2^n equal closed cubes $Q_{j_1 j}, j \in \{1, \ldots, 2^n\}$, and denote by $Q_{j_1 j_2}$ one of the cubes $Q_{j_1 j}, j \in \{1, \ldots, 2^n\}$, for which $A \cap Q_{j_1 j_2} \ne \emptyset$ and there is no finite sub-cover of Ω that covers $A \cap Q_{j_1 j_2}$. Continuing this way, we get the sequence

$$Q_{j_1} \supset Q_{j_1 j_2} \supset \cdots \supset Q_{j_1 j_2 \ldots j_k} \supset \cdots$$

---

4 Heinrich Eduard Heine (16 March 1821–21 October 1881) was a German mathematician. Heine became known for results on special functions and in real analysis. In particular, he authored an important treatise on spherical harmonics and Legendre functions (Handbuch der Kugelfunctionen). He also investigated basic hypergeometric series. He introduced the Mehler–Heine formula. Félix Édouard Justin Émile Borel (7 January 1871–3 February 1956) was a French mathematician and politician. As a mathematician, he was known for his founding work in the areas of measure theory and probability.

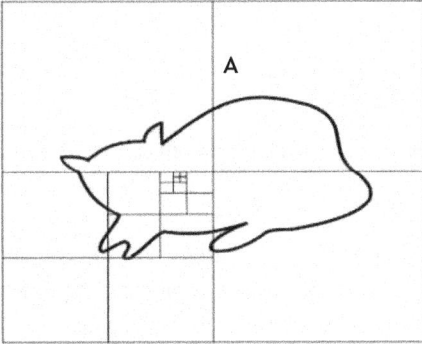

Figure 1.20: The case $n = 2$.

with lengths of their sides

$$\frac{b}{2}, \frac{b}{2^2}, \dots, \frac{b}{2^k}, \dots,$$

respectively. By Theorem 1.25 it follows that there is a unique

$$y \in \bigcap_{l=1}^{\infty} Q_{j_1 \dots j_l}.$$

Let $\epsilon > 0$ be arbitrarily chosen. Take $k \in \mathbb{N}$ such that

$$\frac{b\sqrt{n}}{2^k} < \epsilon,$$

which is possible because

$$\lim_{k \to \infty} \frac{b\sqrt{n}}{2^k} = 0.$$

For any $x \in Q_{j_1 \dots j_k}$, we have

$$d(x,y) \leq \frac{b\sqrt{n}}{2^k}$$
$$< \epsilon,$$

and thus $x \in U(y, \epsilon)$. Since $A \cap Q_{j_1 \dots j_l} \neq \emptyset$, $l \in \mathbb{N}$, in any neighborhood of y, there is at least one element of A. Thus y is an adherent point of A. Because A is closed, we have $y \in A$. Since Ω is a cover of A, there is $\alpha_0 \in I$ such that $y \in A_{\alpha_0}$. We have that A_{α_0} is an open set. Hence there is $\epsilon_1 > 0$ such that $U(y, \epsilon_1) \subset A_{\alpha_0}$, and there is $p \in \mathbb{N}$ such that

$$Q_{j_1 \dots j_p} \subset U(y, \epsilon_1).$$

Thus we get the chain

$$A \bigcap Q_{j_1...j_p} \subset Q_{j_1...j_p} \subset U(y, \epsilon_1) \subset A_{\alpha_0},$$

that is, the cover Ω contains a finite cover of the set $A \bigcap Q_{j_1...j_p}$, which contradicts the construction of the sets $Q_{j_1...j_l}$, $l \in \mathbb{N}$.

2. Suppose that any open cover of A contains a finite open sub-cover. Assume that A is not a compact set. Then there is a sequence $\{x^m\}_{m \in \mathbb{N}}$ that does not contain any convergent subsequence. Hence any point $x \in A$ is not a limit of any subsequence of the sequence $\{x^m\}_{m \in \mathbb{N}}$. Therefore, for any $x \in A$, there is a neighborhood $U(x)$ that contains at most finitely many elements of the sequence $\{x^m\}_{m \in \mathbb{N}}$. Hence

$$\Omega = \{U(x) : x \in A\}$$

is a cover of A. By our assumption there is a finite sub-cover

$$\Omega_0 = \{U(x_1), \ldots, U(x_l)\}$$

that covers A and any element of Ω_0 contains at least finitely many elements of the sequence $\{x^m\}_{m \in \mathbb{N}}$, which is impossible because Ω_0 contains all elements of $\{x^k\}_{k \in \mathbb{N}}$. So A is a compact set. This completes the proof. □

An application of the Heine–Borel theorem is the following result.

Theorem 1.27. *Let A be a compact set. Then for any open cover*

$$\Omega = \{A_\alpha : \alpha \in I\}$$

of A, there is a number $\eta > 0$, called the Lebesgue number, such that if $B \subset A$ and diam$B < \eta$, then there is $\alpha_0 \in I$ for which $B \subset A_{\alpha_0}$.

Proof. Suppose the contrary, that is, assume that there is an open cover

$$\Omega = \{A_\alpha : \alpha \in I\}$$

of the set A such that for any $\eta > 0$, there exists $B \subset A$ such that diam$B < \eta$ and $B \not\subset A_\alpha$. For $\eta = \frac{1}{m}$, $m \in \mathbb{N}$, let

$$B_m = \left\{x \in A : \text{diam}B_m < \frac{1}{m}\right\}$$

be such that $B_m \not\subset A_\alpha$ for all $\alpha \in I$. For any $m \in \mathbb{N}$, take arbitrary $x^m \in B_m$. Thus we have a sequence $\{x^m\}_{m \in \mathbb{N}}$ of elements of A. By the definition of a compact set it follows that there is a subsequence $\{x^{m_k}\}_{k \in \mathbb{N}}$ of the sequence $\{x^m\}_{m \in \mathbb{N}}$ that converges to some $x^0 \in A$, that is,

$$\lim_{k \to \infty} x^{m_k} = x^0. \tag{1.26}$$

Let $\alpha_1 \in I$ be such that $x^0 \in A_{\alpha_1}$. Such α_1 exists because Ω covers A and $x^0 \in A$. Since A_{α_1} is an open set and $x^0 \in A_{\alpha_1}$, there exists $\epsilon > 0$ such that

$$U(x^0, \epsilon) \subset A_{\alpha_1}.$$

Take $k_0 \in \mathbb{N}$ such that

$$\frac{1}{m_{k_0}} < \frac{\epsilon}{2}.$$

Now, using (1.26), we conclude that there exists $k \in \mathbb{N}$ such that $k > k_0$ and

$$|x^{m_k} - x^0| < \frac{\epsilon}{2}. \tag{1.27}$$

For $x \in B_{m_k}$, using $\mathrm{diam} B_{m_k} < \frac{1}{m_k}$ and (1.27), we have

$$\begin{aligned}
|x - x^0| &= |x - x^{m_k} + x^{m_k} - x^0| \\
&\leq |x - x^{m_k}| + |x^{m_k} - x^0| \\
&< \frac{1}{m_k} + \frac{\epsilon}{2} \\
&< \frac{\epsilon}{2} + \frac{\epsilon}{2} \\
&= \epsilon.
\end{aligned}$$

Thus

$$B_{m_k} \subset U(x^0, \epsilon) \subset A_{\alpha_1}.$$

This is a contradiction with our assumption. This completes the proof. □

Now we will give a criterion for the compactness of a set.

Theorem 1.28. *Let A be a compact set. For any $\eta > 0$, define the set*

$$A_\eta = \{x \in \mathbb{R}^n : d(x, A) \leq \eta\}.$$

Then A_η is a compact set.

Proof. Since A is a compact set, by Theorem 1.24 it follows that A is a bounded set. Then there is $\epsilon > 0$ such that

$$A \subset U(0, \epsilon). \tag{1.28}$$

Take arbitrary $x \in A_\eta$. By Theorem 1.21 it follows that there is $y \in A$ such that

$$d(x,y) = d(x,A)$$
$$\leq \eta.$$

By inclusion (1.28) it follows that

$$d(0,y) < \epsilon.$$

Then

$$d(0,x) \leq d(0,y) + d(x,y)$$
$$< \epsilon + \eta.$$

Therefore $x \in U(0, \epsilon + \eta)$. Since $x \in A_\eta$ was arbitrarily chosen and it is an element of $U(0, \epsilon + \eta)$, we arrive at the inclusion

$$A_\eta \subset U(0, \epsilon + \eta).$$

Consequently, A_η is a bounded set. Let now z be an adherent point of A_η. Then, for any $\epsilon_1 > 0$, there is $w \in A_\eta$ such that

$$d(z,w) < \epsilon_1.$$

By Theorem 1.21 it follows that there is $v \in A$ such that

$$d(w,v) = d(w,A)$$
$$\leq \eta.$$

Therefore

$$d(z,A) = \inf_{u \in A} d(z,u)$$
$$\leq d(z,v)$$
$$\leq d(z,w) + d(w,v)$$
$$< \epsilon_1 + \eta.$$

Since the last inequality is valid for all $\epsilon_1 > 0$, we find

$$d(z,A) \leq \eta.$$

Thus $z \in A_\eta$. Because z was an arbitrarily chosen adherent point of A_η and it is an element of A_η, we obtain that all elements of A_η are adherent points. Therefore A_η is closed. Now applying Theorem 1.24, we conclude that A_η is compact. This completes the proof. □

The diameter of a set is an important term, which is used in geometry and differential and integral calculus.

Definition 1.41. The number

$$\sup_{x,y \in A} d(x,y)$$

is called the diameter of A and is denoted by $\operatorname{diam} A$.

We have

$$0 \le \operatorname{diam} A \le \infty$$

and

$$\operatorname{diam} A = \operatorname{diam} \overline{A}.$$

Example 1.44. Consider $U(x^0, r)$. Let $x, y \in U(x^0, r)$. Then

$$d(x, x^0) < r,$$
$$d(y, x^0) < r.$$

Hence

$$d(x,y) \le d(x, x^0) + d(y, x^0)$$
$$< r + r \qquad\qquad (1.29)$$
$$= 2r.$$

For $k \in \mathbb{N}$, $\frac{1}{k} < r$, set

$$x^k = \left(x_1^0 - r + \frac{1}{k}, x_2^0, \ldots, x_n^0 \right),$$
$$y^k = \left(x_1^0 + r - \frac{1}{k}, x_2^0, \ldots, x_n^0 \right).$$

Then

$$d(x^k, x^0) = r - \frac{1}{k},$$
$$d(y^k, x^0) = r - \frac{1}{k}, \quad k \in \mathbb{N}, \quad \frac{1}{k} < r,$$

and

$$d(x^k, y^k) = 2r - \frac{2}{k}, \quad k \in \mathbb{N}, \quad \frac{1}{k} < r.$$

Hence

$$\lim_{k \to \infty} d(x^k, y^k) = 2r.$$

By the last equation and (1.29) we conclude that

$$\operatorname{diam} U(x^0, r) = 2r.$$

Exercise 1.11. Prove that $\operatorname{diam} P(x^0, \frac{a}{2}) = a\sqrt{n}$.

1.6 Multidimensional vector spaces

To a mathematician, the word space does not connote volume but instead refers to a set endowed with some structure. The Euclidean space $\mathbb{R}^n$ comes with two algebraic operations. The first is vector addition.

Definition 1.42. Let $x = (x_1, \ldots, x_n)$, $y = (y_1, \ldots, y_n) \in \mathbb{R}^n$. Then vector sum $x + y$ is defined as follows:

$$x + y = (x_1 + y_1, \ldots, x_n + y_n).$$

Note that the meaning of the "+" sign is now overloaded: on the left of the displayed equality, it denotes the new operation of vector addition, whereas on the right side, it denotes the old addition of real numbers. The multiple meanings of the plus sign should not cause problems since the meaning of "+" is clear from the context, i.e., its clear whether "+" sits between vectors or scalars. An expression such as

$$(1, 2, 3) + 4$$

with the plus sign between a vector and a scalar makes no sense according to our grammar.

The interpretation of vectors as arrows gives a geometric description of vector addition, at least in $\mathbb{R}^2$. To add vectors x and y in $\mathbb{R}^2$, draw them as arrows starting at 0 and then complete the parallelogram P that has x and y as two of its sides. The diagonal of P starting at 0 is then the arrow depicting the vector $x + y$ (see Fig. 1.21). The second operation in $\mathbb{R}^n$ is scalar multiplication.

Definition 1.43. Let $\alpha \in \mathbb{R}$ and $x = (x_1, \ldots, x_n) \in \mathbb{R}^n$. The scalar product of α and x is defined as follows:

$$\alpha x = (\alpha x_1, \ldots, \alpha x_n).$$

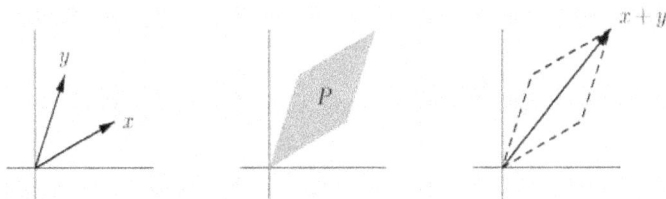

Figure 1.21: The parallelogram law of vector addition.

Then we define

$$x - ay = x + (-a)y \quad \text{and} \quad -x = (-x_1, \ldots, -x_n).$$

The scalar product of a vector x, viewed as an arrow, by a scalar a also has geometric interpretation. It simply stretches x by a factor a. When a is negative, ax turns x around and stretches it in the other direction by $|a|$ (see Fig. 1.22).

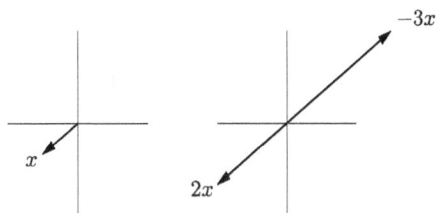

Figure 1.22: Scalar multiplication as stretching.

Example 1.45. Consider $\mathbb{R}^3$. Let

$$x = (-3, 4, 5), \quad y = (-2, 1, 3).$$

We will find

$$x + y, \quad x - 3y, \quad 4y.$$

We have

$$x + y = (-3 - 2, 4 + 1, 5 + 3)$$
$$= (-5, 5, 8).$$

Note that

$$(-3)y = ((-3) \cdot (-2), (-3) \cdot 1, (-3) \cdot 3)$$
$$= (6, -3, -9).$$

Hence

$$x - 3y = x + (-3)y$$
$$= (-3 + 6, 4 - 3, -9 + 5)$$
$$= (3, 1, -4).$$

Next,

$$4y = (4 \cdot (-2), 4 \cdot 1, 4 \cdot 3)$$
$$= (-8, 4, 12).$$

Example 1.46. Consider $\mathbb{R}^5$. Let

$$x = (-4; 1, 0, 3, 2), \quad y = (-1, 0, 2, 3, 1).$$

We will find

$$x - 5y \quad \text{and} \quad 2x + y.$$

We have

$$(-5)y = ((-5) \cdot (-1), (-5) \cdot 0, (-5) \cdot 2, (-5) \cdot 3, (-5) \cdot 1)$$
$$= (5, 0, -10, -15, -5)$$

and

$$2x = (2 \cdot (-4), 2 \cdot 1, 2 \cdot 0, 2 \cdot 3, 2 \cdot 2)$$
$$= (-8, 2, 0, 6, 4).$$

Hence

$$x - 5y = x + (-5)y$$
$$= (-4 + 5, 1 + 0, 0 - 10, 3 - 15, 2 - 5)$$
$$= (1, 1, -10, -12, -3)$$

and

$$2x + y = (-8 - 1, 2 + 0, 0 + 2, 6 + 3, 4 + 1)$$
$$= (-9, 2, 2, 9, 5).$$

Example 1.47. Consider $\mathbb{R}^2$. Let

$$x = (-2, 5), \quad y = (-3, 1).$$

We will find

$$3x - 4y.$$

We have

$$3x = (3 \cdot (-2), 3 \cdot 5)$$
$$= (-6, 15)$$

and

$$(-4)y = ((-4) \cdot (-3), (-4) \cdot 1)$$
$$= (12, -4).$$

Hence

$$3x - 4y = 3x + (-4)y$$
$$= (-6 + 12, 15 - 4)$$
$$= (6, 11).$$

Exercise 1.12. Consider $\mathbb{R}^4$. Let

$$x = (3, -1, 1, 0), \quad y = (2, 3, 1, 1).$$

Find
1. $x + y$.
2. $x - y$.
3. $2x - 4y$.

With these two operations and distinguished element 0, the Euclidean space $\mathbb{R}^n$ satisfies the following algebraic laws. All these algebraic laws are consequences of how "+" and "·" and 0 are defined for $\mathbb{R}^n$ in conjunction with the fact that the real numbers, in turn endowed with "+" and "·" and containing 0 and 1, satisfy field axioms.

Suppose that

$$x = (x_1, \ldots, x_n), \quad y = (y_1, \ldots, y_n), \quad z = (z_1, \ldots, z_n) \in \mathbb{R}^n,$$

and $a, \beta \in \mathbb{R}$. Then we have the following.
1. Addition is commutative:

$$x + y = y + x.$$

Proof. Indeed, using the commutative law for real numbers, we get

$$x + y = (x_1 + y_1, \ldots, x_n + y_n)$$
$$= (y_1 + x_1, \ldots, y_n + x_n)$$
$$= y + x.$$

This completes the proof. □

2. Addition is associative:

$$(x + y) + z = x + (y + z).$$

Proof. Indeed, using the associative law for real numbers, we find

$$(x + y) + z = (x_1 + y_1, \ldots, x_n + y_n) + (z_1, \ldots, z_n)$$
$$= ((x_1 + y_1) + z_1, \ldots, (x_n + y_n) + z_n)$$
$$= (x_1 + (y_1 + z_1), \ldots, x_n + (y_n + z_n))$$
$$= (x_1, \ldots, x_n) + (y_1 + z_1, \ldots, y_n + z_n)$$
$$= x + (y + z).$$

This completes the proof. □

3. 0 is the additive identity:

$$x + 0 = 0 + x = x.$$

Proof. Using the commutative law for real numbers and $0 + x_j = x_j + 0 = x_j, j \in \{1, \ldots, n\}$, we arrive at

$$x + 0 = (x_1 + 0, \ldots, x_n + 0)$$
$$= (x_1, \ldots, x_n)$$
$$= x$$
$$= (0 + x_1, \ldots, 0 + x_n)$$
$$= 0 + x.$$

This completes the proof. □

4. The existence of additive inverses:

$$x + (-x) = 0.$$

Proof. Using that $x_j + (-x_j) = 0, j \in \{1, \dots, n\}$, we find

$$x + (-x) = \left(x_1 + (-x_1), \dots, x_n + (-x_n)\right)$$
$$= (0, \dots, 0)$$
$$= 0.$$

This completes the proof. □

5. 1 is the multiplicative identity:

$$1 \cdot x = x.$$

Proof. Using that $1 \cdot x_j = x_j, j \in \{1, \dots, n\}$, we arrive at

$$1 \cdot x = (1 \cdot x_1, \dots, 1 \cdot x_n)$$
$$= (x_1, \dots, x_n)$$
$$= x.$$

This completes the proof. □

6. Scalar multiplication is associative:

$$\alpha(\beta x) = (\alpha\beta)x.$$

Proof. Using the multiplicative associative law, we find

$$\alpha(\beta x) = \alpha(\beta x_1, \dots, \beta x_n)$$
$$= \left(\alpha(\beta x_1), \dots, \alpha(\beta x_n)\right)$$
$$= \left((\alpha\beta)x_1, \dots, (\alpha\beta)x_n\right)$$
$$= (\alpha\beta)x.$$

This completes the proof. □

7. Scalar multiplication distributes over scalar addition:

$$(\alpha + \beta)x = \alpha x + \beta x.$$

Proof. Using the distributive law for real numbers, we find

$$(\alpha + \beta)x = \left((\alpha + \beta)x_1, \dots, (\alpha + \beta)x_n\right)$$
$$= (\alpha x_1 + \beta x_1, \dots, \alpha x_n + \beta x_n)$$
$$= (\alpha x_1, \dots, \alpha x_n) + (\beta x_1, \dots, \beta x_n)$$
$$= \alpha x + \beta x.$$

This completes the proof. □

8. Scalar multiplication distributes over vector addition:

$$a(x + y) = \alpha x + \beta y.$$

Proof. We have

$$
\begin{aligned}
a(x + y) &= a(x_1 + y_1, \ldots, x_n + y_n) \\
&= (a(x_1 + y_1), \ldots, a(x_n + y_n)) \\
&= (ax_1 + ay_1, \ldots, ax_n + ay_n) \\
&= (ax_1, \ldots, ax_n) + (ay_1, \ldots, ay_n) \\
&= ax + ay.
\end{aligned}
$$

This completes the proof. □

By the above properties it follows that $\mathbb{R}^n$ is a linear vector space. The standard basis of $\mathbb{R}^n$ is the set of vectors

$$\{e_1, e_2, \ldots, e_n\},$$

where

$$e_1 = (1, 0, \ldots, 0), \quad e_2 = (0, 1, \ldots, 0), \quad \ldots, \quad e_n = (0, \ldots, 0, 1).$$

Any element $x = (x_1, \ldots, x_n) \in \mathbb{R}^n$ can be represented in the form

$$
\begin{aligned}
x &= (x_1, x_2, \ldots, x_n) \\
&= (x_1, 0, \ldots, 0) + (0, x_2, \ldots, 0) + \cdots + (0, 0, \ldots, x_n) \\
&= x_1(1, 0, \ldots, 0) + x_2(0, 1, \ldots, 0) + \cdots + x_n(0, 0, \ldots, 1) \\
&= x_1 e_1 + x_2 e_2 + \cdots + x_n e_n,
\end{aligned}
$$

or shortly

$$x = \sum_{j=1}^{n} x_j e_j.$$

This equation shows that any $x \in \mathbb{R}^n$ can be expressed as a linear combination of the standard basis vectors. This expression is unique, for if also

$$x = \sum_{j=1}^{n} x_j' e_j$$

for some scalars $x'_j, j \in \{1, \ldots, n\}$, then the equality says that

$$x = (x'_1, x'_2, \ldots, x'_n),$$

so that $x'_j = x_j, j \in \{1, \ldots, n\}$. The standard basis is handy in that there is a finite set of vectors from which each of the vectors of $\mathbb{R}^n$ can be obtained in exactly one way as a linear combination. However, it is neither the only such set, nor the optimal one.

Definition 1.44. A set of vectors $\{f_j\}$ is a basis of $\mathbb{R}^n$ if any $x \in \mathbb{R}^n$ is uniquely expressible as a linear combination of the f_j.

Example 1.48. We will show that the set

$$\{f_1 = (1, 1), f_2 = (1, -1)\}$$

is a basis of $\mathbb{R}^2$. To see this, take arbitrary $(x, y) \in \mathbb{R}^2$. This vector is expressible as a linear combination of f_1 and f_2 if and only if there are scalars a and b such that

$$(x, y) = a(1, 1) + b(1, -1),$$

which is equivalent to a pair of scalar equations

$$x = a + b,$$
$$y = a - b.$$

We find

$$a = \frac{x + y}{2},$$
$$b = \frac{x - y}{2},$$

and

$$(x, y) = \frac{x + y}{2}(1, 1) + \frac{x - y}{2}(1, -1).$$

These scalars are uniquely determined by the vector (x, y). Thus the set $\{f_1, f_2\}$ is a basis of $\mathbb{R}^2$.

Example 1.49. Consider the set

$$\{f_1 = (1, 2), f_2 = (2, 4)\}.$$

This set is not a basis of $\mathbb{R}^2$ since any linear combination $af_1 + bf_2$ is

$$af_1 + bf_2 = (a + 2b, 2a + 4b),$$

with the second entry equal to two times the first entry. Thus the vector $(1,0)$ is not a linear combination of f_1 and f_2.

The geometric notions of length and angle in $\mathbb{R}^n$ are described in terms of the algebraic notion of inner product.

Definition 1.45. In the linear vector space $\mathbb{R}^n$, we define the inner product $\langle x, y \rangle$ of the elements $x = (x_1, \ldots, x_n), y = (y_1, \ldots, y_n) \in \mathbb{R}^n$ in the following way:

$$\langle x, y \rangle = \sum_{j=1}^{n} x_j y_j.$$

Example 1.50. Let $n = 2$ and

$$x = (-1, 3),$$
$$y = (4, 5).$$

Then

$$\langle x, y \rangle = (-1) \cdot 4 + 3 \cdot 5$$
$$= -4 + 15$$
$$= 11.$$

Example 1.51. Let $n = 3$ and

$$x = \left(-\frac{1}{2}, 2, 4 \right),$$
$$y = (4, -1, 3).$$

Then

$$\langle x, y \rangle = \left(-\frac{1}{2} \right) \cdot 4 + 2 \cdot (-1) + 4 \cdot 3$$
$$= -2 - 2 + 12$$
$$= 8.$$

Example 1.52. Let $n = 4$ and

$$x = \left(-\frac{1}{3}, -\frac{1}{2}, \frac{1}{8}, 1 \right),$$
$$y = (6, 4, -8, -2).$$

Then

$$\langle x,y \rangle = \left(-\frac{1}{3}\right)\cdot 6 + \left(-\frac{1}{2}\right)\cdot 4 + \frac{1}{8}\cdot(-8) + 1\cdot(-2)$$
$$= -2 - 2 - 1 - 2$$
$$= -7.$$

Exercise 1.13. Find $\langle x,y \rangle$, where
1. $n = 3$,

$$x = (-1, 3, 5),$$
$$y = (-4, 1, 0).$$

2. $n = 4$,

$$x = \left(-\frac{1}{2}, 0, 1, 0\right),$$
$$y = (2, -1, 2, 1).$$

3. $n = 5$,

$$x = (-1, 1, -1, 1, -1),$$
$$y = (2, -1, 0, 1, 3).$$

Example 1.53. We will find the parameter $a \in \mathbb{R}$ such that $\langle x,y \rangle = 11$, where

$$x = (a, 1, -1),$$
$$y = (a - 1, 2, 3).$$

We have

$$\langle x,y \rangle = a(a - 1) + 2 - 3$$
$$= a^2 - a - 1$$
$$= 11.$$

Hence

$$a^2 - a - 12 = 0.$$

Thus

$$a_{1,2} = \frac{1 \pm \sqrt{49}}{2}$$
$$= \frac{1 \pm 7}{2},$$

or

$$a_1 = 4, \quad a_2 = -3.$$

Example 1.54. We will find the parameters a and b such that

$$\langle x, y \rangle = 4,$$
$$x - y = (-b, a + 1, 1),$$

where

$$x = (a, -1, 2),$$
$$y = (-2, b, 1).$$

We have

$$\langle x, y \rangle = a \cdot (-2) + (-1) \cdot b + 2$$
$$= -2a - b + 2$$
$$= 4$$

and

$$x - y = (a + 2, -1 - b, 1)$$
$$= (-b, a + 1, 1).$$

Thus we get the system

$$-2a - b = 2,$$
$$a + b = -2.$$

Therefore

$$a = 0 \quad \text{and} \quad b = -2.$$

Example 1.55. We will find the parameters $a, b \in \mathbb{R}$ such that

$$\langle x, y \rangle = -1, \tag{1.30}$$

where

$$x = (a, -1, 2),$$
$$y = \left(b, -ab, -\frac{1}{2} \right).$$

We have

$$\langle x, y \rangle = a \cdot (-b) + (-1) \cdot (-ab) + 2 \cdot \left(-\frac{1}{2}\right)$$
$$= -ab + ab - 1$$
$$= -1.$$

Thus (1.30) holds for all $a, b \in \mathbb{R}$.

Exercise 1.14. Find $a \in \mathbb{R}$ such that $\langle x, y \rangle = -4$, where

$$x = (a + 1, -1, 1),$$
$$y = (a - 2, a, -1).$$

Exercise 1.15. Find the parameters $a, b \in \mathbb{R}$ such that

$$\langle x, y \rangle = -1,$$
$$3x - y = (-1, -2, -5),$$

where

$$x = (a, -b, -1),$$
$$y = (b, -1, 2).$$

Let $x, y, z \in \mathbb{R}^n$ and $\lambda \in \mathbb{R}$. Then the inner product has the following properties.
1. The inner product is positive definite:

$$\langle x, x, \rangle \geq 0, \quad \text{and} \quad \langle x, x \rangle = 0 \quad \text{if and only if} \quad x = 0.$$

Proof. We have

$$\langle x, x \rangle = \sum_{j=1}^{n} x_j^2$$
$$\geq 0$$

and

$$\langle x, x \rangle = 0$$

if and only if

$$\sum_{j=1}^{n} x_j^2 = 0$$

if and only if $x_j = 0, j \in \{1, \ldots, n\}$. This completes the proof. $\square$

2. The inner product is symmetric:

$$\langle x, y \rangle = \langle y, x \rangle.$$

Proof. We have

$$\langle x, y \rangle = \sum_{j=1}^{n} x_j y_j$$

$$= \sum_{j=1}^{n} y_j x_j$$

$$= \langle y, x \rangle.$$

This completes the proof. $\square$

3. The inner product is bilinear:

$$\langle x + y, z \rangle = \langle x, z \rangle + \langle y, z \rangle,$$
$$\langle x, y + z \rangle = \langle x, y \rangle + \langle x, z \rangle$$

and

$$\langle \lambda x, y \rangle = \lambda \langle x, y \rangle,$$
$$\langle x, \lambda y \rangle = \lambda \langle x, y \rangle.$$

Proof. We have

$$x + y = (x_1 + y_1, \dots, x_n + y_n)$$

and

$$\langle x + y, z \rangle = \sum_{j=1}^{n} (x_j + y_j) z_j$$

$$= \sum_{j=1}^{n} (x_j z_j + y_j z_j)$$

$$= \sum_{j=1}^{n} x_j z_j + \sum_{j=1}^{n} y_j z_j$$

$$= \langle x, z \rangle + \langle y, z \rangle.$$

As above, we can prove that

$$\langle x, y + z \rangle = \langle x, y \rangle + \langle x, z \rangle.$$

Next, we have

$$\lambda x = (\lambda x_1, \dots, \lambda x_n),$$
$$\lambda y = (\lambda y_1, \dots, \lambda y_n),$$

and

$$\langle \lambda x, y \rangle = \sum_{j=1}^{n} (\lambda x_j) y_j$$
$$= \sum_{j=1}^{n} \lambda (x_j y_j)$$
$$= \lambda \sum_{j=1}^{n} x_j y_j$$
$$= \sum_{j=1}^{n} x_j (\lambda y_j)$$
$$= \langle x, \lambda y \rangle.$$

We leave to the reader the proof of the equality

$$\langle x, \lambda y \rangle = \lambda \langle x, y \rangle$$

as an exercise. This completes the proof. □

The length of a vector can be represented in terms of the inner product:

$$|x| = \sqrt{\langle x, x \rangle}.$$

If $x = (x_1, \dots, x_n), y = (y_1, \dots, y_n) \in \mathbb{R}^n$, then the Cauchy–Schwarz inequality can be written in terms of the inner product and modulus as follows:

$$|\langle x, y \rangle| \le |x||y|.$$

Note that the absolute value signs mean different things the sides of the Cauchy–Schwarz inequality. On the left side, the quantities x and y are vectors, their inner product $\langle x, y \rangle$ is a scalar, and $|\langle x, y \rangle|$ is its scalar absolute value, whereas on the right side, $|x|$ and $|y|$ are the scalar absolute values of vectors, and $|x||y|$ is their product. Thus the Cauchy–Schwarz inequality says: the size of the inner product of two vectors is at most the product of their sizes.

Definition 1.46. Two vectors $x, y \in \mathbb{R}^n$ are said to be orthogonal, denoted $x \perp y$, if $\langle x, y \rangle = 0$.

Example 1.56. Let

$$x = (-1, 1, 2, 3),$$
$$y = \left(1, 2, -1, \frac{1}{3}\right).$$

We have

$$\langle x, y \rangle = -1 \cdot 1 + 1 \cdot 2 + 2 \cdot (-1) + 3 \cdot \frac{1}{3}$$
$$= 0,$$

and thus $x \perp y$.

Example 1.57. Let

$$x = (-1, 1, 3),$$
$$y = (2, 1, -1).$$

Then

$$\langle x, y \rangle = -1 \cdot 2 + 1 \cdot 1 + 3 \cdot (-1)$$
$$= -2 + 1 - 3$$
$$= -4$$
$$\neq 0,$$

and thus $x \not\perp y$.

Example 1.58. Let

$$x = (a, -1, 2),$$
$$y = (a - 1, 1, a),$$

where $a \in \mathbb{R}$ is such that $\langle x, y \rangle = 0$. Let us find a. We have

$$0 = \langle x, y \rangle$$
$$= a(a - 1) - 1 \cdot 1 + 2 \cdot a$$
$$= a^2 - a - 1 + 2a$$
$$= a^2 + a - 1,$$

and thus

$$a = \frac{-1 \pm \sqrt{5}}{2}.$$

Exercise 1.16. Find $a \in \mathbb{R}$ such that the vectors x and $x + ay$ are orthogonal, where

$$x = (1, 2, 1, 3),$$
$$y = (4, 1, 1, 1).$$

Definition 1.47. Let $x, y \in \mathbb{R}^n$, $x, y \neq 0$. The angle $\phi \in [0, \pi]$ for which

$$\cos \phi = \frac{\langle x, y \rangle}{|x||y|}$$

is said to be the angle between the vectors x and y. Sometimes, it is denoted by $\angle(x, y)$.

By the Cauchy–Schwarz inequality we get

$$
\begin{aligned}
|\cos(\angle(x, y))| &= \left| \frac{\langle x, y \rangle}{|x||y|} \right| \\
&= \frac{|\langle x, y \rangle|}{|x||y|} \\
&\leq \frac{|x||y|}{|x||y|} \\
&= 1.
\end{aligned}
$$

Thus

$$-1 \leq \cos(\angle(x, y)) \leq 1.$$

Example 1.59. Let

$$x = (-1, 1, 2),$$
$$y = (1, 1, 1).$$

Then

$$
\begin{aligned}
\langle x, y \rangle &= (-1) \cdot 1 + 1 \cdot 1 + 2 \cdot 1 \\
&= -1 + 1 + 2 \\
&= 2, \\
|x| &= \sqrt{(-1)^2 + 1^2 + 2^2} \\
&= \sqrt{1 + 1 + 4} \\
&= \sqrt{6}, \\
|y| &= \sqrt{1^2 + 1^2 + 1^2} \\
&= \sqrt{1 + 1 + 1} \\
&= \sqrt{3},
\end{aligned}
$$

and

$$\cos(\angle(x,y)) = \frac{2}{\sqrt{6}\sqrt{3}}$$
$$= \frac{\sqrt{2}}{3}.$$

Exercise 1.17. Let

$$x = (-1,1,0,2),$$
$$y = (1,-1,1,1).$$

Find $\cos(\angle(x,y))$.

Definition 1.48. Let $x \in \mathbb{R}^n$ and $|x| = 1$. Set

$$\cos a_j = \langle x, e_j\rangle, \quad j \in \{1,\ldots,n\}.$$

The cosines $\cos a_j, j \in \{1,\ldots,n\}$, are said to be the directional cosines of the vector x.

We have

$$x = \cos a_1 e_1 + \cdots + \cos a_n e_n.$$

Since $|x| = 1$, we obtain the equality

$$(\cos a_1)^2 + \cdots + (\cos a_n)^2 = 1.$$

1.7 Advanced practical problems

Problem 1.1. Find $d(x,y)$, where
1. $n = 3$ and

$$x = (-1,0,1),$$
$$y = (0,-3,4).$$

2. $n = 2$ and

$$x = (-3,1),$$
$$y = (-1,2).$$

3. $n = 5$ and

$$x = (-1,0,1,2,1),$$
$$y = (3,1,-1,1,1).$$

Problem 1.2. Let $n = 4$, and let

$$x = (-1, 0, a, 1),$$
$$y = (a, -1, 1, 2).$$

Find $a \in \mathbb{R}$ such that
1. $d(x, y) = 3$.
2. $d(x, y) = 5$.
3. $d(x, y) = 9$.

Problem 1.3. Find $\lim_{m \to \infty} x^m$, where
1.

$$\{x^m\}_{m \in \mathbb{N}} = \left\{ \left(m^{\frac{4}{3}} \left(\sqrt[3]{m^2 + 1} - \sqrt[3]{m^2 - 1} \right), \frac{2m + 1}{2m + 3}, 1 \right) \right\}_{m \in \mathbb{N}}.$$

2.

$$\{x^m\}_{m \in \mathbb{N}} = \left\{ \left(\frac{a_m^2 - 6a_m + 8}{a_m^2 - 5a_m + 4}, \frac{b_m^4 + 2b_m^2 - 3}{b_m^2 - 3b_m + 2} \right) \right\}_{m \in \mathbb{N}},$$

where $\lim_{m \to \infty} a_m = 4$, $a_m \neq 1, 4$, $m \in \mathbb{N}$, $\lim_{m \to \infty} b_m = 1$, $b_m \neq 1, 2$, $m \in \mathbb{N}$.

3.

$$\{x^m\}_{m \in \mathbb{N}} = \left\{ \left(\sum_{j=0}^{m} \frac{1}{(2j + 1)(2j + 3)}, \frac{1}{m^2} \sum_{j=1}^{m} j, \frac{1}{1 - a_m} - \frac{3}{1 - a_m^3} \right) \right\}_{m \in \mathbb{N}},$$

where $\lim_{m \to \infty} a_m = 1$, $a_m \neq 1$, $m \in \mathbb{N}$.

4.

$$\{x^m\}_{m \in \mathbb{N}} = \left\{ \left(\sum_{j=0}^{m} \frac{1}{(2j + 1)(2j + 3)(2j + 5)}, \prod_{j=2}^{m} \frac{j^3 - 1}{j^3 + 1}, \prod_{j=2}^{m} \left(1 - \frac{1}{1 + 2 + \cdots + j} \right) \right) \right\}_{m \in \mathbb{N}}.$$

5.

$$\{x^m\}_{m \in \mathbb{N}} = \left\{ \left(\frac{1}{m^3} \sum_{j=1}^{m} j^2, \frac{\sqrt[3]{a_m} + 1}{\sqrt[5]{a_m} + 1}, \left(1 - \frac{2}{m} \right)^m, \left(\frac{m^2 - 4m + 3}{m^2 + 3m + 2} \right) \right) \right\}_{m \in \mathbb{N}},$$

where $\lim_{m \to \infty} a_m = -1$, $a_m \neq -1$, $m \in \mathbb{N}$.

Problem 1.4. Prove that $\lim_{m \to \infty} x^m = \infty$, where
1.

$$\{x^m\} = \left\{ \left(m + 1, e^m, \frac{1}{m + 2} \right) \right\}_{m \in \mathbb{N}}.$$

2.

$$\{x^m\} = \{ (m^2, e^{-m}, m + 1, e^{-m^3}) \}_{m \in \mathbb{N}}.$$

3.

$$\{x^m\} = \left\{ \left(\frac{1}{m + 1}, \frac{2}{1 + m^2}, e^{m^2} \right) \right\}_{m \in \mathbb{N}}.$$

4.
$$\{x^m\} = \{(e^{-m^2}, 1, 2m)\}_{m\in\mathbb{N}}.$$

5.
$$\{x^m\} = \left\{\left(e^{\frac{m^2+1}{m+2}}, -m, \frac{1}{m+2}, 3, 8\right)\right\}_{m\in\mathbb{N}}.$$

Problem 1.5. Find the adherent points of the set
$$B = \left\{x = (x_1, x_2) \in \mathbb{R}^2 : x_2 = \sin\left(\frac{1}{x_1}\right)\right\}$$
that do not belong to B.

Problem 1.6. Construct a set whose points are isolated but the set of its limit points is not empty.

Problem 1.7. Prove that the set of all isolated points of a set A is at most countable.

Problem 1.8. Construct a set $A \subset \mathbb{R}^n$ satisfying the following conditions:
1. All points of A are isolated.
2. The set A has no any limit points.
3. $\inf_{x,y\in A} d(x,y) = 0$.

Problem 1.9. Let $A \subset B \subset \mathbb{R}^n$. Prove that $\overline{A} \subset \overline{B}$.

Problem 1.10. Let $f \in C(\mathbb{R})$, and let $y_0, z_0 \in \mathbb{R}$ be such that $z_0 < y_0$. Prove that the following sets of $\mathbb{R}$ are open:
1.
$$\{x \in \mathbb{R} : f(x) < y_0\}.$$

2.
$$\{x \in \mathbb{R} : z_0 < f(x) < y_0\}.$$

Problem 1.11. Let $f \in C(\mathbb{R})$ and $y_0 \in \mathbb{R}$. Prove that the sets
$$B = \{x \in \mathbb{R} : f(x) \leq y_0\},$$
$$C = \{x \in \mathbb{R} : f(x) \geq y_0\}$$
are closed sets.

Problem 1.12. Let $f \in C(\mathbb{R})$ and $y_0, z_0 \in \mathbb{R}$, $z_0 \leq y_0$. Prove that
$$B = \{x \in \mathbb{R} : z_0 \leq f(x) \leq y_0\}$$
is a closed set.

Problem 1.13. Let $F \subset \mathbb{R}$ be a closed set, and let $f \in C(\mathbb{R})$. Prove that $f(F)$ is a closed set of $\mathbb{R}$.

Problem 1.14. Let $\{\epsilon_n\}_{n\in\mathbb{N}}$ be a sequence of real numbers such that

$$0 < \epsilon_1 < \epsilon_2 < \cdots < \epsilon_n < \cdots.$$

Check if
1. $\sum_{k=1}^{\infty} Q^n(x, \epsilon_k)$ is an open set.
2. $\sum_{k=1}^{\infty} Q^n(x, \epsilon_k)$ is a closed set.

Problem 1.15. Let $\{\epsilon_n\}_{n\in\mathbb{N}}$ be a sequence of real numbers such that

$$0 < \epsilon_1 < \epsilon_2 < \cdots < \epsilon_n < \cdots.$$

Check if $\sum_{k=1}^{\infty} S^{n-1}(x, \epsilon_k)$ is a closed set.

Problem 1.16. Let $A_j \subset \mathbb{R}^n, j \in \mathbb{N}$. Prove
1. $\overline{\bigcup_{j=1}^{m} E_j} = \bigcup_{j=1}^{m} \overline{E_j}$.
2. $\overline{\bigcup_{j=1}^{\infty} E_j} \supset \bigcup_{j=1}^{\infty} \overline{E_j}$.

Problem 1.17. Give an example of a closed set different from the closure of the set of all its interior points.

Problem 1.18. Give an example of an open set different from the set of all interior points of its closure.

Problem 1.19. Find $d(A, B)$ if
1.

$$A = \{(1, \ldots, 1)\},$$
$$B = \{(x_1, 0, \ldots, 0)\}.$$

2.

$$A = \{x = (x_1, x_2) \in \mathbb{R}^2 : x_2 = x_1^2\},$$
$$B = \{x = (x_1, x_2) : x_2 = x_1 - 2\}.$$

3.

$$A = \{x = (x_1, x_2) \in \mathbb{R}^2 : x_1^2 + 4x_2^2 = 4\},$$
$$B = \{(x_1, x_2) \in \mathbb{R}^2 : x_1 + 2\sqrt{3}x_2 = 8\}.$$

Problem 1.20. Determine if the set $A \subset \mathbb{R}^2$ is connected, linear connected, open, and a domain, where
1. $A = \{x = (x_1, x_2) \in \mathbb{R}^2 : x_1^2 + x_2^2 = 0\}$.
2. $A = \{x = (x_1, x_2) \in \mathbb{R}^2 : x_1^2 + x_2^2 < 1\} \bigcup \{x = (x_1, x_2) \in \mathbb{R}^2 : (x_1 - 2)^2 + x_2^2 < 1\}$.
3. $A = \{x = (x_1, x_2) \in \mathbb{R}^2 : x_1^2 + x_2^2 \le 1\} \bigcup \{x = (x_1, x_2) \in \mathbb{R}^2 : (x_1 - 2)^2 + x_2^2 < 1\}$.

Problem 1.21. Check if the set A is a domain in $\mathbb{R}^3$, where
1. $A = \{x = (x_1, x_2, x_3) \in \mathbb{R}^3 : x_1^2 + x_2^2 < 1, x_3 = 0\}$.

2. $A = \{x = (x_1, x_2, x_3) \in \mathbb{R}^3 : x_1^2 + 2x_2^2 + 3x_3^2 < 4\}$.
3. $A = \{x = (x_1, x_2, x_3) \in \mathbb{R}^3 : x_1^2 + x_2^2 - x_3^2 < 1\}$.

Problem 1.22. Find diam A, where
1. $A = \{x = (x_1, x_2, x_3) \in \mathbb{R}^3 : x_1^2 + x_2^2 + x_3^2 < 9\}$.
2. $A = \{x = (x_1, x_2) \in \mathbb{R}^2 : 4x_1^2 - 3x_2^2 = 2\}$.

Problem 1.23. A simplex in $\mathbb{R}^n$ is a set of the form

$$\left\{ x \in \mathbb{R}^n : x = \sum_{j=1}^{n+1} a_j a^j, \ a_j \geq 0, \ j \in \{1, \ldots, n\}, \ \sum_{j=1}^{n+1} a_j = 1 \right\},$$

where $a^j \in \mathbb{R}^n, j \in \{1, \ldots, n\}$. Prove that any simplex is a convex set.

Problem 1.24. Consider $\mathbb{R}^4$. Find $d(x, y)$, where
1. $x = (2, 4, 2, 4, 2), y = (6, 4, 4, 4, 6)$.
2. $x = (2, 4, 2, 4, 2), y = (5, 7, 5, 7, 2)$.
3. $x = (6, 4, 4, 4, 6), y = (5, 7, 5, 7, 2)$.

Problem 1.25. Consider $\mathbb{R}^5$. Find $d(x, y)$, where

$$x = (1, 2, 3, 2, 1), \quad y = (3, 4, 0, 4, 3).$$

Problem 1.26. Consider $\mathbb{R}^4$. Find $a \in \mathbb{R}$ such that $d(x, y) = \sqrt{15}$, where

$$x = (-2, -1, a, 1), \quad y = (-1, 1, 0, 2).$$

Problem 1.27. Consider $\mathbb{R}^3$. Let

$$x = (-3, 4, 1), \quad y = (-1, 2, 1).$$

Find
1. $x + y$.
2. $x - y$.
3. $3x - y$.

Problem 1.28. Find $\langle x, y \rangle$, where
1. $n = 2$,

$$x = (-4, -1),$$
$$y = \left(\frac{1}{4}, 0 \right).$$

2. $n = 5$,

$$x = (-1, 2, 1, 0, 1),$$
$$y = (-3, 1, -1, 1, 1).$$

3. $n = 3$,

$$x = \left(-\frac{1}{6}, 2, -1\right),$$
$$y = \left(-\frac{1}{3}, 1, 1\right).$$

Problem 1.29. Find $a \in \mathbb{R}$ such that $\langle x, y \rangle = 23$, where

$$x = (2a, a + 1, -1),$$
$$y = (a, a - 1, 3).$$

Problem 1.30. Find $a, b \in \mathbb{R}$ such that

$$\langle x, y \rangle = -6,$$
$$x + y = (4, 0, 1),$$

where

$$x = (a, b, -1),$$
$$y = (a, -b, 2).$$

Problem 1.31. Find $a \in \mathbb{R}$ such that the vectors x and $x + ay$ are orthogonal, where

$$x = (1, 2, \dots, n),$$
$$y = (n, n - 1, \dots, 1), \quad n > 1.$$

Problem 1.32. Find $a \in \mathbb{R}$ such that $|x| = 6$, where

$$x = (a - 2, 2, 2, 0, a).$$

Problem 1.33. Let

$$x = (1, -1, -1, 4),$$
$$y = (2, 1, -1, 0).$$

Find $\cos(\angle(x, y))$.

2 Limit and continuous functions of several variables

In this chapter, functions of several variables are introduced. Limits of functions of several variables are defined, and some of their properties are deduced. Continuous functions of several variables are defined and investigated. Properties of continuous functions on compact sets are deduced. Uniform continuity is introduced and explored.

2.1 Definition for a function of several variables

Multivariable functions are fundamental in machine learning algorithms. They are used to model complex relationships in data where multiple input features influence the output. Functions of several variables play a key role in regression analysis, classification, and clustering tasks. In graphics programming, functions of multiple variables are employed to define and manipulate three-dimensional objects. For example, functions may represent surfaces or volumes, and their properties are crucial in rendering realistic graphics. Functions of several variables are extensively used in optimization problems. In computer science, optimization algorithms aim to find the best solution based on various criteria. These algorithms often involve optimizing a multivariable function representing the objective or cost. In the analysis of computer networks, functions of several variables can model complex relationships between network parameters. This is essential for understanding network performance, optimizing routing algorithms, and detecting anomalies. Functions of multiple variables play an important role in database design and query optimization. They are used to model relationships between tables and aid in the optimization of queries involving multiple parameters. In computer-aided design software, functions of several variables are used to represent and manipulate three-dimensional shapes. This is crucial for architects, engineers, and designers working on complex projects. Functions of several variables are applied in image and signal processing to represent and analyze multidimensional data. For instance, functions may be used to filter images, compress data, or perform transformations in various domains. In computer vision applications, functions of multiple variables are employed to model the relationships between different visual features. This is integral to tasks such as object recognition, tracking, and scene understanding. Computer simulations often involve functions with multiple variables to model the behavior of complex systems. This is applicable in fields like physics simulations, environmental modeling, and financial simulations.

Definition 2.1. Let A be a subset of $\mathbb{R}^n$. A function f of n variables is a rule that assigns each n tuple $(x_1, \ldots, x_n)$ in A a value $f(x_1, \ldots, x_n)$ in $\mathbb{R}$. The set A is called the domain of f, and the set of all outputs of f is called the range of f.

https://doi.org/10.1515/9783112218082-002

Example 2.1. Let

$$z = f(x,y) = x - y^2.$$

We will evaluate $f(1,2)$ and $f(-2,1)$ and we will find the domain and range of f. Using the definition $f(x,y) = x - y^2$, we have

$$f(1,2) = 1 - 2^2$$
$$= -3$$

and

$$f(-2,1) = -2 - 1^2$$
$$= -3.$$

The domain of f is not specified. We take it to be all possible pairs in $\mathbb{R}^2$ for which the function f is defined. In this example, f is defined for all pairs (x,y), and thus the domain of f is $\mathbb{R}^2$. The outputs of f can be made as large or small as possible. Any real number r can be the output of f. Therefore the range of f is $\mathbb{R}$.

Example 2.2. Let

$$f(x,y) = \sqrt{1 - \frac{x^2}{9} - \frac{y^2}{4}}.$$

We will find the domain and range of the function f. The domain is the set of all pairs (x,y) allowable as inputs of f. Because of the square root, we need (x,y) to be such that

$$1 - \frac{x^2}{9} - \frac{y^2}{4} \geq 0$$

or

$$\frac{x^2}{9} + \frac{y^2}{4} \leq 1.$$

This inequality describes the interior of an ellipse shown in Fig. 2.1. We can represent the domain A graphically with the figure. In the given situation, we can write

$$A = \left\{ (x,y) \in \mathbb{R}^2 : \frac{x^2}{9} + \frac{y^2}{4} \leq 1 \right\}.$$

The range of f is the set of all possible output values. The square root ensures that all outputs are ≥ 0. Since the x and y terms are squared, then subtracted, inside the square root, the largest output value comes at $x = 0, y = 0$, i. e., $f(0,0) = 1$. Thus the range A of the given function is the interval $[0,1]$.

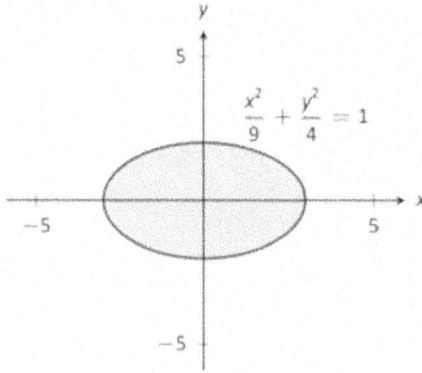

Figure 2.1: The interior of the ellipse $\frac{x^2}{9} + \frac{y^2}{4} \leq 1$.

Exercise 2.1. Find the domain of the function f defined as

1.

$$f(x,y) = \frac{1}{\sqrt{x^2 - 16(y^2 + 1)}}, \quad (x,y) \in \mathbb{R}^2.$$

2.

$$f(x,y) = \sqrt{1 - x^2 - y^2}, \quad (x,y) \in \mathbb{R}^2.$$

3.

$$f(x,y) = \frac{1}{\sqrt{1 - x^2 - y^2}}, \quad (x,y) \in \mathbb{R}^2.$$

4.

$$f(x,y) = \log(x^2 + y^2 - 1), \quad (x,y) \in \mathbb{R}^2.$$

5.

$$f(x,y) = \log(y^2 - 4x + 8), \quad (x,y) \in \mathbb{R}^2.$$

Example 2.3. Consider the function

$$f(x,y,z) = \frac{2x^2 + 3z - \sin y}{2x + 3y - z}.$$

We will evaluate the function f at $(1, 0, -1)$, and we will find its domain and range. By the definition of f we find

$$f(1, 0, -1) = \frac{2 \cdot 1^2 + 3 \cdot (-1) - \sin 0}{2 \cdot 1 + 3 \cdot 0 - (-1)}$$

$$= -\frac{1}{3}.$$

As the domain of f is not specified, we take it to be the set of all triples (x, y, z) for which the function f is defined. Since we cannot divide by 0, the domain A of the function f is

$$A = \{(x,y,z) \in \mathbb{R}^3 : 2x + 3y - z \neq 0\}.$$

Notice that the set of all points in $\mathbb{R}^3$ that are not in A form the plane that passes through the origin with the normal vector $(2,3,-1)$. We determine that the range of f is $\mathbb{R}$, i.e., all real numbers are possible outputs of f. There is no set way of establishing this. To get numbers near 0, we can take $y = 0$ and $z \approx -\frac{2}{3}x^2$. To get arbitrarily large numbers, we can take $z \approx 2x + 3y$.

Exercise 2.2. Find the domain of the function f, where

1.
$$f(x,y,z) = \log(1 - x - y - z), \quad (x,y,z) \in \mathbb{R}^3.$$

2.
$$f(x,y,z) = \sqrt{1 - |x| - |y| - |z|}, \quad (x,y,z) \in \mathbb{R}^3.$$

3.
$$f(x,y,z) = \frac{1}{\sqrt{x}} + \frac{1}{\sqrt{y}}, \quad (x,y,z) \in \mathbb{R}^3.$$

4.
$$f(x,y,z) = \frac{1}{\sqrt{1 - x^2 - y^2 - z^2}}, \quad (x,y,z) \in \mathbb{R}^3.$$

5.
$$f(x,y,z) = \sqrt{16 - x^2 - z^2}, \quad (x,y,z) \in \mathbb{R}^3.$$

Example 2.4. We will find the domain of the function

$$f(x_1,\ldots,x_n) = \arcsin\left(2 - \sum_{j=1}^{n} x_j^2\right), \quad (x_1,\ldots,x_n) \in \mathbb{R}^n.$$

By the definition of the function arcsin it follows that the function f is defined if

$$-1 \leq 2 - \sum_{j=1}^{n} x_j^2 \leq 1, \quad (x_1,\ldots,x_n) \in \mathbb{R}^n,$$

or

$$1 \leq \sum_{j=1}^{n} x_j^2 \leq 3, \quad (x_1,\ldots,x_n) \in \mathbb{R}^n.$$

Thus the domain of the function f is

$$\left\{(x_1,\ldots,x_n) \in \mathbb{R}^n : 1 \leq \sum_{j=1}^{n} x_j^2 \leq 3\right\}.$$

Exercise 2.3. Find the domain of the function f, where

1.

$$f(x_1, \ldots, x_4) = \frac{1}{x_1^2 + x_2^2 + x_3^2 + x_4^2 - 1}, \quad (x_1, \ldots, x_4) \in \mathbb{R}^4.$$

2.

$$f(x_1, \ldots, x_4) = \sqrt{1 - x_1^2} + \sqrt{4 - x_2^2} + \sqrt{9 - x_3^2} + \sqrt{16 - x_4^2}, \quad (x_1, \ldots, x_4) \in \mathbb{R}^4.$$

3.

$$f(x_1, \ldots, x_n) = \sqrt{\log\left(2 - \sum_{j=1}^{n} x_j^2\right)}, \quad (x_1, \ldots, x_n) \in \mathbb{R}^n.$$

4.

$$f(x_1, \ldots, x_n) = \sum_{j=1}^{n} \sin x_j + 1, \quad (x_1, \ldots, x_n) \in \mathbb{R}^n.$$

5.

$$f(x_1, \ldots, x_n) = \sqrt{1 - x_1} + \sum_{j=1}^{n} x_j, \quad (x_1, \ldots, x_n) \in \mathbb{R}^n.$$

Exercise 2.4. Find the range of the function f, where

1.

$$f(x, y) = \sqrt{2 + x + y - x^2 - 2xy - y^2}, \quad (x, y) \in \mathbb{R}^2.$$

2.

$$f(x, y) = \log_y x + \log_x y, \quad (x, y) \in \mathbb{R}^2.$$

3.

$$f(x, y) = x^2 - xy + y^2,$$

where

$$(x, y) \in \{(x, y) \in \mathbb{R}^2 : |x| + |y| = 1\}.$$

Example 2.5. We will find a function $z = f(x, y)$ such that z is the area of a rhombus, x is its perimeter, and y is the sum of the lengths of its diagonals. Let $a > 0$ be the length of the side of the rhombus, let p and q be the lengths of the diagonals of the rhombus, and let α be the measure of the vertex angle of the rhombus. Then

$$p = 2a \cos\left(\frac{\alpha}{2}\right),$$

$$q = 2a \sin\left(\frac{\alpha}{2}\right),$$

and

$$x = 4a,$$

$$y = p + q,$$
$$z = \frac{pq}{2}.$$

To find explicitly the function z, we will express p and q in terms of x and y. By the second equation of the last system we obtain

$$y = \frac{x}{2}\left(\sin\left(\frac{a}{2}\right) + \cos\left(\frac{a}{2}\right)\right),$$

and therefore

$$\sin\left(\frac{a}{2}\right) + \cos\left(\frac{a}{2}\right) = 2\frac{y}{x}.$$

Employing the basic trigonometric formula

$$\left(\sin\left(\frac{a}{2}\right)\right)^2 + \left(\cos\left(\frac{a}{2}\right)\right)^2 = 1,$$

we find

$$2\left(\sin\left(\frac{a}{2}\right)\right)^2 - 4\frac{y}{x}\sin\left(\frac{a}{2}\right) + 4\frac{y^2}{x^2} - 1 = 0.$$

We consider the last equation as a quadratic equation with respect to $\sin(\frac{a}{2})$. For its roots, we have the following representations:

$$\left(\sin\left(\frac{a}{2}\right)\right)_{1,2} = \frac{2\frac{y}{x} \pm \sqrt{2 - 4\frac{y^2}{x^2}}}{2}, \quad x > 0, \quad y > 0, \quad \frac{y^2}{x^2} < \frac{1}{2},$$

and then

$$\left(\cos\left(\frac{a}{2}\right)\right)_{1,2} = \frac{2\frac{y}{x} \mp \sqrt{2 - 4\frac{y^2}{x^2}}}{2}, \quad x > 0, \quad y > 0, \quad \frac{y^2}{x^2} < \frac{1}{2}.$$

Therefore

$$f(x,y) = \frac{1}{2}pq$$
$$= a^2\frac{4y^2 - x^2}{x^2}, \quad x > 0, \quad y > 0, \quad \frac{1}{4} < \frac{y^2}{x^2} < \frac{1}{2}.$$

Example 2.6. We will find the function $t = f(x,y,z)$, where t is the area of a triangle, and x, y, z are the lengths of its sides. By the Heron formula we have

$$f(x,y,z) = \frac{1}{2}\sqrt{(x + y + z)(x + y - z)(y + z - x)(x + z - y)},$$

where

$$x > 0, \quad y > 0, \quad z > 0, \quad x < y + z, \quad y < x + z, \quad z < x + y.$$

Example 2.7. We will find the function $z = f(x,y)$, where z is the volume of a right circular cone, x is its slant length, and y is its height. By the definition of x and y we find $x > 0, y > 0$. Moreover, denoting by r the radius of the base of the cone, we have

$$r = \sqrt{x^2 - y^2}, \quad x > y.$$

Then

$$f(x,y) = \frac{\pi}{3}y\sqrt{x^2 - y^2}, \quad x > 0, \quad y > 0, \quad x > y.$$

Exercise 2.5. Find the function $t = f(x,y,z)$, where t is the area of an isosceles trapezoid, x and y are the lengths of its bases, and z is the length of its legs.

Example 2.8. We will find the function f for which

$$f(x+y, x-y) = y(x+y), \quad (x,y) \in \mathbb{R}^2.$$

For this aim, we introduce new variables

$$u_1 = x + y,$$
$$u_2 = x - y, \quad (x,y) \in \mathbb{R}^2.$$

We express the old variables via them as follows:

$$x = \frac{u_1 + u_2}{2},$$
$$y = \frac{u_1 - u_2}{2}, \quad (u_1, u_2) \in \mathbb{R}^2.$$

Thus, for the function f, we get the expression

$$f(u_1, u_2) = u_1 \frac{u_1 - u_2}{2}, \quad (u_1, u_2) \in \mathbb{R}^2.$$

Example 2.9. We will find the function

$$f(x,y,z) = g\left(\frac{x}{z}, \frac{x}{y} - x^3 y\right), \quad (x,y,z) \in \mathbb{R}^3, \quad x > 0, \quad z > 0,$$

if

$$f(x,x,z) = x^3 z, \quad (x,y,z) \in \mathbb{R}^3, \quad x > 0, \quad z > 0.$$

By this condition and the definition of the function f we find

$$x^3 z = f(x, x, z)$$

$$= g\left(\frac{x}{z}, 1 - x^4\right), \quad (x, y, z) \in \mathbb{R}^3, \quad x > 0, \quad z > 0.$$

We introduce the following new variables:

$$u_1 = \frac{x}{z},$$

$$u_2 = 1 - x^4, \quad (x, y, z) \in \mathbb{R}^3, \quad x > 0, \quad z > 0.$$

The old variables x and y can be expressed via the new variables as follows:

$$x = \sqrt[4]{1 - u_2},$$

$$z = \frac{\sqrt[4]{1 - u_2}}{u_1}, \quad u_2 \leq 1, \quad u_1 > 0.$$

Thus

$$g(u_1, u_2) = \sqrt[4]{(1 - u_2)^3} \, \frac{\sqrt[4]{1 - u_2}}{u_1}$$

$$= \frac{1 - u_2}{u_1}, \quad (u_1, u_2) \in \mathbb{R}^2, \quad u_2 \leq 1, \quad u_1 > 0,$$

and

$$f(x, y, z) = g\left(u_1, \frac{\sqrt[4]{1 - u_2}}{y} - \sqrt[4]{(1 - u_2)^3} y\right)$$

$$= \frac{z}{x} - \frac{z}{y} + x^2 yz, \quad (x, y, z) \in \mathbb{R}^3, \quad x > 0, \quad z > 0.$$

Example 2.10. We will find the function

$$f(x, y) = \phi(x + 2\sqrt{y}) + \psi(x - 2\sqrt{y})$$

for $(x, y) \in A = \{(x, y) \in \mathbb{R}^2 : x \geq 0, \, y \geq 0\}$ that satisfies the following conditions:

$$f(x, 0) = \sin x,$$

$$f\left(x, \frac{x^2}{4}\right) = \cos x - 1, \quad x > 0,$$

and $\phi, \psi \in C(\mathbb{R})$. Employing these conditions and the definition of the function f, we arrive at the system

$$\sin x = \phi(x) + \psi(x),$$
$$\cos x - 1 = \phi(2x) + \psi(0), \quad x \geq 0.$$

From the last system we express ϕ and ψ:

$$\psi(x) = \sin x - \cos\left(\frac{x}{2}\right) + 1 + \psi(0),$$

$$\phi(x) = \cos\left(\frac{x}{2}\right) - 1 - \psi(0), \quad x \geq 0.$$

Consequently,

$$f(x,y) = \phi(x + 2\sqrt{y}) + \psi(x - 2\sqrt{y})$$
$$= \sin(x - 2\sqrt{y}) - 2\sin\left(\frac{x}{2}\right)\sin(\sqrt{y}), \quad (x,y) \in A.$$

Exercise 2.6. Find the function f, where
1. $f = f(x,y)$, and

$$f\left(xy, \frac{y}{x}\right) = x^2 - y^2, \quad (x,y) \in \mathbb{R}^2.$$

2.
$$f(x,y) = \sqrt{y} + \phi(\sqrt{x} - 1),$$
$$f(x,1) = x, \quad (x,y) \in \mathbb{R}^2,$$

and $\phi \in C(\mathbb{R})$.

3.
$$f(x,y) = x\phi\left(\frac{y}{x}\right),$$
$$f(1,y) = \sqrt{1 + y^2}, \quad (x,y) \in \mathbb{R}^2,$$

and $\phi \in C(\mathbb{R})$.

4.
$$f(x,y,z) = g\left(\frac{z-x}{xz}, \frac{z^2}{e^y - z}\right), \quad (x,y,z) \in \mathbb{R}^3, \quad x > 0, \quad z > 0,$$

$g \in C(\mathbb{R}^2)$, and

$$f(x, \log x, z) = \left(1 - \frac{z}{x}\right)^2, \quad (x,y,z) \in \mathbb{R}^3, \quad x > 0, \quad z > 0.$$

Definition 2.2. Let $f : X \to Y$. The set

$$\{(x_1,\ldots,x_n,f(x_1,\ldots,x_n)) : (x_1,\ldots,x_n) \in X\}$$

is called the graph of the function f (see Fig. 2.2).

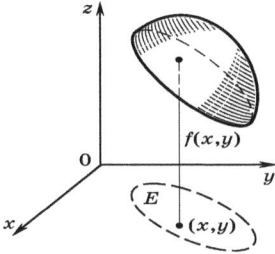

Figure 2.2: Graph of a function of two variables.

Suppose we wish to graph the function $z = f(x,y)$. This function has two indepen-dent variables x and y and one dependent variable z. Recall that when graphing a func-tion $y = f(x)$ of one variable, we use the Cartesian plane. We are able to graph any ordered pair (x,y) in the plane, and every point in the plane has an ordered pair (x,y) associated with it. With a function of two variables, each ordered pair (x,y) in the do-main of the function is mapped to a real number z. Therefore the graph of the function f consists of ordered triples (x,y,z). The graph of a function $z = f(x,y)$ of two variables is called a surface.

To understand more completely the concept of plotting a set of ordered triples to obtain a surface in the three-dimensional space, imagine the (x,y) coordinate system laying flat. Then any point in the domain of the function f has a unique z value associ-ated with it. If z is positive, then the graphed point is located above the (x,y)-plane. If z is negative, then the graphed point is located below the (x,y)-plane. The set of all the graphed points becomes the two-dimensional surface that is the graph of the function f. We will illustrate this with the following examples.

Example 2.11. We will create the graph of the function

$$z = g(x,y) = \sqrt{9 - x^2 - y^2}, \quad x^2 + y^2 \le 9, \quad (x,y) \in \mathbb{R}^2.$$

When $x^2 + y^2 = 9$, then $g(x,y) = 0$. Therefore any point on the circle of radius 3 centered at the origin in the (x,y)-plane is mapped to $z = 0$ in $\mathbb{R}^3$. If $x^2 + y^2 = 8$, then $g(x,y) = 1$, and thus any point on the circle of radius $2\sqrt{2}$ centered at the origin in the (x,y)-plane is mapped to $z = 1$ in $\mathbb{R}^3$. As $x^2 + y^2$ gets closer to zero, the value of z approaches 3. If $x^2 + y^2 = 0$, then $g(x,y) = 3$. This is the origin in the (x,y)-plane. If $x^2 + y^2$ equals any other value between 0 and 9, then $g(x,y)$ equals some other constant between 0 and 3. The surface described by this function is the hemisphere centered at the origin with radius 3, as it is shown in Fig. 2.3.

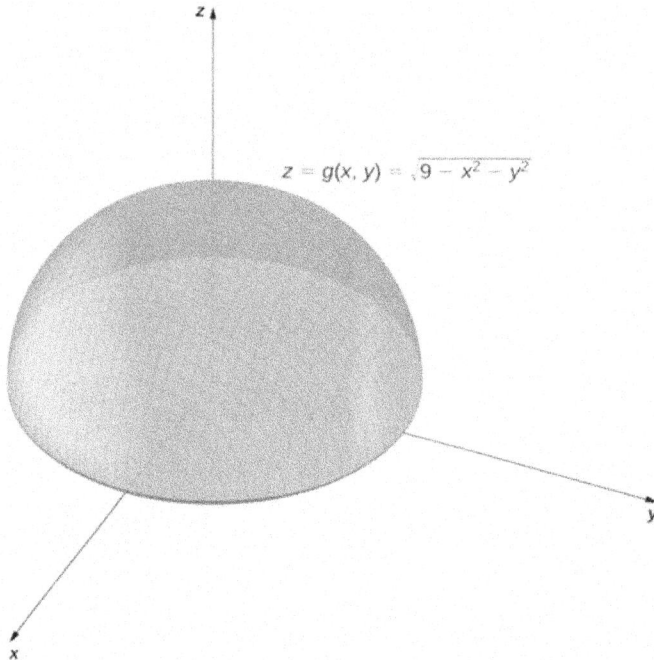

Figure 2.3: The graph of $z = \sqrt{9 - x^2 - y^2}$.

Example 2.12. Consider the function

$$z = g(x,y) = x^2 + y^2, \quad (x,y) \in \mathbb{R}^2.$$

Setting this expression to various values starting at zero, we obtain circles of increasing radius. The minimum value of $g(x,y)$ is zero attained when $x = y = 0$. When $x = 0$, the function becomes $z = y^2$, and when $y = 0$, the function becomes $z = x^2$. These are cross-sections of the graph, and they are parabolas. The graph of $g(x,y)$ is the paraboloid shown in Fig. 2.4.

Example 2.13. Consider the function

$$z = g(x,y) = 16 - (x - 3)^2 - (y - 2)^2.$$

This function is a polynomial function of two variables. The domain of the function is

$$(x - 3)^2 + (y - 2)^2 \leq 16.$$

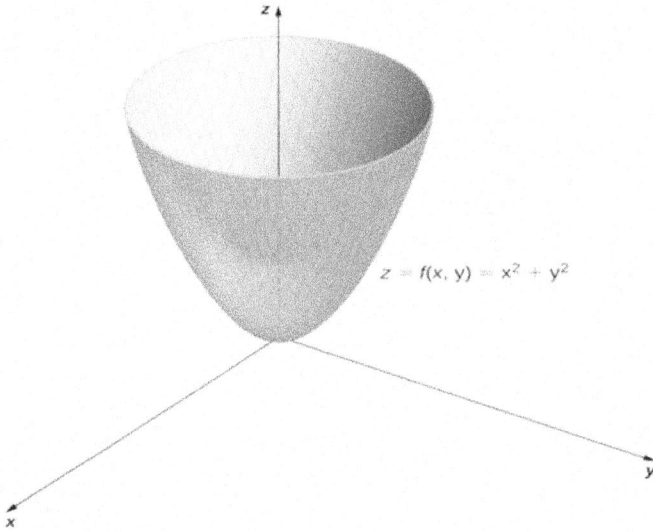

Figure 2.4: The graph of the paraboloid $z = x^2 + y^2$.

Thus the domain of the function $g(x,y)$ is the disc of radius 4 centered at $(3,2)$. For $x = 3$ and $y = 2$, $g(x,y) = 16$. For any $z < 16$, we can solve the equation

$$16 - (x-3)^2 - (y-2)^2 = z,$$

or

$$(x-3)^2 + (y-2)^2 = 16 - z.$$

This equation describes the circle with radius $\sqrt{16-z}$ centered at $(3,2)$. Therefore the range of the function $g(x,y)$ is

$$\{z \in \mathbb{R} : z \le 16\}.$$

The graph of $g(x,y)$ is also a paraboloid shown in Fig. 2.5.

Another way of visualizing a function is through level sets, i. e., the set of points in the domain of a function where the function is constant. The nice part of level sets is that they live in the same domains as the domain of the function.

Definition 2.3. Let $c \in \mathbb{R}$. The set of all points $x \in \mathbb{R}^n$ for which $f(x) = c$ is said to be the level set of f corresponding to c.

Example 2.14. We will find the level sets of the function

$$f(x,y) = \sqrt{9 - x^2 - y^2}, \quad (x,y) \in \mathbb{R}^2.$$

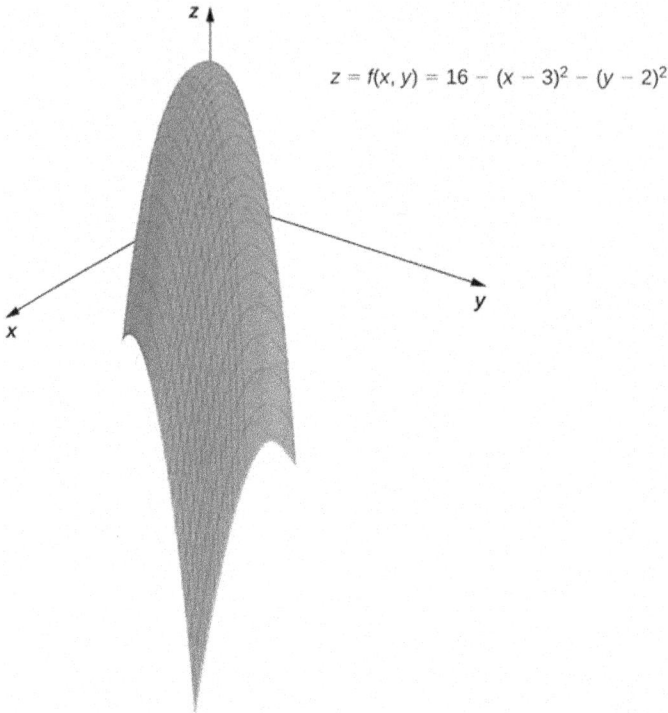

Figure 2.5: The graph of the paraboloid $z = 16 - (x - 3)^2 - (y - 2)^2$.

Its graph is shown in Fig. 2.3. Let $c \geq 0$. Then

$$\sqrt{9 - x^2 - y^2} = c, \quad (x, y) \in \mathbb{R}^2,$$

or

$$9 - x^2 - y^2 = c^2, \quad (x, y) \in \mathbb{R}^2,$$

or

$$x^2 + y^2 = 9 - c^2, \quad (x, y) \in \mathbb{R}^2,$$

are the level sets of the considered function. Figure 2.6 shows the level sets of the considered function for $c = 0$, $c = 1$, and $c = 2$.

Example 2.15. We will find the level sets of the function

$$f(x, y) = \sqrt{8 + 8x - 4y - 4x^2 - 4y^2}, \quad (x, y) \in \mathbb{R}^2.$$

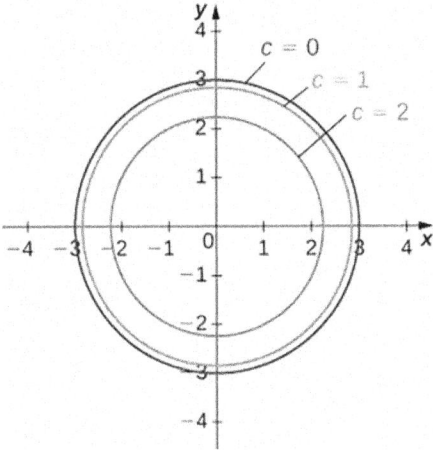

Figure 2.6: The level sets of the function in Example 2.14 for $c = 0$, $c = 1$, and $c = 2$.

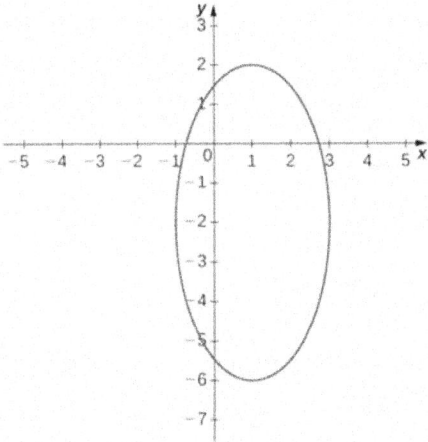

Figure 2.7: The graph of the function in Example 2.15.

Its graph is shown in Fig. 2.7. Let $c \geq 0$. Then

$$\sqrt{8 + 8x - 4y - 4x^2 - 4y^2} = c, \quad (x, y) \in \mathbb{R}^2,$$

whereupon

$$8 + 8x - 4y - 4x^2 - 4y^2 = c^2, \quad (x, y) \in \mathbb{R}^2,$$

and

$$4(x - 1)^2 + (2y + 1)^2 = 13 - c^2, \quad (x, y) \in \mathbb{R}^2,$$

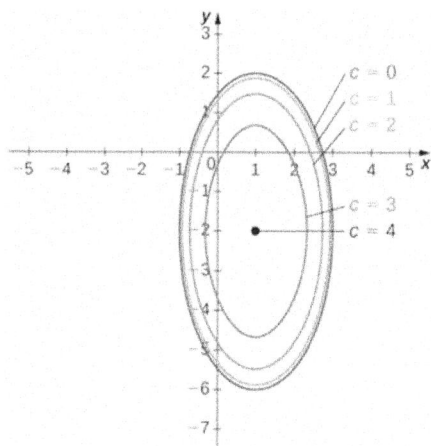

Figure 2.8: The level sets of the function in Example 2.15 for $c = 0$, $c = 1$, $c = 2$, $c = 3$, and $c = 4$.

are the level sets of the considered function. Figure 2.8 shows the level sets of the considered function for $c = 0$, $c = 1$, $c = 2$, $c = 3$, and $c = 4$.

Example 2.16. Consider the function

$$f(x,y,z) = z^2 - x^2 + y^2.$$

Its level sets are the surfaces

$$z^2 - x^2 + y^2 = c,$$

where c are real constants. Figure 2.9 shows the level surface for $c = -2$, and Fig. 2.10 shows the level surface for $c = 2$.

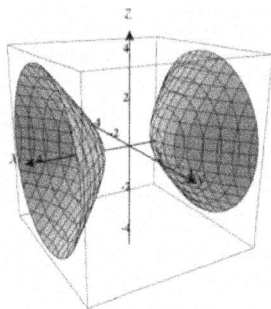

Figure 2.9: Level surface for $c = -2$.

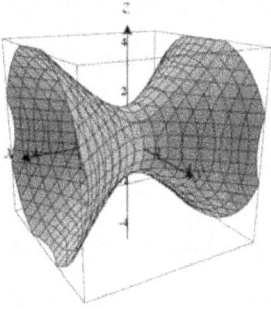

Figure 2.10: Level surface for $c = 2$.

Exercise 2.7. Find the level sets of the following functions:

1.
$$f(x,y) = y - x, \quad (x,y) \in \mathbb{R}^2.$$

2.
$$f(x,y) = \frac{1}{x^2 + y^2}, \quad (x,y) \in \mathbb{R}^2.$$

3.
$$f(x,y) = \log(1 - x^2 - y^2), \quad (x,y) \in \mathbb{R}^2.$$

4.
$$f(x,y) = \sqrt{36 - 4x^2 - 9y^2}, \quad (x,y) \in \mathbb{R}^2.$$

5.
$$f(x,y) = \frac{1}{\sqrt{x^2 - y^2}}, \quad (x,y) \in \mathbb{R}^2.$$

Exercise 2.8. Find the level sets of the following functions:

1.
$$f(x,y,z) = e^{x+2y+3z}, \quad (x,y,z) \in \mathbb{R}^3.$$

2.
$$f(x,y,z) = \frac{1}{x^2 + y^2 + z^2 + 2x}, \quad (x,y,z) \in \mathbb{R}^3.$$

3.
$$f(x,y,z) = \frac{2z}{x^2 + y^2 + z^2}, \quad (x,y,z) \in \mathbb{R}^3.$$

4.
$$f(x,y,z) = \log \frac{1 + \sqrt{x^2 + y^2 + z^2}}{1 - \sqrt{x^2 + y^2 + z^2}}, \quad (x,y,z) \in \mathbb{R}^3.$$

5.
$$f(x,y,z) = \arcsin \sqrt{\frac{x^2 + y^2}{z^2}}, \quad (x,y,z) \in \mathbb{R}^3.$$

For a given set $A \subset \mathbb{R}^n$, by $\mathcal{F}(A)$ we denote the set of all functions $f : A \to \mathbb{R}$. This set forms a vector space over $\mathbb{R}$, whose points are functions, under the following operations:
1. $+ : \mathcal{F}(A) \times \mathcal{F}(A) \to \mathcal{F}(A)$, defined by

$$(f + g)(x) = f(x) + g(x) \quad \text{for } x \in A.$$

2. $\cdot : \mathbb{R} \times \mathcal{F}(A) \to \mathcal{F}(A)$, defined by

$$(a \cdot f)(x) = a \cdot f(x) \quad \text{for } x \in A.$$

Here the signs "+" and "$\cdot$" are overloaded. On the left, they denote operations in $\mathcal{F}(A)$, and on the right, they denote operations on $\mathbb{R}$ defined in the previous chapter. The origin in $\mathcal{F}(A)$ is the zero function

$$0 : A \to \mathbb{R}$$

defined by

$$0(x) = 0 \quad \text{for all } x \in A.$$

There are various classes of functions of several variables. One of them is the class of homogeneous functions. The concept of a homogeneous function was originally introduced for functions of several variables. With the definition of vector spaces at the end of the nineteenth century, the concept has been naturally extended to functions between vector spaces, since a tuple of variable values can be considered as a coordinate vector. Homogeneous functions are useful for solving problems involving physics and engineering concepts. In many cases, it is possible to simplify complex equations by using the techniques that come with working with homogeneous functions.

Definition 2.4. A function $f : X \to Y$ is said to be a homogeneous function with homogeneity degree $\alpha \in \mathbb{R}$ if for all $\lambda \in \mathbb{R}$ and $x = (x_1, \dots, x_n) \in \mathbb{R}^n$ such that $(\lambda x_1, \dots, \lambda x_n) \in X$, we have

$$f(\lambda x_1, \dots, \lambda x_n) = \lambda^\alpha f(x_1, \dots, x_n).$$

If for all $\lambda \in \mathbb{R}$ and $x = (x_1, \dots, x_n) \in X$ such that $(\lambda x_1, \dots, \lambda x_n) \in X$, we have

$$f(\lambda x_1, \dots, \lambda x_n) = |\lambda|^\alpha f(x_1, \dots, x_n),$$

then we say that f is a homogeneous function of positive homogeneity degree α. The function f is said be a locally homogeneous function of homogeneity degree α if it is a homogeneous function of homogeneity degree α in a neighborhood of each point of X.

Example 2.17. Let

$$f(x,y) = \frac{\sqrt{x^2 + y^2}}{x},$$

where

$$(x,y) \in A = \{(x,y) \in \mathbb{R}^2 : x > 0\}.$$

Take $\lambda \in \mathbb{R}$ and $(x,y) \in A$ such that $(\lambda x, \lambda y) \in A$. Then $\lambda > 0$, and

$$f(\lambda x, \lambda y) = \frac{\sqrt{(\lambda x)^2 + (\lambda y)^2}}{\lambda x}$$

$$= \lambda^0 f(x,y).$$

Thus $f : A \to \mathbb{R}$ is a homogeneous function of degree 0.

Example 2.18. Let

$$f(x,y,z) = \sqrt[6]{x^6 z^2 + 2x^3 y^4 z + xy^2 z^5},$$

where

$$(x,y,z) \in A = \{(x,y,z) \in \mathbb{R}^3 : xz > 0\}.$$

Take $\lambda \in \mathbb{R}$ and $(x,y,z) \in A$ such that $(\lambda x, \lambda y, \lambda z) \in A$. Then

$$f(\lambda x, \lambda y, \lambda z) = \sqrt[6]{(\lambda x)^6 (\lambda z)^2 + 2(\lambda x)^3 (\lambda y)^4 (\lambda z) + (\lambda x)(\lambda y)^2 (\lambda z)^5}$$

$$= \sqrt[6]{\lambda^8 x^6 z^2 + 2\lambda^8 x^3 y^4 z + \lambda^8 xy^2 z^5}$$

$$= \lambda^{\frac{4}{3}} f(x,y,z).$$

Thus $f : A \to \mathbb{R}$ is a homogeneous function of degree $\frac{4}{3}$.

Example 2.19. Let

$$f(x_1, \ldots, x_n) = \left| \prod_{j=1}^{n} x_j \right|^3, \quad (x_1, \ldots, x_n) \in \mathbb{R}^n.$$

Take $\lambda \in \mathbb{R}$ and $(x_1, \ldots, x_n) \in \mathbb{R}^n$. Then $(\lambda x_1, \ldots, \lambda x_n) \in \mathbb{R}^n$, and

$$f(\lambda x_1, \ldots, \lambda x_n) = \left| \prod_{j=1}^{n} \lambda x_j \right|^3$$

$$= |\lambda|^3 f(x_1, \ldots, x_n).$$

Thus $f : \mathbb{R}^n \to \mathbb{R}$ is a homogeneous function of positive homogeneity degree 3.

Exercise 2.9. Find the degree of homogeneity or the degree of positive homogeneity of the following functions:

1.
$$f(x,y) = \sqrt{x^2 + y^2}, \quad (x,y) \in \mathbb{R}^2.$$

2.
$$f(x,y) = x^{\sqrt{2}} \sin\frac{y}{x} + y^{\sqrt{2}} \cos\frac{x}{y},$$

where

$$(x,y) \in A = \{(x,y) \in \mathbb{R}^2 : x > 0, \ y > 0\}.$$

3.
$$f(x_1, \ldots, x_n) = \sum_{j=1}^{n-1} x_j^{10}(\log x_{j+1} - \log x_j),$$

where

$$(x_1, \ldots, x_n) \subset A = \{(y_1, \ldots, y_n) \subset \mathbb{R}^n : y_j > 0, \ j \subset \{1, \ldots, n\}\}.$$

Now we will give an example of a function that is not homogeneous.

Example 2.20. Consider the function

$$f(x,y) = \begin{cases} x^4 y & \text{if } y \geq 0, \\ y^5 & \text{if } y < 0. \end{cases}$$

Take arbitrary $(x,y) \in \mathbb{R}^2$. Consider the following cases for y.
1. Let $y > 0$. Then there exists $\epsilon > 0$ such that $y > \epsilon > 0$. Consider $U((x,y), \epsilon)$ and let $(y_1, y_2) \in U((x,y), \epsilon)$. Then $y_2 > 0$, and $y_1^2 + y_2^2 \neq 0$. Take

$$\lambda = \frac{xy_1 + yy_2 \pm \sqrt{(y_1^2 + y_2^2)\frac{\epsilon^2}{4} - (xy_2 + yy_1)^2}}{y_1^2 + y_2^2}. \tag{2.1}$$

Then

$$\lambda^2(y_1^2 + y_2^2) - 2\lambda(xy_1 + yy_2) + x^2 + y^2 - \frac{\epsilon^2}{4} = 0, \tag{2.2}$$

whereupon

$$(x - \lambda y_1)^2 + (y - \lambda y_2)^2 = \frac{\epsilon^2}{4}$$

$$< \epsilon^2.$$

Therefore $(\lambda y_1, \lambda y_2) \in U((x,y), \epsilon)$, and

$$f(\lambda y_1, \lambda y_2) = (\lambda y_1)^4 (\lambda y_2)$$
$$= \lambda^5 f(y_1, y_2).$$

2. Let $y < 0$. Let also (y_1, y_2) and $\epsilon > 0$ be as in case 1, and let λ be as in (2.1). Then we have that (2.2) holds and $(\lambda y_1, \lambda y_2) \in U((x,y), \epsilon)$. Hence

$$f(\lambda y_1, \lambda y_2) = \lambda^5 y_2^5.$$

3. Let $y = 0$. Take $\epsilon > 0$ and consider

$$U((x,y), \epsilon) = (x - \epsilon, x + \epsilon).$$

Let $(y_1, y_2) \in U((x,y), \epsilon)$. Then there is $\lambda \in \mathbb{R}$ such that $(\lambda y_1, \lambda y_2) \in U((x,y), \epsilon)$ and

$$f(y_1, y_2) = f(\lambda y_1, \lambda y_2)$$
$$= 0$$
$$= \lambda^5 f(y_1, y_2).$$

Thus $f : \mathbb{R}^2 \to \mathbb{R}$ is a locally homogeneous function of degree 5. Suppose $f : \mathbb{R}^2 \to \mathbb{R}$ is a homogeneous function of degree 5. Consider $U((2,0), 4)$. Then $(\frac{3}{2}, \frac{1}{2}), (-\frac{3}{2}, -\frac{1}{2}) \in U((2,0), 4)$,

$$f\left(\frac{3}{2}, \frac{1}{2}\right) = \left(\frac{3}{2}\right)^4 \frac{1}{2}$$
$$= \frac{81}{32},$$

and

$$f\left(-\frac{3}{2}, -\frac{1}{2}\right) = \left(-\frac{1}{2}\right)^5$$
$$= -\frac{1}{32}$$
$$\neq (-1)^5 f\left(\frac{3}{2}, \frac{1}{2}\right).$$

This is a contradiction. Therefore $f : \mathbb{R}^2 \to \mathbb{R}$ is not a homogeneous function.

Exercise 2.10. Prove that the function

$$f(x,y,z) = \begin{cases} x^3y^5z & \text{if } z \geq 0, \\ x^3y^6 & \text{if } z < 0 \end{cases}$$

is a locally homogeneous function but not a homogeneous function.

Exercise 2.11. Let $f,g : X \rightarrow Y$ be homogeneous functions of degree k. Prove that $f \pm g$ is a homogeneous function of degree k.

Exercise 2.12. Let $f,g : X \rightarrow Y$ be homogeneous functions of degrees k_1 and k_2, respectively. Prove that
1. fg is a homogeneous function of degree $k_1 k_2$.
2. $\frac{f}{g}$ is a homogeneous function of degree $\frac{k_1}{k_2}$.

Exercise 2.13. Let f and g be homogeneous functions of positive degree k. Prove that $f \pm g$ is a homogeneous function of positive degree k.

Exercise 2.14. Let $f,g : X \rightarrow Y$ be homogeneous functions of positive degrees k_1 and k_2, respectively. Prove that
1. fg is a homogeneous function of positive degree $k_1 k_2$.
2. $\frac{f}{g}$ is a homogeneous function of positive degree $\frac{k_1}{k_2}$.

Exercise 2.15. Let $f : X \rightarrow \mathbb{R}$ be a homogeneous function of degree k, where

$$X = \{(x_1,\ldots,x_{n-1},x_n) \in \mathbb{R}^n : x_n \neq 0\}.$$

Prove that f can be represented in the form

$$f(x_1,\ldots,x_n) = x_n^k F\left(\frac{x_1}{x_n},\ldots,\frac{x_{n-1}}{x_n}\right), \quad (x_1,\ldots,x_n) \in X.$$

2.2 Limit of a function

In mathematics, the limit of a function is a fundamental concept in calculus and analysis concerning the behavior of that function near a particular input, which may or may not be in the domain of the function. Limits are used to find the maximum or minimum values of a function, which is important in optimization problems in fields like economics, engineering, and physics. For example, finding the dimensions that maximize the volume of a box or the angle that minimizes the time it takes.

After defining a new kind of functions, functions of several variables, we apply calculus ideas to it. This section investigates what is meant by the "limit of a function of several variables".

Suppose that $X \subset \mathbb{R}^n$, $Y \subset \mathbb{R}$, $a \in Y$, $x^0 \in \mathbb{R}^n$ or $x^0 = \infty$, $a \in \mathbb{R}$ or $a = \pm\infty$, and $f : X \to Y$ is a given function. In the case $n = 1$, it is possible that $x^0 = -\infty$.

Definition 2.5. We will say that the point a is the limit of the function f as $x \to x^0$ if for any sequence $\{x^k\}_{k\in\mathbb{N}} \subset X$ that converges to x^0, the sequence $\{f(x^k)\}_{k\in\mathbb{N}}$ converges to a, i. e.,

$$\lim_{k\to\infty} f(x^k) = a.$$

We will write

$$\lim_{x\to x^0} f(x) = a, \quad \text{or} \quad f(x) \to_{x\to x^0} a, \quad \text{or} \quad f(x) \to a \quad \text{as} \quad x \to x^0.$$

In terms of neighborhood of a point, the above definition can be formulated as follows.

Definition 2.6. We will say that the point a is the limit of the function f as $x \to x^0$ if for any neighborhood $U(a)$ of the point a, there is a neighborhood $U(x^0)$ of the point x^0 such that

$$f(X \cap U(x^0)) \subset U(a).$$

Now we can give the following definition for a limit of a function in terms of the distance between two points.

Definition 2.7. We will say that the point a is the limit of the function f as $x \to x^0$ if for any $\epsilon > 0$, there is $\delta = \delta(\epsilon) > 0$ such that the inequality $d(x, x^0) < \delta$ implies the inequality

$$|f(x) - a| < \epsilon.$$

By the above definitions it follows that the term "limit of a function at x^0" is equivalent to the term "limit of the function $|f(x) - a|$ when $a \in \mathbb{R}$" and "limit of the function $|f(x)|$ when $a = \infty$", i. e.,

$$\lim_{x\to x^0} f(x) = a \quad \Longleftrightarrow \quad \lim_{x\to x^0} |f(x) - a| = 0, \quad a \in \mathbb{R}^n,$$

and

$$\lim_{x\to x^0} f(x) = \infty \quad \Longleftrightarrow \quad \lim_{x\to x^0} |f(x)| = \infty, \quad a = \infty.$$

The limit of the function f is unique because the convergent sequence $\{f(x^k)\}_{k\in\mathbb{N}}$ has a unique limit.

Computing limits using these definitions is rather cumbersome. The following theorem allows us to evaluate limits much more easily.

Theorem 2.1. *Let $f, g : X \to Y$, $x^0 \in \mathbb{R}^n$, $a, b \in \mathbb{R}$, and*

$$\lim_{x \to x^0} f(x) = a,$$

$$\lim_{x \to x^0} g(x) = b. \tag{2.3}$$

Then we have

1. $\lim_{x \to x^0} (f(x) \pm g(x)) = a \pm b.$
2. $\lim_{x \to x^0} (f(x) g(x)) = ab.$
3. $\lim_{x \to x^0} \frac{f(x)}{g(x)} = \frac{a}{b}$, *provided that $g(x^0) \neq 0$ and $b \neq 0$.*

Proof. Let $\epsilon > 0$. By (2.3) and the definition of a limit, it follows that there is $\delta = \delta(\epsilon) > 0$ such that the inequality

$$d(x, x^0) < \epsilon \tag{2.4}$$

implies the inequalities

$$|f(x) - a| < \epsilon,$$

$$|g(x) - b| < \epsilon, \tag{2.5}$$

or, equivalently,

$$|a| - \epsilon < |f(x)| < |a| + \epsilon,$$

$$|b| - \epsilon < |g(x)| < |b| + \epsilon. \tag{2.6}$$

Now suppose that $x \in X$ satisfies (2.4).

1. Applying (2.5) and the triangle inequality, we obtain

$$\begin{aligned} |(f(x) \pm g(x)) - (a \pm b)| &= |(f(x) - a) \pm (g(x) - b)| \\ &\leq |f(x) - a| + |g(x) - b| \\ &< \epsilon + \epsilon \\ &= 2\epsilon. \end{aligned}$$

2. Employing inequalities (2.5) and (2.6), we find

$$\begin{aligned} |f(x)g(x) - ab| &= |f(x)g(x) - ag(x) + ag(x) - ab| \\ &= |(f(x) - a)g(x) + a(g(x) - b)| \\ &\leq |f(x) - a||g(x)| + |a||g(x) - b| \\ &< \epsilon(|b| + \epsilon) + \epsilon|a| \\ &= \epsilon(|a| + |b| + \epsilon). \end{aligned}$$

3. Let $g(x^0) \neq 0$ and $b \neq 0$. Then there are small enough $\epsilon > 0$ and $\delta = \delta(\epsilon) > 0$ such that $\epsilon < |b|$ and $|g(x)| > |b| - \epsilon$ for $x \in \mathbb{R}^n$ for which (2.4) holds. Applying (2.5) and (2.6), we find

$$\left| \frac{f(x)}{g(x)} - \frac{a}{b} \right| = \left| \frac{bf(x) - ag(x)}{bg(x)} \right|$$
$$= \frac{|bf(x) - ab + ab - ag(x)|}{|b||g(x)|}$$
$$\leq \frac{|b||f(x) - a| + |a||g(x) - b|}{|b||g(x)|}$$
$$< \epsilon \frac{|a| + |b|}{|b|(|b| - \epsilon)}.$$

This completes the proof.   □

We will illustrate the above properties of the limit of a function with the following examples.

Example 2.21. We will find the limit

$$\lim_{(x,y)\to(1,\pi)} \left(\frac{y}{x} + \cos(xy) \right).$$

The last theorem allows us to simply evaluate

$$\frac{y}{x} + \cos(xy) \quad \text{for } x = 1 \text{ and } y = \pi.$$

If an indeterminate form is returned, then we have to do more work to evaluate the limit. Otherwise, the result is the limit. Therefore

$$\lim_{(x,y)\to(1,\pi)} \left(\frac{y}{x} + \cos(xy) \right) = \frac{\pi}{1} + \cos \pi$$
$$= \pi - 1.$$

Example 2.22. We will find the limit

$$\lim_{(x,y)\to(\infty,\infty)} \left(xy \sin\left(\frac{\pi}{xy} \right) \right).$$

For this aim, we will use the representation

$$xy \sin\left(\frac{\pi}{xy} \right) = \frac{\sin(\frac{\pi}{xy})}{\frac{1}{xy}}$$

and the well-known limit

$$\lim_{x \to 0} \frac{\sin x}{x} = 1.$$

Therefore

$$\lim_{(x,y) \to (\infty,\infty)} \left(xy \sin\left(\frac{\pi}{xy}\right) \right) = \lim_{(x,y) \to (\infty,\infty)} \frac{\sin\left(\frac{\pi}{xy}\right)}{\frac{1}{xy}}$$

$$= \pi \lim_{(x,y) \to (\infty,\infty)} \frac{\sin\left(\frac{\pi}{xy}\right)}{\frac{\pi}{xy}}$$

$$= \pi.$$

In the space, there are infinite directions from which x may approach x_0. In fact, we do not have to restrict ourselves to approaching x_0 from a particular direction, but rather we can approach that point along a path that is not a straight line.

Example 2.23. Let

$$f(x,y) = \frac{x^2 y^2}{x^2 + y^2}.$$

We will find $\lim_{(x,y) \to (0,0)} f(x,y)$. We see that along any line

$$y = mx, \quad m \in \mathbb{R},$$

the limit is zero, i. e.,

$$\lim_{x \to 0} f(x, mx) = \lim_{x \to 0} \frac{m^2 x^4}{x^2 + m^2 x^2}$$

$$= \frac{m^2}{1 + m^2} \lim_{x \to 0} x^2$$

$$= 0.$$

But this is not sufficient to prove that the limit exists and that it is 0. To prove that the limit is 0, take arbitrary $\epsilon > 0$. We will find $\delta > 0$ such that if

$$\sqrt{(x - 0)^2 + (y - 0)^2} < \delta,$$

then

$$|f(x,y) - 0| < \epsilon.$$

Fix $\delta \in (0, \sqrt{\epsilon})$. Observe that

$$\frac{y^2}{x^2 + y^2} \leq 1 \quad \text{for all } (x,y) \neq (0,0)$$

and that if $\sqrt{x^2 + y^2} < \delta$, then $x^2 < \delta^2$. Let $\sqrt{x^2 + y^2} < \delta$. Consider $|f(x,y) - 0|$. We have

$$|f(x,y) - 0| = \left| \frac{x^2 y^2}{x^2 + y^2} - 0 \right|$$

$$= x^2 \frac{y^2}{x^2 + y^2}$$

$$< \delta^2$$

$$< \epsilon.$$

Thus

$$\lim_{(x,y) \to (0,0)} \frac{x^2 y^2}{x^2 + y^2} = 0.$$

The case where the limit does not exist is often easier to deal with. We can often pick two paths along which the limit is different.

Example 2.24. We will evaluate

$$\lim_{(x,y) \to (0,0)} \frac{xy}{x^2 + 4y^2}.$$

Along the lines $y = mx$, $m \in \mathbb{R}$, the resulting limit is

$$\lim_{(x,mx) \to (0,0)} \frac{3x(mx)}{x^2 + 4(mx)^2} = \lim_{x \to 0} \frac{3mx^2}{x^2(1 + 4m^2)}$$

$$= \frac{3m}{1 + 4m^2}.$$

Along different lines we get different limit values. For instance, for $m = 0$, the limit is 0, whereas for $m = 1$, the limit is $\frac{3}{5}$. This means that the limit of the function does not exist.

Example 2.25. Let

$$f(x,y) = \frac{\sin(2xy)}{x + 3y}.$$

We will evaluate $\lim_{(x,y) \to (0,0)} f(x,y)$. First, consider the limits along the lines $y = mx$, $m \in \mathbb{R}$, as it was done in the previous example. We have

$$\lim_{(x,mx) \to (0,0)} f(x, mx) = \lim_{x \to 0} \frac{\sin(2x(mx))}{x + 3mx}$$

$$= \lim_{x \to 0} \frac{\sin(2mx^2)}{(1 + 3m)x}.$$

By applying the L'Hôpital rule, we can see that this limit is 0, except for $m \neq -\frac{1}{3}$, i. e., along the line $y = -\frac{1}{3}x$. This line is not in the domain of the function f. So, along any line $y = mx$, $m \in \mathbb{R}$, in the domain of the function f, we have

$$\lim_{(x,mx)\to(0,0)} f(x,mx) = 0.$$

Now we consider the limit along the path $y = -\frac{1}{3}\sin x$, and we find

$$\lim_{(x,-\frac{1}{3}\sin x)\to(0,0)} f(x,y) = \lim_{x\to 0} \frac{\sin(-\frac{2}{3}x \sin x)}{x - \sin x}.$$

The L'Hôpital rule gives

$$\lim_{(x,-\frac{1}{3}\sin x)\to(0,0)} f(x,y) = \lim_{x\to 0} \frac{-\frac{2}{3}\cos(-\frac{2}{3}x \sin x)(\sin x + x \cos x)}{1 - \cos x} \quad (= \text{``}\frac{0}{0}\text{''})$$

$$= \lim_{x\to 0} \frac{-\frac{4}{9}\sin(-\frac{2}{3}x \sin x)(\sin x + x \cos x)^2 - \frac{2}{3}\cos(-\frac{2}{3}x \sin x)(2\cos x - x \sin x)}{\sin x}$$

$$= \text{``}\frac{-\frac{4}{3}}{0}\text{''},$$

and thus the limit does not exist. Since the limit is not the same along any path to $(0,0)$, we see that $\lim_{(x,y)\to(0,0)} f(x,y)$ does not exist.

Exercise 2.16. Find

1.

$$\lim_{(x,y)\to(0,0)} \frac{x^2 + y}{\sqrt{x^2 + y + 9} - 3}.$$

2.

$$\lim_{(x,y)\to(0,0)} \frac{\sin(y - x^2)}{y - x^2}.$$

3.

$$\lim_{(x,y)\to(0,2)} \frac{\sin(xy)}{x}.$$

4.

$$\lim_{(x,y)\to(0,1)} (1 + x)^{\frac{1}{x+x^2y}}.$$

5.

$$\lim_{(x,y)\to(\infty,\infty)} \frac{\sqrt{x^2 + y^2 + 6} + \sqrt{x^2 + y^2}}{\sqrt{x^4 + y^4 + 2(1 + x^2 y^2)} - \sqrt{x^2 + y^2}}.$$

Exercise 2.17. Find

1.

$$\lim_{x\to 0}\lim_{y\to 0}\frac{xy}{x^2+y^2}, \quad \lim_{y\to 0}\lim_{x\to 0}\frac{xy}{x^2+y^2}, \quad \lim_{(x,y)\to(0,0)}\frac{xy}{x^2+y^2}.$$

2.

$$\lim_{x\to 0}\lim_{y\to 0}\frac{y^2-x^2}{x^2+y^2}, \quad \lim_{y\to 0}\lim_{x\to 0}\frac{y^2-x^2}{x^2+y^2}, \quad \lim_{(x,y)\to(0,0)}\frac{y^2-x^2}{x^2+y^2}.$$

3.

$$\lim_{x\to 0}\lim_{y\to 0}\frac{x^2y+xy^2}{x^2-xy+y^2}, \quad \lim_{y\to 0}\lim_{x\to 0}\frac{x^2y+xy^2}{x^2-xy+y^2}, \quad \lim_{(x,y)\to(0,0)}\frac{x^2y+xy^2}{x^2-xy+y^2}.$$

4.

$$\lim_{x\to\infty}\lim_{y\to\infty}\frac{x^3+xy^2}{x^2+y^4}, \quad \lim_{y\to\infty}\lim_{x\to\infty}\frac{x^3+xy^2}{x^2+y^4}, \quad \lim_{(x,y)\to(\infty,\infty)}\frac{x^3+xy^2}{x^2+y^4}.$$

5.

$$\lim_{x\to\infty}\lim_{y\to\infty}\sin\left(\frac{\pi y^2}{x^2+3y^2}\right), \quad \lim_{y\to\infty}\lim_{x\to\infty}\sin\left(\frac{\pi y^2}{x^2+3y^2}\right), \quad \lim_{(x,y)\to(\infty,\infty)}\sin\left(\frac{\pi y^2}{x^2+3y^2}\right).$$

Exercise 2.18. Find

$$\lim_{X\ni(x,y)\to(0,0)}\frac{x^2\sin y+y^2\sin x}{x^4+y^2},$$

where

$$X=\{(x,y)\in\mathbb{R}^2 : x=\alpha t,\ y=\beta t,\ \alpha^2+\beta^2\neq 0,\ t\in\mathbb{R}\}.$$

2.3 Continuous functions

Continuity is one of the core concepts of calculus and mathematical analysis, where arguments and values of functions are real or complex numbers. In mathematics, a continuous function is a function such that a small variation of the argument induces a small variation of the value of the function. In other words, a function is continuous if arbitrarily small changes in its value can be ensured by restricting to sufficiently small changes of its argument. A discontinuous function is a function that is not continuous. Until the nineteenth century, mathematicians largely relied on intuitive notions of continuity and considered only continuous functions. The epsilon-delta definition of a limit was introduced to formalize the definition of continuity.

Applications of continuity play a pivotal role in various disciplines such as mathematics, engineering, and physics, by ensuring that functions behave predictably without abrupt changes. This principle is essential in error minimization in computational models, the smooth operation of electronic circuits, and in the study of fluid dynamics, where it guarantees the steady flow of fluids. Grasping the concept of continuity enables a deeper understanding of the world around us, making it a fundamental tool for problem solving in numerous scientific fields.

Continuity, like convergence, is typographically indistinguishable in $\mathbb{R}$ and $\mathbb{R}^n$. As for a function of one variable, the continuity of functions of several variables means that the graph of the function does not have any breaks, holes, jumps, and so on. We define the continuity of a function of several variables in a similar way as we did for a function of one variable. Suppose that $X \subset \mathbb{R}^n$, $Y \subset \mathbb{R}$, $x^0 \in X$, and $f : X \rightarrow Y$.

Definition 2.8. The function f is said to be continuous at the point x^0 if

$$\lim_{x \to x^0} f(x) = f(x^0).$$

This definition can be reformulated in terms of sequences.

Definition 2.9. The function f is said to be continuous at x^0 if for any sequence $\{x^k\}_{k \in \mathbb{N}}$ for which $\lim_{k \to \infty} x^k = x^0$, we have

$$\lim_{k \to \infty} f(x^k) = f(x^0).$$

Now we will represent a definition for a continuous function in terms of neighborhoods.

Definition 2.10. The function f is said to be continuous at x^0 if for any neighborhood $U(f(x^0))$ of the point $f(x^0)$, there is a neighborhood $U(x^0)$ of the point x^0 such that

$$f(X \cap U(x^0)) \subset U(f(x^0)).$$

Definition 2.11. The function f is said to be continuous in the set X if it is continuous at all its points. The set of all continuous functions on X will be denoted by $\mathcal{C}(X)$.

Example 2.26. Let

$$f(x,y) = \begin{cases} \frac{2\cos y \sin x}{x} & \text{if } x \neq 0, \\ 2\cos y & \text{if } x = 0. \end{cases}$$

We will check if f is continuous at the point $(0,0)$ and if f is continuous everywhere. To determine if f is continuous at $(0,0)$, we have to compare $\lim_{(x,y) \to (0,0)} f(x,y)$ and $f(0,0)$. By the definition of the function f we get

$$f(0,0) = 2\cos 0$$
$$= 2.$$

Substituting 0 for x and y into the expression

$$\frac{2\cos y \sin x}{x}$$

returns to the indeterminate form "$\frac{0}{0}$". Thus we have to do more work to compute this limit. For this aim, consider two related limits

$$\lim_{(x,y)\to(0,0)} (2\cos y) \quad \text{and} \quad \lim_{(x,y)\to(0,0)} \frac{\sin x}{x}.$$

The first limit does not contain x, and since $\cos y$ is a continuous function, we find

$$\lim_{(x,y)\to(0,0)} (2\cos y) = 2\lim_{y\to 0} \cos y$$
$$= 2\cos 0$$
$$= 2.$$

The second limit does not contain y, and

$$\lim_{(x,y)\to(0,0)} \frac{\sin x}{x} = \lim_{x\to 0} \frac{\sin x}{x}$$
$$= 1.$$

Finally, Theorem 2.1 states that we can combine these two limits as follows:

$$\lim_{(x,y)\to(0,0)} \frac{2\cos y \sin x}{x} = \left(\lim_{(x,y)\to(0,0)} (2\cos y)\right)\left(\lim_{(x,y)\to(0,0)} \frac{\sin x}{x}\right)$$
$$= 2\cdot 1$$
$$= 2.$$

Thus we get

$$\lim_{(x,y)\to(0,0)} f(x,y) = f(0,0),$$

and we conclude that f is a continuous function at $(0,0)$. A similar analysis shows that f is continuous at all points of $\mathbb{R}^2$. As long as $x \neq 0$, we can evaluate the limit directly. For $x = 0$, the limit is $2\cos y$. Therefore we can say that f is a continuous function everywhere. A graph of f is given in Fig. 2.11. Notice that it has no breaks, holes, jumps, and so on.

Example 2.27. We will show that the function

$$f(x,y) = \frac{x - 2y}{2x + y + 1}$$

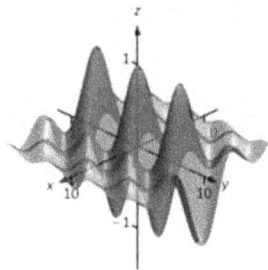

Figure 2.11: The graph of the function f in Example 2.26.

is a continuous function at the point $(1, -1)$. First, we will check if $f(1, -1)$ exists. This is true because the domain of the function f consists of all ordered pairs for which the denominator is nonzero, i. e.,

$$2x + y + 1 \neq 0.$$

The point $(1, -1)$ satisfies this condition, and furthermore,

$$f(1, -1) = \frac{1 - 2(-1)}{2 \cdot 1 + (-1) + 1}$$
$$= \frac{3}{2}.$$

Now we will show that the limit $\lim_{(x,y) \to (1,-1)} f(x, y)$ exists. This is also true because using the quotient law for limits, we get

$$\lim_{(x,y) \to (1,-1)} f(x, y) = \frac{\lim_{(x,y) \to (1,-1)} (x - 2y)}{\lim_{(x,y) \to (1,-1)} (2x + y + 1)}$$
$$= \frac{3}{2}.$$

Therefore

$$\lim_{(x,y) \to (1,-1)} f(x, y) = f(1, -1),$$

and the function f is a continuous function at the point $(1, -1)$.

Example 2.28. Let

$$f(x, y, z) = \frac{x^2 y + z}{2x - 3y + z}.$$

We will show that the function f is a continuous function at the point $(1, -1, 0)$. Because the domain of the function f consists of all ordered triples (x, y, z) for which

$$2x - 3y + z \neq 0,$$

we have that the point $(1, -1, 0)$ satisfies this condition, and thus $f(1, -1, 0)$ exists, and

$$f(1, -1, 0) = \frac{1^2(-1) + 0}{2 \cdot 1 - 3(-1) + 0}$$

$$= -\frac{1}{5}.$$

For the limit $\lim_{(x,y,z) \to (1,-1,0)} f(x, y, z)$, applying the quotient law, we find

$$\lim_{(x,y,z) \to (1,-1,0)} f(x, y, z) = \frac{\lim_{(x,y,z) \to (1,-1,0)} (x^2 y + z)}{\lim_{(x,y,z) \to (1,-1,0)} (2x - 3y + z)}$$

$$= -\frac{1}{5}.$$

Therefore

$$\lim_{(x,y,z) \to (1,-1,0)} f(x, y, z) = f(1, -1, 0),$$

so that the function f is a continuous function at the point $(1, -1, 0)$.

Example 2.29. The functions

$$f(x, y) = \frac{1}{x^2 + y^2} \quad \text{and} \quad g(x, y) = \sin\left(\frac{1}{\sqrt{x^2 + y^2}}\right)$$

have infinite discontinuities at the origin (see Fig. 2.12). Each of the directional limits of f tends to $+\infty$ as (x, y) approaches $(0, 0)$, and thus $\lim_{(x,y) \to (0,0)} g(x, y)$ does not exist. The function g represents the surface obtained by revolving the function $y = \sin\frac{1}{x}$ about the y-axis.

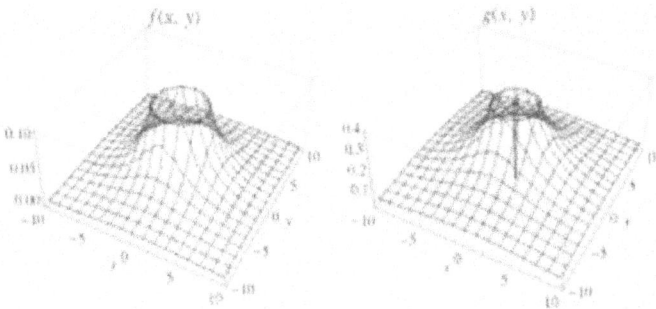

Figure 2.12: The functions f and g in Example 2.12.

Example 2.30. The functions

$$h(x,y) = \frac{1}{(x-y)^2} \quad \text{and} \quad k(x,y) = \frac{1}{x-y}$$

have infinite discontinuities along the entire line $y = x$ (see Fig. 2.13). The limit of h approaches $+\infty$ from both sides of that line. The limits of the values of k disagree on either side of that line.

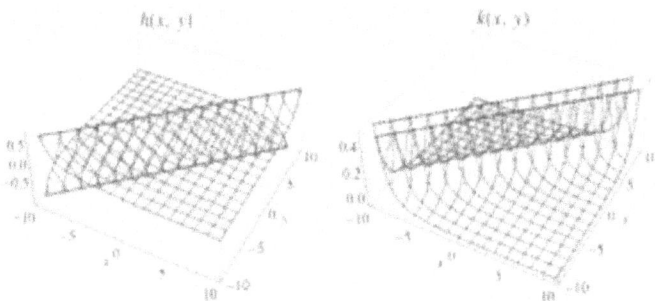

Figure 2.13: The functions h and k in Example 2.13.

Exercise 2.19. Find $a \in \mathbb{R}$ such that the function

$$f(x,y) = \begin{cases} \frac{x^2-y^2}{x^2+y^2} & \text{if } x^2 + y^2 \neq 0, \\ a & \text{if } x^2 + y^2 = 0 \end{cases}$$

1. is a continuous function at $(0,0)$ along the curve

$$X = \{(x,y) \in \mathbb{R}^2 : y = a\sqrt{x},\ x \geq 0,\ a \neq 0\}.$$

2. is a continuous function at $(0,0)$.

The following theorem enables us to combine continuous functions and to get other continuous functions. Its proof follows directly from Theorem 2.1 and the definition of a continuous function.

Theorem 2.2. *Let f and g be continuous functions at $x^0 \in X$, and let $c \in \mathbb{R}$. Then so are cf, $f+g, f-g$, and fg. In addition, $\frac{f}{g}$ is a continuous function at x_0, provided that $g(x_0) \neq 0$.*

Composition of functions is a process or operation that combines two or more functions together into a single function. Let us have a look at the definition of composition of functions.

Definition 2.12. Suppose that $X \subset \mathbb{R}^n$, $Y, Z \subset \mathbb{R}$, and

$$g : X \rightarrow Y, \quad f : Y \rightarrow Z.$$

The composition $f \circ g$ of the functions f and g is defined by

$$f \circ g(x) = f(g(x)), \quad x \in X.$$

We say that the function $y = g(x)$ is a change of the independent variable of the function $f(y)$.

Figure 2.14 shows the representation of the composition of functions. The order of functions is important while dealing with the composition of functions since $f \circ g(x)$ is not equal to $g \circ f(x)$. Another consequence of the definition of continuity is as follows.

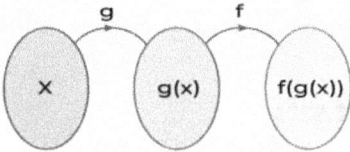

Figure 2.14: Composition of functions.

Theorem 2.3. *Let $X \subset \mathbb{R}^n$, $Y, Z \subset \mathbb{R}$, $x^0 \in X$, and $f : X \rightarrow Y$ and $g : Y \rightarrow Z$ be functions so that the limits*

$$\lim_{x \to x^0} f(x) = a \tag{2.7}$$

and

$$\lim_{y \to a} g(y) = b \tag{2.8}$$

exist for some $a, b \in \mathbb{R}$. Then there exists the limit

$$\lim_{x \to x^0} g \circ f(x) = \lim_{x \to x^0} g(f(x)) = b.$$

Proof. Let $\{x^k\}_{k \in \mathbb{N}} \subset X$ be such that

$$\lim_{k \to \infty} x^k = x^0.$$

Applying (2.7), we get

$$\lim_{k \to \infty} f(x^k) = a.$$

Now using that $\{f(x^k)\}_{k\in\mathbb{N}} \subset Y$ and applying (2.8), we get

$$\lim_{k\to\infty} g(f(x^k)) = b.$$

This completes the proof. $\qquad\qquad\qquad\qquad\qquad\qquad\qquad\qquad\qquad\qquad\qquad\square$

Corollary 2.1. *Assume that $X \subset \mathbb{R}^n$, $Y, Z \subset \mathbb{R}$, $x^0 \in X$, $f : X \to Y$ is a continuous function at x^0, and $g : Y \to Z$ is a continuous function at $f(x^0)$. Then $g \circ f$ is a continuous function at x^0.*

Proof. By Theorem 2.3 we get

$$\lim_{x\to x^0} g(f(x)) = \lim_{y\to f(x^0)} g(y)$$
$$= g(f(x^0)).$$

This completes the proof. $\qquad\qquad\qquad\qquad\qquad\qquad\qquad\qquad\qquad\qquad\qquad\square$

2.4 Continuous functions and compact sets

In calculus, and especially in mathematical analysis, a fundamental role is played by the space of continuous functions on compact sets.

The next theorem is the main result in this section. It states that the continuous image of a compact set is a compact set.

Theorem 2.4. *Let $K \subset \mathbb{R}^n$ be a compact set, and let $f : K \to \mathbb{R}$ be a continuous function. Then $f(K)$ is a compact set.*

Proof. To prove that $f(K)$ is compact set in $\mathbb{R}$, we take a sequence $\{y^k\}_{k\in\mathbb{N}} \subset f(K)$, and we will prove that it contains a subsequence that converges in $f(K)$. Since $\{y^k\}_{k\in\mathbb{N}} \subset f(K)$, we have

$$y^k = f(x^k), \quad k \in \mathbb{N},$$

for some $x^k \in K$, $k \in \mathbb{N}$. As K is a compact set in $\mathbb{R}^n$, any of its sequences contain a subsequence that converges in K. Therefore there is a subsequence $\{x^{k_m}\}_{m\in\mathbb{N}}$ of the sequence $\{x^k\}_{k\in\mathbb{N}}$ that converges to some point $x^0 \in K$, i. e.,

$$\lim_{m\to\infty} x^{k_m} = x^0.$$

Since f is a continuous function on K, we get

$$\lim_{m\to\infty} y^{k_m} = \lim_{m\to\infty} f(x^{k_m})$$
$$= f(x^0) \in f(K).$$

Let $y = f(x^0)$. Thus $\{y^{k_m}\}_{m\in\mathbb{N}}$ is a subsequence of the sequence $\{y^k\}_{k\in\mathbb{N}}$ that converges to $y \in f(K)$. Because $\{y^k\}_{k\in\mathbb{N}}$ was an arbitrarily chosen sequence of $f(K)$, we obtain that any sequence of $f(K)$ contains a convergent subsequence in $f(K)$. Therefore $f(K)$ is a compact set. This completes the proof. □

This theorem can be used to prove the compactness of various sets.

Example 2.31. Let $a = (a_1,\ldots,a_n), b = (b_1,\ldots,b_n) \in \mathbb{R}^n$. Then the set

$$\{a + tb : t \in [0,1]\}$$

is a closed segment of $\mathbb{R}^n$. This set is a compact set because it is a continuous image of the compact interval $[0,1]$ in $\mathbb{R}$ under the map $f : \mathbb{R} \rightarrow \mathbb{R}^n$ given by

$$f(t) = a + tb, \quad t \in [0,1].$$

Exercise 2.20. Prove that the closed solid ellipsoid in $\mathbb{R}^3$,

$$\left\{(x,y,z) \in \mathbb{R}^3 : \frac{x^2}{a^2} + \frac{y^2}{b^2} + \frac{z^2}{c^2} \leq 1\right\},$$

is compact.

A topological property of sets is a property that is preserved under continuity. Theorem 2.4 says that compactness is a topological property. Neither the property of closedness nor of boundedness is topological in itself. That is, the continuous image of a closed set need not be closed, and the continuous image of a bounded set need not be bounded.

The following result is a corollary of Theorem 2.4.

Theorem 2.5. *Let $K \subset \mathbb{R}^n$ be a compact set, and let $f : K \rightarrow \mathbb{R}$ be a continuous function. Then $f(K)$ attains the largest and least values at some points of K.*

Proof. Indeed, by Theorem 2.4 we have that $f(K)$ is a compact set in $\mathbb{R}$. By the properties of the compact sets in $\mathbb{R}$ we obtain that $f(K)$ attains the maximum and minimum in $\mathbb{R}$. This completes the proof. □

A homomorphism is the mathematical tool for succinctly expressing precise structural correspondences. A definition of a homomorphism reads as follows.

Definition 2.13. Let $X \subset \mathbb{R}^n$ and $f : X \rightarrow \mathbb{R}$ be such that
1. $f : X \rightarrow f(X)$ is a bijection.
2. $f \in C(X)$.
3. $f^{-1} \in C(f(X))$.

Then f is called a homomorphism on X.

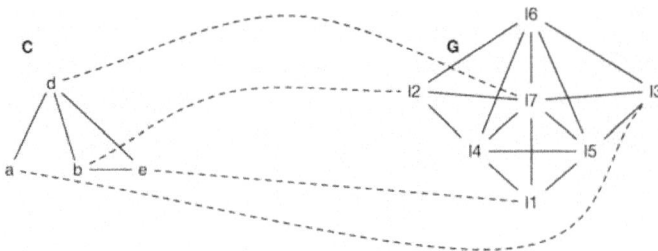

Figure 2.15: A homomorphism.

Figure 2.15 shows a homomorphism. A sufficient condition for a function to be a homomorphism is as follows.

Theorem 2.6. *Let $K \subset \mathbb{R}^n$ be a compact set, and let $f : K \to f(K)$ be a bijection. Then f is a homomorphism on K.*

Proof. To show that f^{-1} is continuous at each point $y^0 \in f(K)$, we fix a sequence $\{y^k\}_{k \in \mathbb{N}} \subset f(K)$ that converges to y^0. Then there are a point $x^0 \in K$ and a sequence $\{x^k\}_{k \in \mathbb{N}} \subset K$ for which y^0 and y^k, $k \in \mathbb{N}$, are the images of x^0 and x^k, $k \in \mathbb{N}$, respectively, under the map f. We have to show that the sequence $\{x^k\}_{k \in \mathbb{N}}$ converges to x^0. Seeking a contradiction, suppose the contrary. Then there is $\epsilon > 0$ such that for any $m \in \mathbb{N}$, there is $k_m \in \mathbb{N}$, $k_m > m$, such that

$$|x^{k_m} - x^0| > \epsilon. \tag{2.9}$$

As K is a compact set in $\mathbb{R}^n$, it follows that there is a subsequence $\{x^{k_{m_l}}\}_{l \in \mathbb{N}}$ of the sequence $\{x^{k_m}\}_{m \in \mathbb{N}}$ that converges to some $x^1 \in K$, and since (2.9) holds, we have the inequality

$$|x^{k_m} - x^0| > \epsilon, \quad m \in \mathbb{N}.$$

Taking the limit of both sides of this inequality as $m \to \infty$ and using that $x^{k_m} \to x^1$ as $m \to \infty$, we find

$$\epsilon \le |x^1 - x^0|.$$

Therefore $x^1 \neq x^0$, and since $f : K \to f(K)$ is a bijection, we get $f(x^1) \neq f(x^0)$. However, the subsequence $\{y^{k_{m_l}}\}_{l \in \mathbb{N}}$ must have the same limit as the sequence $\{y^k\}_{k \in \mathbb{N}}$, i. e.,

$$\lim_{l \to \infty} y^{k_{m_l}} = y^0.$$

Since f is a continuous function on K, we obtain

$$y^0 = \lim_{l \to \infty} y^{k_{m_l}}$$

$$= \lim_{l \to \infty} f(x^{k_{m_l}})$$
$$= f(x^1),$$

which is a contradiction. Thus the sequence $\{x^k\}_{k \in \mathbb{N}}$ converges to x^0, and f^{-1} is a continuous function on $f(K)$. This completes the proof. □

Example 2.32. For any fixed $n \in \mathbb{N}$, define $f : [0, \infty) \to \mathbb{R}$ by

$$f(x) = x^n.$$

Then f is one-to-one, as it is strictly increasing and continuous, being a monomial. Thus, applying Theorem 2.6, we have that f^{-1}, the nth root function, is continuous on any interval $[a^n, b^n]$, $0 \le a \le b$, and hence, it is continuous on $[0, \infty)$.

2.5 Uniform continuity

Uniform continuity is a mathematical concept that describes the behavior of a function when its input and output values are close to each other. A function is uniformly continuous if for any small change in the input there is a corresponding small change in the output. Uniform continuity is a stronger condition than continuity. While a continuous function only requires small changes in input to result in small changes in output at a specific point, uniform continuity requires this behavior to hold for the entire domain of the function.

Uniform continuity has many applications in mathematics and other fields, such as physics and engineering. It is used to prove the existence and uniqueness of solutions to differential equations, to study the convergence of series and integrals, and to analyze the behavior of functions in optimization problems.

Let $X \subset \mathbb{R}^n$.

Definition 2.14. The function $f : X \to \mathbb{R}$ is said to be uniformly continuous if for any $\epsilon > 0$, there is $\delta > 0$ such that for all $x^1, x^2 \in X$ for which

$$d(x^1, x^2) < \delta,$$

we have

$$d(f(x^1), f(x^2)) < \epsilon.$$

Clearly, if a function f is uniformly continuous on X, then it is continuous on X. The converse is not true in general.

Definition 2.15. For a function $f : X \to \mathbb{R}$, we define its oscillation as

$$\omega(f, X) = \operatorname{diam} f(X).$$

Using this definition, we can reformulate the definition of uniform continuity of a function as follows.

Definition 2.16. The function $f : X \rightarrow \mathbb{R}$ is said to be uniformly continuous if for any $\epsilon > 0$, there is $\delta > 0$ such that for any $Y \subset X$ for which diam $Y < \delta$, we have

$$\omega(f, Y) < \epsilon.$$

Example 2.33. Consider

$$f(x, y) = 2x + 4y + 5, \quad (x, y) \in \mathbb{R}^2.$$

Take arbitrary $\epsilon > 0$. Let $\delta = \frac{\epsilon}{8}$. Then, for $x^1 = (x_1^1, x_2^1), x^2 = (x_1^2, x_2^2) \in \mathbb{R}^2$ for which

$$|x^1 - x^2| < \delta,$$

we have

$$|x_1^1 - x_2^1| < \frac{\epsilon}{8},$$
$$|x_1^2 - x_2^2| < \frac{\epsilon}{8},$$

and

$$\begin{aligned}
|f(x_1^1, x_2^1) - f(x_1^2, x_2^2)| &= |2x_1^1 + 4x_2^1 + 5 - 2x_1^2 - 4x_2^2 - 5| \\
&= |2(x_1^1 - x_1^2) + 4(x_2^1 - x_2^2)| \\
&\le 2|x_1^1 - x_1^2| + 4|x_2^1 - x_2^2| \\
&< 2\frac{\epsilon}{8} + 4\frac{\epsilon}{8} \\
&= \frac{\epsilon}{4} + \frac{\epsilon}{2} \\
&< \frac{\epsilon}{2} + \frac{\epsilon}{2} \\
&= \epsilon.
\end{aligned}$$

Thus f is uniformly continuous on $\mathbb{R}^2$.

Example 2.34. Consider the function

$$f(x, y) = \log(x^2 + y^2), \quad (x, y) \in X = \{(y_1, y_2) \in \mathbb{R}^2 : y_1^2 + y_2^2 \ge 1\}.$$

Let $\epsilon > 0$ be arbitrarily chosen. Take $\delta > 0$ such that

$$1 + \delta < e^{\frac{\epsilon}{2}}.$$

Let $x = (x_1, x_2), y = (y_1, y_2) \in X$ be such that

$$d(x,y) < \delta.$$

Without loss of generality, suppose that

$$x_1^2 x_2^2 \geq y_1^2 + y_2^2.$$

Then

$$d(x,0) \leq d(x,y) + d(y,0)$$
$$< \delta + d(y,0),$$

i. e.,

$$\sqrt{x_1^2 + x_2^2} < \sqrt{y_1^2 + y_2^2} + \delta.$$

Hence

$$|f(x_1,x_2) - f(y_1,y_2)| = |\log(x_1^2 + x_2^2) - \log(y_1^2 + y_2^2)|$$
$$= \left|\log \frac{x_1^2 + x_2^2}{y_1^2 + y_2^2}\right|$$
$$= \log \frac{x_1^2 + x_2^2}{y_1^2 + y_2^2}$$
$$= 2\log \frac{\sqrt{x_1^2 + x_2^2}}{\sqrt{y_1^2 + y_2^2}}$$
$$< 2\log \frac{\delta + \sqrt{y_1^2 + y_2^2}}{\sqrt{y_1^2 + y_2^2}}$$
$$= 2\log\left(1 + \frac{\delta}{\sqrt{y_1^2 + y_2^2}}\right)$$
$$\leq 2\log(1 + \delta)$$
$$< 2\log e^{\frac{\epsilon}{2}}$$
$$= 2 \cdot \frac{\epsilon}{2}$$
$$= \epsilon.$$

Thus f is uniformly continuous on X.

Example 2.35. Consider the function

$$f(x,y) = \sin \frac{\pi}{1 - x^2 - y^2}, \quad (x,y) \in X = \{(y_1,y_2) \in \mathbb{R}^2 : y_1^2 + y_2^2 < 1\}.$$

For $m \in \mathbb{N}$, define

$$x^m = (x_1^m, x_2^m)$$
$$= \left(\sqrt{1 - \frac{1}{2m}} \cos \alpha, \sqrt{1 - \frac{1}{2m}} \sin \alpha \right),$$
$$y^m = (y_1^m, y_2^m)$$
$$= \left(\sqrt{1 - \frac{2}{1 + 4m}} \cos \alpha, \sqrt{1 - \frac{2}{1 + 4m}} \sin \alpha \right),$$

where $\alpha \in [0, 2\pi)$. For any $m \in \mathbb{N}$, we have

$$(x_1^m)^2 + (x_2^m)^2 = \left(\sqrt{1 - \frac{1}{2m}} \cos \alpha \right)^2 + \left(\sqrt{1 - \frac{1}{2m}} \sin \alpha \right)^2$$
$$= \left(1 - \frac{1}{2m} \right)(\cos \alpha)^2 + \left(1 - \frac{1}{2m} \right)(\sin \alpha)^2$$
$$= \left(1 - \frac{1}{2m} \right)((\cos \alpha)^2 + (\sin \alpha)^2)$$
$$= 1 - \frac{1}{2m}$$
$$< 1$$

and

$$(y_1^m)^2 + (y_2^m)^2 = \left(\sqrt{1 - \frac{2}{1 + 4m}} \cos \alpha \right)^2 + \left(\sqrt{1 - \frac{2}{1 + 4m}} \sin \alpha \right)^2$$
$$= \left(1 - \frac{2}{1 + 4m} \right)(\cos \alpha)^2 + \left(1 - \frac{2}{1 + 4m} \right)(\sin \alpha)^2$$
$$= \left(1 - \frac{2}{1 + 4m} \right)((\cos \alpha)^2 + (\sin \alpha)^2)$$
$$= 1 - \frac{2}{1 + 4m}$$
$$< 1,$$

i. e., $x^m, y^m \in X, m \in \mathbb{N}$. Next,

$$d(x^m, y^m) = \sqrt{(x_1^m - y_1^m)^2 + (x_2^m - y_2^m)^2}$$
$$= \sqrt{\left(\sqrt{1 - \frac{1}{2m}} - \sqrt{1 - \frac{2}{1 + 4m}} \right)^2 (\cos \alpha)^2 + \left(\sqrt{1 - \frac{1}{2m}} - \sqrt{1 - \frac{2}{1 + 4m}} \right)^2 (\sin \alpha)^2}$$
$$= \left| \sqrt{1 - \frac{1}{2m}} - \sqrt{1 - \frac{2}{1 + 4m}} \right|$$

$$= \frac{\left|\left(\sqrt{1-\frac{1}{2m}} - \sqrt{1-\frac{2}{1+4m}}\right)\left(\sqrt{1-\frac{1}{2m}} + \sqrt{1-\frac{2}{1+4m}}\right)\right|}{\sqrt{1-\frac{1}{2m}} + \sqrt{1-\frac{2}{1+4m}}}$$

$$= \frac{\left|1 - \frac{1}{2m} - 1 + \frac{2}{1+4m}\right|}{\sqrt{1-\frac{1}{2m}} + \sqrt{1-\frac{2}{1+4m}}}$$

$$= \frac{\left|\frac{1+4m-4m}{2m(1+4m)}\right|}{\sqrt{1-\frac{1}{2m}} + \sqrt{1-\frac{2}{1+4m}}}$$

$$= \frac{1}{2m(1+4m)\left(\sqrt{1-\frac{1}{2m}} + \sqrt{1-\frac{2}{1+4m}}\right)}$$

$$\to 0 \quad \text{as} \quad m \to \infty,$$

and

$$|f(x_1^m, x_2^m) - f(y_1^m, y_2^m)| = \left|\sin\frac{\pi}{\frac{1}{2m}} - \sin\frac{\pi}{\frac{2}{1+4m}}\right|$$

$$= \left|\sin(2m\pi) - \sin\left((2m+1)\frac{\pi}{2}\right)\right|$$

$$= 1, \quad m \in \mathbb{N}.$$

Thus f is not uniformly continuous on X.

Exercise 2.21. Check if the following functions are uniformly continuous on X:
1. $f(x,y) = \arcsin\frac{y}{x}$, $(x,y) \in X = \{(y_1, y_2) \in \mathbb{R}^2 : |y_2| < |y_1|\}$.
2. $f(x,y) = 5x + 4y + 12$, $(x,y) \in X = \mathbb{R}^2$.

Another characteristic of a function is the modulus of continuity.

Definition 2.17. Let $f : X \to \mathbb{R}$. The function

$$\omega(\delta, f, X) = \sup_{|x^1-x^2|\leq\delta} |f(x^1) - f(x^2)|, \quad x^1, x^2 \in X, \quad \delta > 0,$$

is called the modulus of continuity of the function f.

A sufficient condition for uniform continuity reads as follows.

Theorem 2.7. *The function $f : X \to \mathbb{R}$ is uniformly continuous if and only if*

$$\lim_{\delta \to 0} \omega(\delta, f, X) = 0. \tag{2.10}$$

Proof. 1. Let f be uniformly continuous on X. Take $\epsilon > 0$ arbitrarily. Then there is $\delta_\epsilon > 0$ such that the inequality

$$|x^1 - x^2| < \delta_\epsilon$$

implies the inequality

$$|f(x^1) - f(x^2)| < \frac{\epsilon}{2}.$$

Let $\delta \in (0, \delta_\epsilon)$. Then

$$w(\delta, f, X) = \sup_{|x^1 - x^2| \le \delta} |f(x^1) - f(x^2)|$$

$$\le \frac{\epsilon}{2}$$

$$< \epsilon.$$

Since $\epsilon > 0$ was arbitrarily chosen, we get (2.10).

2. Suppose that (2.10) holds. Take $\epsilon > 0$ arbitrarily. Then there is $\delta_\epsilon > 0$ such that if $\delta \in (0, \delta_\epsilon)$, then

$$\omega(\delta, f, X) < \epsilon.$$

Hence, if $\delta \in (0, \delta_\epsilon)$ and $x^1, x^2 \in X$, $|x^1 - x^2| < \delta$, then we have

$$|f(x^1) - f(x^2)| \le \omega(\delta, f, X)$$

$$< \epsilon.$$

Thus f is uniformly continuous on X. This completes the proof. □

Example 2.36. We will find the modulus of continuity of the function

$$f(x, y) = x^2 + y^2, \quad (x, y) \in \mathbb{R}^2.$$

Take $\delta > 0$ arbitrarily. Then

$$\omega(\delta, f, \mathbb{R}^2) = \sup_{|x^1 - x^2| \le \delta} |f(x^1) - f(x^2)|$$

$$= \sup_{|x^1 - x^2| \le \delta} |(x_1^1)^2 + (x_2^1)^2 - (x_1^2)^2 - (x_2^2)^2|$$

$$\ge (x_1^1)^2 + (x_2^1)^2 - \left(x_1^1 - \frac{\delta}{\sqrt{2}}\right)^2 - \left(x_2^1 - \frac{\delta}{\sqrt{2}}\right)^2$$

$$= (x_1^1)^2 + (x_2^1)^2 - (x_1^1)^2 + \sqrt{2}x_1^1 \delta - \frac{\delta^2}{2}$$

$$- (x_2^1)^2 + \sqrt{2}x_2^1 \delta - \frac{\delta^2}{2}$$

$$= \sqrt{2}\delta\left(x_1^1 + x_2^1 - \frac{\delta}{\sqrt{2}}\right)$$

for all $x^1 = (x_1^1, x_2^1) \in \mathbb{R}^2$ and $x^2 = (x_1^2, x_2^2) \in \mathbb{R}^2$. Since

$$\lim_{\substack{x_1^1 \to \infty \\ x_2^1 \to \infty}} \sqrt{2}\delta\left(x_1^1 + x_2^1 - \frac{\delta}{\sqrt{2}}\right) = \infty,$$

we find

$$\omega(\delta, f, \mathbb{R}^2) = \infty.$$

Example 2.37. Consider the function

$$f(x, y) = x^2 + y^2, \quad (x, y) \in X,$$

where

$$X = \{(x, y) \in \mathbb{R}^2 : 0 \le x \le 1, \ 0 \le y \le 1\}.$$

Take $\delta > 0$ arbitrarily. Let $x^1 = (x_1^1, x_2^1)$ and $x^2 = (x_1^2, x_2^2)$ be such that

$$0 \le x_1^2 - \delta \le x_1^1 \le x_1^2 \le 1,$$
$$0 \le x_2^2 - \delta \le x_2^1 \le x_2^2 \le 1.$$

Then

$$0 \le x_1^2 - x_1^1 \le \delta,$$
$$0 \le x_2^2 - x_2^1 \le \delta,$$

i. e., $x^1, x^2 \in X$. Next,

$$
\begin{aligned}
\left| f(x_1^2, x_2^2) - f(x_1^1, x_2^1) \right| &= f(x_1^2, x_2^2) - f(x_1^1, x_2^1) \\
&= \left(x_1^2\right)^2 + \left(x_2^2\right)^2 - \left(x_1^1\right)^2 - \left(x_1^2\right)^2 \\
&= \left(x_1^2\right)^2 - \left(x_1^1\right)^2 + \left(x_2^2\right)^2 - \left(x_2^1\right)^2 \\
&\le \left(x_1^2\right)^2 - \left(x_1^2 - \delta\right)^2 + \left(x_2^2\right)^2 - \left(x_2^2 - \delta\right)^2 \\
&= \left(x_1^2\right)^2 - \left(x_1^2\right)^2 + 2x_1^2\delta - \delta^2 + \left(x_2^2\right)^2 - \left(x_2^2\right)^2 + 2x_2^2\delta - \delta^2 \\
&= 2x_1^2\delta + 2x_2^2\delta - 2\delta^2 \\
&\le 2\delta + 2\delta - 2\delta^2 \\
&= 4\delta - 2\delta^2.
\end{aligned}
$$

Therefore

$$\omega(\delta, f, X) \le 4\delta - 2\delta^2. \tag{2.11}$$

Let now

$$y^1 = (y_1^1, y_2^1)$$
$$= (1 - \delta, 1 + \delta),$$
$$y^2 = (y_1^2, y_2^2)$$
$$= (1, 1).$$

Then

$$|f(y_1^2, y_2^2) - f(y_1^1, y_2^1)| = f(y_1^2, y_2^2) - f(y_1^1, y_2^1)$$
$$= (y_1^2)^2 + (y_1^2)^2 - (y_1^1)^2 - (y_1^1)^2$$
$$= 1^2 + 1^2 - (1 - \delta)^2 - (1 - \delta)^2$$
$$= 1 + 1 - 1 + 2\delta - \delta^2 - 1 + 2\delta - \delta^2$$
$$= 4\delta - 2\delta^2.$$

Therefore

$$\omega(\delta, f, X) = \sup_{|z^1 - z^2| \le \delta} |f(z^1) - f(z^2)|$$
$$\ge |f(y_1^1, y_2^1) - f(y_1^2, y_2^2)|$$
$$= 4\delta - 2\delta^2.$$

By the last inequality and inequality (2.11) we find

$$\omega(\delta, f, X) = 4\delta - 2\delta^2.$$

Note that

$$\lim_{\delta \to 0} \omega(\delta, f, X) = \lim_{\delta \to 0} (4\delta - 2\delta^2)$$
$$= 0.$$

Hence and Theorem 2.7 we conclude that f is uniformly continuous on X.

Example 2.38. Consider the function

$$f(x, y) = \sin \frac{1}{x} + \sin \frac{1}{y}, \quad (x, y) \in X,$$

where

$$X = \{(x, y) \in \mathbb{R}^2 : x \ne 0, y \ne 0\}.$$

Let $\delta > 0$ be arbitrarily chosen. Then

$$\omega(\delta, f, X) = \sup_{|x^1 - x^2| \le \delta} |f(x_1^1, x_2^1) - f(x_1^2, x_2^2)|$$

$$= \sup_{|x^1 - x^2| \le \delta} \left| \sin \frac{1}{x_1^1} + \sin \frac{1}{x_2^1} - \sin \frac{1}{x_1^2} - \sin \frac{1}{x_2^2} \right|$$

$$\le \sup_{|x^1 - x^2| \le \delta} \left(\left| \sin \frac{1}{x_1^1} \right| + \left| \sin \frac{1}{x_2^1} \right| + \left| \sin \frac{1}{x_1^2} \right| + \left| \sin \frac{1}{x_2^2} \right| \right)$$

$$\le 1 + 1 + 1 + 1$$

$$= 4,$$

i. e.,

$$\omega(\delta, f, X) \le 4. \tag{2.12}$$

Let

$$y^1 = (y_1^1, y_2^1)$$

$$= \left(\frac{1}{\frac{\pi}{2} + 2m\pi}, \frac{1}{\frac{\pi}{2} + 2m\pi} \right),$$

$$y^2 = (y_1^2, y_2^2)$$

$$= \left(\frac{1}{3\frac{\pi}{2} + 2m\pi}, \frac{1}{3\frac{\pi}{2} + 2m\pi} \right),$$

where $m \in \mathbb{N}$ is chosen so that

$$\frac{1}{\frac{\pi}{2} + 2m\pi} \le \frac{\delta}{2\sqrt{2}}.$$

Then

$$\frac{1}{3\frac{\pi}{2} + 2m\pi} \le \frac{1}{\frac{\pi}{2} + 2m\pi}$$

$$\le \frac{\delta}{2\sqrt{2}}$$

and

$$|y_1^1 - y_1^2| \le |y_1^1| + |y_2^1|$$

$$= \frac{1}{\frac{\pi}{2} + 2m\pi} + \frac{1}{3\frac{\pi}{2} + 2m\pi}$$

$$\le \frac{\delta}{2\sqrt{2}} + \frac{\delta}{2\sqrt{2}}$$

$$= \frac{\delta}{\sqrt{2}},$$

$$|y_2^1 - y_2^2| \le |y_2^1| + |y_2^2|$$

$$= \frac{1}{\frac{\pi}{2} + 2m\pi} + \frac{1}{3\frac{\pi}{2} + 2m\pi}$$

$$\le \frac{\delta}{2\sqrt{2}} + \frac{\delta}{2\sqrt{2}}$$

$$= \frac{\delta}{\sqrt{2}}.$$

Hence

$$|y^1 - y^2| = \sqrt{(y_1^1 - y_1^2)^2 + (y_2^1 - y_2^2)^2}$$

$$\le \sqrt{\frac{\delta^2}{2} + \frac{\delta^2}{2}}$$

$$= \sqrt{\delta^2}$$

$$= \delta.$$

Then

$$\omega(\delta, f, X) \ge |f(y_1^1, y_2^1) - f(y_1^2, y_2^2)|$$

$$= \left| \sin\frac{1}{y_1^1} + \sin\frac{1}{y_2^1} - \sin\frac{1}{y_1^2} - \sin\frac{1}{y_2^2} \right|$$

$$= \left| \sin\left(\frac{\pi}{2} + 2m\pi\right) + \sin\left(\frac{\pi}{2} + 2m\pi\right) - \sin\left(3\frac{\pi}{2} + 2m\pi\right) - \sin\left(3\frac{\pi}{2} + 2m\pi\right) \right|$$

$$= |1 + 1 + 1 + 1|$$

$$= 4.$$

Hence by (2.12) we get

$$\omega(\delta, f, X) = 4.$$

Exercise 2.22. Find the modulus of continuity of the following functions:
1.

$$f(x, y) = \frac{1}{x} + \frac{1}{y}, \quad (x, y) \in X,$$

where

$$X = \{(x, y) \in \mathbb{R}^2 : 0 < x < 1, \ 0 < y < 1\}.$$

2.

$$f(x, y) = \sqrt{x} + \sqrt{y}, \quad (x, y) \in X,$$

where

$$X = \{(x,y) \in \mathbb{R}^2 : x \geq 0, \ y \geq 0\}.$$

3.

$$f(x,y) = \frac{1}{\sqrt{x}} + \frac{1}{\sqrt{y}}, \quad (x,y) \in X,$$

where

$$X = \{(x,y) \in \mathbb{R}^2 : x > 0, \ y > 0\}.$$

4.

$$f(x,y) = 2x + 2y - 1, \quad (x,y) \in X,$$

where $X = \mathbb{R}^2$.

2.6 Properties of continuous functions

In the previous sections, we saw that a continuous map does not necessarily take open sets and closed sets to open sets and closed sets, respectively. In this section, we consider some cases where this is possible. We will start with the following useful theorem needed for the proof of the main results in this section.

Theorem 2.8. *Let X be a linearly connected set in $\mathbb{R}^n$, let $f \in C(X)$, and let $f(a) = A$ and $f(b) = B$ for some $a, b \in X$ and $A < B$. Then, for any $C \in [A, B]$, there is $c \in X$ such that $f(c) = C$.*

Proof. Since X is linearly connected, there is a continuous curve

$$\Gamma = \{x(t) : t \in [p,q]\} \subset X$$

such that

$$x(p) = a,$$
$$x(q) = b.$$

Set

$$F(t) = f(x(t)), \quad t \in [p,q].$$

Then $F \in C([p,q])$, and

$$\begin{aligned} F(p) &= f(x(p)) \\ &= f(a) \\ &= A, \end{aligned}$$

$$F(q) = f(x(q))$$
$$= f(b)$$
$$= B.$$

Hence and the Cauchy mean value theorem it follows that there is $c_1 \in [p, q]$ such that

$$F(c_1) = C.$$

Let

$$x(c_1) = c.$$

Then

$$f(c) = C.$$

This completes the proof. □

The next result is an auxiliary result for the main results in this section, which gives a necessary and sufficient condition for the continuity of a function at a point.

Theorem 2.9. *Let X be an open set in $\mathbb{R}^n$, and let $f : X \to \mathbb{R}$. The function f is a continuous function at $x^0 \in X$ if for any neighborhood V of $f(x^0)$, there is a neighborhood U of x^0 such that*

$$f(U) \subset V, \quad U \subset X. \tag{2.13}$$

Proof. 1.   Let f be continuous at x^0. Then by the definition of a continuous function at a point it follows that for any neighborhood V of $f(x^0)$, there is a neighborhood $U(x^0)$ of x^0 such that

$$f(X \cap U(x^0)) \subset V.$$

Since $U(x^0)$ is a neighborhood of x^0, we have that $U(x^0)$ is an open set in $\mathbb{R}^n$. Hence $X \cap U(x^0)$ is an open set. Set $U = X \cap U(x^0)$. Then we get (2.13).
2.   The converse statement follows directly from the definition of a continuous function at a point. This completes the proof. □

The next result is one of the main results in this section, which gives a necessary and sufficient condition for a continuous map to take open sets to open sets.

Theorem 2.10. *Let X be an open set in $\mathbb{R}^n$, and let $f : X \to \mathbb{R}$. Then $f \in C(X)$ if and only if for any open set V in $\mathbb{R}$, the set $f^{-1}(V)$ is an open set in X.*

Proof. 1.   Let $f \in C(X)$, and let V be an open set in $\mathbb{R}$. Let $x \in f^{-1}(V)$. Since V is an open set, we have that it is a neighborhood of $f(x)$. Now, using that f is continuous at x and Theorem 2.9, it follows that there is a neighborhood U of x such that $f(U) \subset V$, and then $U \subset f^{-1}(V)$. Because $x \in f^{-1}(V)$ was arbitrarily chosen and we have found its neighborhood that is contained in $f^{-1}(V)$, we conclude that $f^{-1}(V)$ is an open set.
2.   Suppose that for any open set V in $\mathbb{R}$, the set $f^{-1}(V)$ is an open set. Take any $x \in X$. Then for any neighborhood V of $f(x)$, we have that $U = f^{-1}(V)$ is an open set and it is a neighborhood of x. Since $f(U) = V$, we conclude that f is continuous at x. Because $x \in X$ was arbitrarily chosen and f is continuous at it, we obtain that $f \in C(X)$. This completes the proof.   □

Now we give a necessary and sufficient condition for a continuous map to take closed sets to closed sets.

Theorem 2.11. *Let X be a closed set in $\mathbb{R}^n$, and let $f : X \to \mathbb{R}$. Then $f \in C(X)$ if and only if for any closed set V of $\mathbb{R}$, the set $f^{-1}(V)$ is closed in X.*

Proof. 1.   Let $f \in C(X)$, and let $Y \subset \mathbb{R}$ be a closed set in $\mathbb{R}$. Let also x be an adherent point of $f^{-1}(Y) \subset X$. Then

$$x \in \overline{f^{-1}(Y)} \subset \overline{X} = X.$$

Because x is an adherent point of $f^{-1}(Y)$, there is a sequence $\{x^m\}_{m \in \mathbb{N}} \subset f^{-1}(Y)$ such that

$$\lim_{m \to \infty} x^m = x.$$

We have that

$$\{y^m = f(x^m)\}_{m \in \mathbb{N}} \subset Y,$$

and using that $f \in C(X)$, we find

$$\lim_{m \to \infty} y^m = \lim_{m \to \infty} f(x^m)$$
$$= f(x).$$

Because Y is closed and $y^m \in Y$, $m \in \mathbb{N}$, we get that $y = f(x) \in Y$. Hence $x \in f^{-1}(Y)$. Since x was an arbitrarily chosen adherent point of $f^{-1}(Y)$ and $x \in f^{-1}(Y)$, we conclude that $f^{-1}(Y)$ is a closed set.
2.   Let for any closed set Y of $\mathbb{R}$, the set $f^{-1}(Y)$ be a closed set in X. Take $x^0 \in X$ arbitrarily. Let also V be a neighborhood of $f(x^0)$ and $f(x^0) \in W$. Since V is an open set in $\mathbb{R}$, we have that $\mathbb{R}\backslash V$ is a closed set in $\mathbb{R}$. By Theorem 2.10 we have that $f^{-1}(V)$ is

an open set, and thus $f^{-1}(\mathbb{R}\backslash V)$ is a closed set. Hence $W = \mathbb{R}^n\backslash f^{-1}(\mathbb{R}\backslash V)$ is an open set. Set $y^0 = f(x^0)$. We have

$$x^0 = f^{-1}(y^0) \in f^{-1}(V).$$

Because $V \cap (\mathbb{R}\backslash V) = \emptyset$, we get

$$f^{-1}(V) \bigcap f^{-1}(\mathbb{R}\backslash V) = \emptyset.$$

Hence $x^0 \notin f^{-1}(\mathbb{R}\backslash V)$, and thus $x^0 \in W$. Moreover, $f(W \cap X) \subset V$. Note that if $x \in W \cap X$, then $f(x) \notin \mathbb{R}\backslash V$ and $x \in \mathbb{R}^n\backslash f^{-1}(\mathbb{R}\backslash V)$. Thus $f(x) \in V$. Therefore f is continuous at x^0. Because $x^0 \in X$ was arbitrarily chosen, we conclude that $f \in C(X)$. This completes the proof. □

2.7 Advanced practical problems

Problem 2.1. Find the function $z = f(x,y)$, where z is the volume of a right circular cone, x is its length, and y is its circle circumference.

Problem 2.2. Find the domain of the function f, where

1.

$$f(x,y) = \log(3x + y - 3) + \frac{\log(3 - x)}{\sqrt{3x - 2y + 6}}, \quad (x,y) \in \mathbb{R}^2.$$

2.

$$f(x,y) = \log(x^2 + 4y^2 - 2x - 3), \quad (x,y) \in \mathbb{R}^2.$$

3.

$$f(x,y) = \log \frac{x^2 + y^2 - x}{2x - x^2 - y^2}, \quad (x,y) \in \mathbb{R}^2.$$

4.

$$f(x,y) = \sqrt{1 - |x| - |y|}, \quad (x,y) \in \mathbb{R}^2.$$

5.

$$f(x,y) = \frac{\log x \log y}{\sqrt{1 - x - y}}, \quad (x,y) \in \mathbb{R}^2.$$

6.

$$f(x,y) = \sqrt{\log(2 - x^2 - y^2)}, \quad (x,y) \in \mathbb{R}^2.$$

7.

$$f(x,y) = \sqrt{(x^2 + y^2 - 1)(2 - x^2 - y^2)}, \quad (x,y) \in \mathbb{R}^2.$$

Problem 2.3. Find the domain of the function f, where

1.
$$f(x,y,z) = \frac{1}{\sqrt{z - x^2 - y^2}}, \quad (x,y,z) \in \mathbb{R}^3.$$

2.
$$f(x,y,z) = \log(2z^2 - 6x^2 - 3y^2 - 6), \quad (x,y,z) \in \mathbb{R}^3.$$

3.
$$f(x,y,z) = \sqrt{x^2 + 2y^2 + 2yz + z^2 - 2y + 1}, \quad (x,y,z) \in \mathbb{R}^3.$$

4.
$$f(x,y,z) = \sqrt{16 - x^2 - y^2 - z^2} \log(x^2 + y^2 + z^2 - 4), \quad (x,y,z) \in \mathbb{R}^3.$$

5.
$$f(x,y,z) = \sqrt{2(x^2 + y^2 + z^2) - (x^2 + y^2 + z^2)^2 - 1}, \quad (x,y,z) \in \mathbb{R}^3.$$

6.
$$f(x,y,z) = \frac{\log x + \log z}{\sqrt{y-1}} + \log(5 - x - y - z), \quad (x,y,z) \in \mathbb{R}^3.$$

7.
$$f(x,y,z) = \sqrt{x} + \sqrt{y} + \sqrt{z} + \sqrt{1 - x - y} + \sqrt{3x + y - 3z}, \quad (x,y,z) \in \mathbb{R}^3.$$

8.
$$f(x,y,z) = \frac{\log(z^2 - x^2 - y^2)}{\sqrt{1 - x^2 - y^2 - z^2}}, \quad (x,y,z) \in \mathbb{R}^3.$$

9.
$$f(x,y,z) = \frac{\log(x^2 + y^2 + z^2 - 4)}{\sqrt{4 - x^2 - y^2}}, \quad (x,y,z) \in \mathbb{R}^3.$$

10.
$$f(x,y,z) = \sqrt{z - xy} + \sqrt{1 - z - \sqrt{x^2 + y^2}}, \quad (x,y,z) \in \mathbb{R}^3.$$

Problem 2.4. Find the domain of the function f, where

1.
$$f(x_1,\ldots,x_n) = \sum_{j=1}^n \sqrt{1 - |x_j|}, \quad (x_1,\ldots,x_n) \in \mathbb{R}^n.$$

2.
$$f(x_1,\ldots,x_n) = \sqrt{1 - \sum_{j=1}^n x_j} + \sum_{j=1}^n \sqrt{x_j}, \quad (x_1,\ldots,x_n) \in \mathbb{R}^n.$$

3.
$$f(x_1,\ldots,x_n) = \log\left(1 - \sum_{j=1}^n (x_j - j)^2\right), \quad (x_1,\ldots,x_n) \in \mathbb{R}^n.$$

4.
$$f(x_1,\ldots,x_n) = \log\left(\sum_{j=1}^n x_j^2 + \sum_{i<j,j=1}^n x_i x_j\right), \quad (x_1,\ldots,x_n) \in \mathbb{R}^n.$$

5.
$$f(x_1, \ldots, x_n) = \sum_{j=1}^{n} \arcsin(x_j - j), \quad (x_1, \ldots, x_n) \in \mathbb{R}^n.$$

6.
$$f(x_1, \ldots, x_n) = \sum_{j=1}^{n} \arccos(jx_j), \quad (x_1, \ldots, x_n) \in \mathbb{R}^n.$$

7.
$$f(x_1, \ldots, x_n) = \sum_{j=1}^{n} \sqrt{j + x_j}, \quad (x_1, \ldots, x_n) \in \mathbb{R}^n.$$

8.
$$f(x_1, \ldots, x_n) = \sum_{j=1}^{n} \sqrt[3]{j^2 + x_j^2}, \quad (x_1, \ldots, x_n) \in \mathbb{R}^n.$$

9.
$$f(x_1, \ldots, x_n) = \sum_{j=1}^{n} \sqrt[3]{4 - x_j^3}, \quad (x_1, \ldots, x_n) \in \mathbb{R}^n.$$

10.
$$f(x_1, \ldots, x_n) = \sum_{j=1}^{n} \frac{2 + x_j}{3 - x_j}, \quad (x_1, \ldots, x_n) \in \mathbb{R}^n.$$

Problem 2.5. Find the codomain of the function f, where

1.
$$f(x,y) = \log(4x^2 + 2y^2 - 4xy + 12x - 12y + 21), \quad (x,y) \in \mathbb{R}^2.$$

2.
$$f(x,y) = e^{2xy} - e^{xy} + 2, \quad (x,y) \in \mathbb{R}^2.$$

3.
$$f(x,y) = 3\sin\left(\frac{y}{x}\right) + 8\left(\sin\left(\frac{y}{2x}\right)\right)^2, \quad (x,y) \in \mathbb{R}^2.$$

4.
$$f(x,y) = 3\sin(x - y) + 6\sin(x + y) + 4\cos(x - y) + 8(x + y), \quad (x,y) \in \mathbb{R}^2.$$

5.
$$f(x,y) = \arccos\frac{1 + x^2 y^2}{2xy}, \quad (x,y) \in \mathbb{R}^2.$$

6.
$$f(x,y) = x^2 + y^2 - 12x + 16y + 25,$$

where

$$(x,y) \in \{(y_1, y_2) \in \mathbb{R}^2 : y_1^2 + y_2^2 = 25\}.$$

7.
$$f(x,y) = \log(2x^2 + 3y^2),$$

where

$$(x,y) \in \{(y_1, y_2) \in \mathbb{R}^2 : y_1 + y_2 = 2,\ y_1 \geq 0,\ y_2 \geq 0\}.$$

Problem 2.6. Find the function f, where

1.

$$f(x,y) = \phi(xy) + \sqrt{xy}\,\psi\!\left(\frac{y}{x}\right), \quad (x,y) \in \mathbb{R}^2,$$

if

$$f(1,y) = 1,$$
$$f(x,x) = x, \quad (x,y) \in \mathbb{R}^2,$$

and $\phi \in C(\mathbb{R})$.

2.

$$f(x,y) = \phi(xy) + \psi\!\left(\frac{y}{x}\right), \quad (x,y) \in \mathbb{R}^2,$$

if

$$f(x,1) = \sin\!\left(\frac{\pi x}{2}\right),$$
$$f(x,x) = 1, \quad x \in \mathbb{R},$$

and $\phi, \psi \in C(\mathbb{R})$.

3.

$$f(x,y) = \phi(x) + \psi(y + e^x), \quad (x,y) \in \mathbb{R}^2,$$

if

$$f(0,y) = y^2,$$
$$f(x, -e^x) = x^2 + 1, \quad (x,y) \in \mathbb{R}^2,$$

and $\phi, \psi \in C(\mathbb{R})$.

4.

$$f(x,y,z) = g\!\left(\frac{x}{z}, \frac{y}{z} - x\right), \quad (x,y,z) \in \mathbb{R}^3, \quad x > 0, \quad z > 0,$$

if

$$f(x,x,z) = x, \quad (x,y,z) \in \mathbb{R}^3, \quad x > 0, \quad z > 0.$$

5.
$$f(x,y,z) = g(x^2 z, 2y^2 z - z^4), \quad (x,y,z) \in \mathbb{R}^3, \quad x > 0, \quad z > 0,$$

if

$$f(x,x,z) = \frac{z^3}{x^2}, \quad (x,y,z) \in \mathbb{R}^3, \quad x > 0, \quad z > 0.$$

Problem 2.7. Find the level sets of the following functions:

1.
$$f(x,y) = \sqrt{y - \sin x}, \quad (x,y) \in \mathbb{R}^2.$$

2.
$$f(x,y) = \log x - \log(\sin(y)), \quad (x,y) \in \mathbb{R}^2.$$

3.
$$f(x,y) = \arcsin \frac{y}{x}, \quad (x,y) \in \mathbb{R}^2.$$

4.
$$f(x,y) = \arctan \frac{2y}{x^2 + y^2 - 1}, \quad (x,y) \in \mathbb{R}^2.$$

Problem 2.8. Find the set levels of the following functions:

1.
$$f(x,y,z) = \log(1 - |x| - |y| - |z|), \quad (x,y,z) \in \mathbb{R}^3.$$

2.
$$f(x,y,z) = \frac{z}{x + y + z - 1}, \quad (x,y,z) \in \mathbb{R}^3.$$

3.
$$f(x,y,z) = \log(z^2 - x^2 - y^2), \quad (x,y,z) \in \mathbb{R}^3.$$

Problem 2.9. Find the homogeneity degree or the positive homogeneity degree of the following functions:

1.
$$f(x_1, x_2, x_3, x_4) = \sqrt{x_1^2 + x_2^2 + x_3^2 - x_4^2},$$

where

$$(x_1, x_2, x_3, x_4) \in A = \{(y_1, y_2, y_3, y_4) \in \mathbb{R}^4 : y_1^2 + y_2^2 + y_3^2 \geq y_4^2\}.$$

2.
$$f(x_1, \ldots, x_n) = \left| \sum_{j=1}^{n} x_j \right|, \quad (x_1, \ldots, x_n) \in \mathbb{R}^n.$$

3.
$$f(x_1, x_2, x_3, x_4) = \frac{x_1 x_2 + x_3 x_4}{x_1^4 + x_2^4 + x_3^4 + x_4^4},$$

where

$$(x_1, x_2, x_3, x_4) \in A = \{(y_1, y_2, y_3, y_4) \in \mathbb{R}^4 : (y_1, y_2, y_3, y_4) \neq (0, 0, 0, 0)\}.$$

Problem 2.10. Prove that the function

$$f(x, y, z) = \begin{cases} xy^7 z^3 & \text{if } z \geq 0, \\ z^{11} & \text{if } z < 0 \end{cases}$$

is a locally homogeneous function but not a homogeneous function.

Problem 2.11. Find

1.
$$\lim_{(x,y) \to (0,0)} \frac{xy^2(x^2 + y^2)}{1 - \cos(x^2 + y^2)}.$$

2.
$$\lim_{(x,y) \to (0,0)} \frac{\sqrt{2} - \sqrt{1 + \cos(x^2 + y^2)}}{(\tan(x^2 + y^2))^2}.$$

3.
$$\lim_{(x,y) \to (0,0)} \frac{1 - \sqrt[3]{(\sin x)^4 + (\cos y)^4}}{\sqrt{x^2 + y^2}}.$$

4.
$$\lim_{(x,y) \to (0,0)} \sqrt{x^2 + y^2} \log(x^2 + y^2).$$

5.
$$\lim_{(x,y) \to (0,0)} (x^2 + y^2)^{x^2 y^2}.$$

6.
$$\lim_{(x,y) \to (0,0)} (1 + xy^2)^{\frac{1}{x^2 + y^2}}.$$

7.
$$\lim_{(x,y) \to (0,0)} (1 + xy)^{\frac{1}{x^2 + y^2}}.$$

8.
$$\lim_{(x,y) \to (0,0)} (1 + xy)^{\frac{1}{|x| + |y|}}.$$

9.
$$\lim_{(x,y) \to (0,0)} (\cos \sqrt{x^2 + y^2})^{-\frac{1}{x^2 + y^2}}.$$

10.
$$\lim_{(x,y) \to (0,2)} \frac{\sin(xy)}{x}.$$

11.
$$\lim_{(x,y) \to (\infty,\infty)} (\sqrt{4(x^4 + y^4) + 13(x^2 + y^2) + 8x^2 y^2 - 7} - 2(x^2 + y^2)).$$

Problem 2.12. Find

1.

$$\lim_{x\to0}\lim_{y\to0}\frac{x^2y^2}{x^2y^2+(x-y)^2}, \quad \lim_{y\to0}\lim_{x\to0}\frac{x^2y^2}{x^2y^2+(x-y)^2}, \quad \lim_{(x,y)\to(0,0)}\frac{x^2y^2}{x^2y^2+(x-y)^2}.$$

2.

$$\lim_{x\to0}\lim_{y\to0}\frac{x^8+x^5+x^4+y^4+y^5-y^8}{x^4+y^4}, \quad \lim_{y\to0}\lim_{x\to0}\frac{x^8+x^5+x^4+y^4+y^5-y^8}{x^4+y^4},$$

$$\lim_{(x,y)\to(0,0)}\frac{x^8+x^5+x^4+y^4+y^5-y^8}{x^4+y^4}.$$

3.

$$\lim_{x\to0}\lim_{y\to0}\left(x+y\sin\frac1x\right), \quad \lim_{y\to0}\lim_{x\to0}\left(x+y\sin\frac1x\right), \quad \lim_{(x,y)\to(0,0)}\left(x+y\sin\frac1x\right).$$

4.

$$\lim_{x\to0}\lim_{y\to0}\left(x\sin\frac1y+y\sin\frac1x\right), \quad \lim_{y\to0}\lim_{x\to0}\left(x\sin\frac1y+y\sin\frac1x\right),$$

$$\lim_{(x,y)\to(0,0)}\left(x\sin\frac1y+y\sin\frac1x\right).$$

5.

$$\lim_{x\to0}\lim_{y\to0}\left(\frac yx\tan\frac{x}{x+y}\right), \quad \lim_{y\to0}\lim_{x\to0}\left(\frac yx\tan\frac{x}{x+y}\right), \quad \lim_{(x,y)\to(0,0)}\left(\frac yx\tan\frac{x}{x+y}\right).$$

6.

$$\lim_{x\to0}\lim_{y\to0}\log_{1+x}(1+x+y), \quad \lim_{y\to0}\lim_{x\to0}\log_{1+x}(1+x+y), \quad \lim_{(x,y)\to(0,0)}\log_{1+x}(1+x+y).$$

7.

$$\lim_{x\to\infty}\lim_{y\to\infty}(x^2+y^2)^a e^{-x^2-y^2}, \quad \lim_{y\to\infty}\lim_{x\to\infty}(x^2+y^2)^a e^{-x^2-y^2}, \quad \lim_{(x,y)\to(\infty,\infty)}(x^2+y^2)^a e^{-x^2-y^2}.$$

Problem 2.13. Let $n=2, f:X\to Y, x^0=(x_1^0,x_2^0)\in\mathbb{R}^2$, and let $\delta_1,\delta_2>0$ be such that

$$\{(x_1,x_2):|x_1-x_1^0|<\delta_1, |x_2-x_2^0|<\delta_2\}\subset X.$$

Suppose that

$$\lim_{(x_1,x_2)\to(x_1^0,x_2^0)}f(x_1,x_2)=a$$

for some $a\in\mathbb{R}$ and that for all $x_2\in(x_1^0-\delta_2,x_2^0+\delta_2), x_2\neq x_2^0$, there exists the limit $\lim_{x_1\to x_1^0}f(x_1,x_2)$. Prove that

$$\lim_{x_2\to x_2^0}\lim_{x_1\to x_1^0}f(x_1,x_2)=a.$$

Problem 2.14. Find

$$\lim_{\substack{(x,y)\in X \\ t\to\infty}}f(x,y),$$

where

$$X = \{(x,y) \in \mathbb{R}^2 : x = t\cos\phi, \ y = t\sin\phi, \ \phi \in [0, 2\pi), \ t \to \infty\}$$

and

1.

$$f(x,y) = e^{\frac{xy^2}{x^2+y^2}}, \quad (x,y) \in X.$$

2.

$$f(x,y) = e^{x^2-y^2}\sin(2xy), \quad (x,y) \in X.$$

3.

$$f(x,y) = \frac{xy}{\sqrt{x^2+y^2}}\log\left(\frac{1}{x} + e^{\frac{1}{y}}\right), \quad (x,y) \in X.$$

Problem 2.15. Find

$$\lim_{X \ni (x,y) \to (0,0)} f(x,y),$$

where

$$X = \{(x,y) \in \mathbb{R}^2 : x = at^m, \ y = \beta t^p, \ a^2 + \beta^2 \neq 0, \ m,p \in \mathbb{N}, \ t > 0\},$$

and

$$f(x,y) = \begin{cases} \dfrac{ye^{-\frac{1}{x^2}}}{y^2 + e^{-\frac{2}{x^2}}} & \text{if } x \neq 0, \\ 0 & \text{if } x = 0. \end{cases}$$

Problem 2.16. Find $a \in \mathbb{R}$ such that the function

$$f(x,y) = \begin{cases} \dfrac{x^2 y}{x^4 + y^2} & \text{if } x^2 + y^2 \neq 0, \\ a & \text{if } x^2 + y^2 = 0 \end{cases}$$

1. is a continuous function at $(0,0)$ along the curve

$$X = \{(x,y) \in \mathbb{R}^2 : x = at, \ y = \beta t, \ a^2 + \beta^2 \neq 0, \ t \in \mathbb{R}\}.$$

2. is a continuous function at $(0,0)$ along the curve

$$X = \{(x,y) \in \mathbb{R}^2 : y = ax^2\}.$$

3. is a continuous function at $(0,0)$.

Problem 2.17. Find $a, b \in \mathbb{R}$ such that the function

$$f(x,y) = \begin{cases} a & \text{if } x^2 + y^2 = 0, \\ \sqrt{5(x^2 + y^2) - 4 - (x^2 + y^2)^2} & \text{if } 1 \le x^2 + y^2 \le 4, \\ b & \text{if } x^2 + y^2 \ge 4 \end{cases}$$

is a continuous function in its domain.

Problem 2.18. Check if the function

$$f(x,y) = \begin{cases} 2\arctan(\frac{1}{x^2-y^2}) & \text{if } x + y \ne 0, \\ \pi & \text{if } x + y = 0 \end{cases}$$

is a continuous function in its domain.

Problem 2.19. Find $a, b \in \mathbb{R}$ such that the function

$$f(x_1,\ldots,x_n) = \begin{cases} a & \text{if } \sum_{j=1}^n x_j^2 = 0, \\ \frac{\sum_{j=1}^n x_j^2}{\log(1-\sum_{j=1}^n x_j^2)} & \text{if } 0 < \sum_{j=1}^n x_j^2 < 1, \\ b & \text{if } \sum_{j=1}^n x_j^2 \ge 1 \end{cases}$$

is a continuous function in $\mathbb{R}^n$.

Problem 2.20. Prove that the function

$$f(x_1,\ldots,x_n) = \sum_{j=1}^n |x_j|, \quad (x_1,\ldots,x_n) \in \mathbb{R}^n,$$

is a continuous function in $\mathbb{R}^n$.

Problem 2.21. Prove that the function

$$f(x_1,\ldots,x_n) = \max_{j \in \{1,\ldots,n\}} |x_j|, \quad (x_1,\ldots,x_n) \in \mathbb{R}^n,$$

is a continuous function in $\mathbb{R}^n$.

Problem 2.22. Let $X \subset \mathbb{R}^n$. Prove that the function

$$f(x) = d(x, X), \quad x \in \mathbb{R}^n,$$

is a continuous function in $\mathbb{R}^n$.

Problem 2.23. Let $X \subset \mathbb{R}^n$, and let $f : X \to \mathbb{R}$ be continuous at $x^0 \in X$ such that

$$f(x^0) > c \tag{2.14}$$

for some $c \in \mathbb{R}$. Prove that there is a neighborhood $U(x^0)$ of x^0 such that

$$f(x) > c \quad \text{for all } x \in U(x^0).$$

Problem 2.24. Let $X \subset \mathbb{R}^n$, and let $f : X \to \mathbb{R}$ be continuous at $x^0 \in X$ such that

$$f(x^0) < c$$

for some $c \in \mathbb{R}$. Prove that there is a neighborhood $U(x^0)$ of x^0 such that

$$f(x) < c \quad \text{for all } x \in U(x^0).$$

Problem 2.25. Prove that the level sets of a continuous function are closed sets.

Problem 2.26. Let $f : \mathbb{R}^n \to \mathbb{R}, f \in C(\mathbb{R}^n)$, and $c \in \mathbb{R}$. Prove that
1. the set

$$X = \{x \in \mathbb{R}^n : f(x) < c\}$$

is an open set.
2. the set

$$Y = \{x \in \mathbb{R}^n : f(x) \le c\}$$

is a closed set.

Problem 2.27. Let $X \subset \mathbb{R}^n$ be a closed set, let $f : X \to \mathbb{R}, f \in C(X)$, and let $c \in \mathbb{R}$. Prove that the set

$$Y = \{x \in X : f(x) \ge c\}$$

is a closed set.

Problem 2.28. Let $n = 2$, let $X \subset \mathbb{R}^2$ be a domain, and let $f : X \to \mathbb{R}$ be such that $f(x,y)$ is continuous in x and continuous in y uniformly with respect to x. Prove that $f \in C(X)$.

Problem 2.29. Let $n = 2$, let $X \subset \mathbb{R}^2$ be a domain, and let $f : X \to \mathbb{R}$ be such that $f(x,y)$ is continuous in x and

$$|f(x,y^1) - f(x,y^2)| \le L|y^1 - y^2|$$

for all $(x,y^1), (x,y^2) \in X$ and some $L > 0$. Prove that $f \in C(X)$.

Problem 2.30. Let $n = 2$, let $X \subset \mathbb{R}^2$, and let $f : X \to \mathbb{R}$ be such that $f(x,y)$ is continuous in x and continuous and monotonic in y. Prove that $f \in C(X)$.

Problem 2.31. Check if the following functions are uniformly continuous on X:

1. $f(x,y) = 10x - 2y - 5$, $(x,y) \in X = \mathbb{R}^2$.
2. $f(x,y) = \sqrt{x^2 + y^2}$, $(x,y) \in X = \mathbb{R}^2$.
3. $f(x,y) = \cos\frac{\pi}{4-x^2-y^2}$, $(x,y) \in X = \{(y_1,y_2) \in \mathbb{R}^2 : y_1^2 + y_2^2 < 4\}$.
4.

$$f(x,y) = \begin{cases} \sqrt{x^2 + y^2} \sin\frac{1}{\sqrt{x^2+y^2}} & \text{if } x^2 + y^2 > 0, \\ 0 & \text{if } (x,y) = (0,0), \end{cases}$$

$X = \mathbb{R}^2$.

Problem 2.32. Find the modulus of continuity of the following functions:

1.

$$f(x,y) = \cos x + \cos y, \quad (x,y) \in X,$$

where $X = \mathbb{R}^2$.

2.

$$f(x,y) = \log x + \log y, \quad (x,y) \in X,$$

where

$$X = \{(x,y) \in \mathbb{R}^2 : x \geq 1, \ y \geq 1\}.$$

3.

$$f(x,y) = x^3 + y^3, \quad (x,y) \in X,$$

where $X = \mathbb{R}^2$.

4.

$$f(x,y) = \frac{1}{x^2} + \frac{1}{y^2}, \quad (x,y) \in X,$$

where

$$X = \{(x,y) \in \mathbb{R}^2 : 0 < x < 1, \ 0 < y < 1\}.$$

Problem 2.33. Let $X \subset \mathbb{R}^n$, let $f : X \to \mathbb{R}$, and let $\omega(\delta,f,X)$, $\delta > 0$, be the modulus of continuity of f.

1. Prove that if $\delta_1 < \delta_2$, then

$$\omega(\delta_1,f,X) \leq \omega(\delta_2,f,X).$$

2. Prove that if f is bounded on X, then $\omega(\delta,f,X)$ is bounded.
3. Prove that if X is bounded and f is unbounded on X, then $\omega(\delta,f,X) = \infty$.

Problem 2.34. Let $X \subset \mathbb{R}^n$, and let $f, g : X \to \mathbb{R}$ be uniformly continuous on X. Prove that $\alpha f + \beta g$ is uniformly continuous on X for all $\alpha, \beta \in \mathbb{R}$.

Problem 2.35. Let $X \subset \mathbb{R}^n$ be bounded, and let $f, g : X \to \mathbb{R}$ be uniformly continuous. Prove that fg is uniformly continuous on X.

3 Partial derivatives, partial differentials, and differentiable functions

In this chapter, we introduce partial derivatives of first- and higher-order and differentials of functions of several variables. Some of their properties are deduced. Criteria for differentiability of functions of several variables are deduced. The gradient of a function is defined and explored. Directional derivatives are defined and investigated.

3.1 Definition for partial derivatives

In mathematics, the derivative is a fundamental tool that quantifies the sensitivity of change of a function output with respect to its input. The derivative of a function of a single variable at a chosen input value, when it exists, is the slope of the tangent line to the graph of the function at that point. The tangent line is the best linear approximation of the function near that input value. For this reason, the derivative is often described as the instantaneous rate of change, the ratio of the instantaneous change in the dependent variable to that of the independent variable. The process of finding a derivative is called differentiation.

A partial derivative of a function of several variables is its derivative with respect to one of those variables, with the others held constant. Partial derivatives are used in vector calculus and differential geometry.

Definition 3.1. Suppose that $x^0 = (x_1^0, \ldots, x_n^0) \in \mathbb{R}^n$ and $f = f(x_1, \ldots, x_n)$ is a function defined in a neighborhood of x^0. Then the partial derivative of f with respect to the variable $x_j, j \in \{1, \ldots, n\}$, at the point x^0 is defined as the number

$$\frac{\partial f}{\partial x^0}(x_1^0, \ldots, x_n^0) = \frac{df}{dx_j}(x_1^0, \ldots, x_{j-1}^0, x_j, x_{j+1}^0, \ldots, x_n^0)|_{x_j = x_j^0}$$

$$= \lim_{h \to 0} \frac{f(x_1^0, \ldots, x_{j-1}^0, x_j^0 + h, x_{j+1}^0, \ldots, x_n^0) - f(x_1^0, \ldots, x_{j-1}^0, x_j^0, x_{j+1}^0, \ldots, x_n^0)}{h}.$$

Sometimes, the partial derivative of f with respect to $x_j, j \in \{1, \ldots, n\}$, at the point x^0 is denoted by $f_{x_j}(x^0)$ or $f_{x_j}(x_1^0, \ldots, x_n^0)$.

A geometrical interpretation of partial derivatives is as follows. Consider a surface $x = f(x, y), (x, y) \in \mathbb{R}^2$, and let it intersect the vertical plane $y = y_0$, where y_0 is a given constant (see Fig. 3.1). As a result, we get the curve of intersection, denoted by AP_0B. The partial derivative $\frac{\partial f}{\partial x}(x_0, y_0)$ is the slope of the tangent P_0T to this curve at the point (x_0, y_0, z_0). The partial derivative $\frac{\partial f}{\partial y}(x_0, y_0)$ is the slope of the tangent Q_0T to the curve of intersection BQ_0C of the surface with a vertical plane $x = x_0$, where x_0 is a given constant (see Fig. 3.2).

https://doi.org/10.1515/9783112218082-003

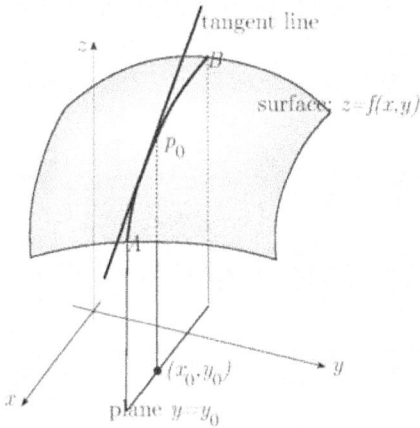

Figure 3.1: Tangent to a surface in the x-direction.

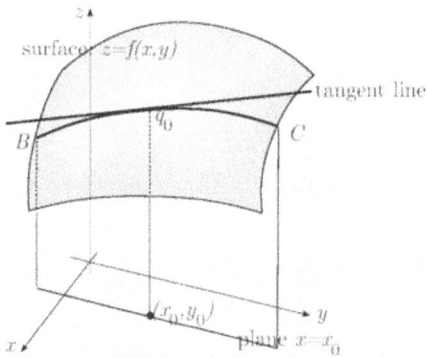

Figure 3.2: Tangent to a surface in the y-direction.

Example 3.1. Let us find the partial derivatives of the function

$$f(x,y) = x^3 + y^3 - 3xy, \quad (x,y) \in \mathbb{R}^2.$$

We have

$$f_x(x,y) = 3x^2 - 3y,$$
$$f_y(x,y) = 3y^2 - 3x, \quad (x,y) \in \mathbb{R}^2.$$

Example 3.2. Let us find the partial derivatives of the function

$$f(x,y,z) = \frac{z}{x} + \frac{x}{z} + y, \quad (x,y,z) \in \mathbb{R}^3, \quad x \neq 0, \quad z \neq 0.$$

For $(x, y, z) \in \mathbb{R}^3$, $x \neq 0$, $z \neq 0$, we have

$$f_x(x, y, z) = -\frac{z}{x^2} + \frac{1}{z},$$
$$f_y(x, y, z) = 1,$$
$$f_z(x, y, z) = \frac{1}{x} - \frac{x}{z^2}.$$

Example 3.3. Let us find the partial derivatives of the function

$$f(x_1, \ldots, x_n) = \sum_{j=1}^{n} (\sin x_j)^2, \quad (x_1, \ldots, x_n) \in \mathbb{R}^n.$$

We have

$$f_{x_j}(x_1, \ldots, x_n) = 2 \sin x_j \cos x_j$$
$$= \sin(2x_j), \quad j \in \{1, \ldots, n\}, \quad (x_1, \ldots, x_n) \in \mathbb{R}^n.$$

Exercise 3.1. Find the partial derivatives of the following functions:
1.

$$f(x, y) = \frac{x(x - y)}{y^2}, \quad (x, y) \in \mathbb{R}^2, \quad y \neq 0.$$

2.

$$f(x, y, z) = xy + xz + yz, \quad (x, y, z) \in \mathbb{R}^3.$$

3.

$$f(x_1, \ldots, x_n) = e^{\sum_{j=1}^{n} x_j^2}, \quad (x_1, \ldots, x_n) \in \mathbb{R}^n.$$

Example 3.4. Let us find $f_x(1, 1)$ and $f_y(1, 1)$, where

$$f(x, y) = (2x + y)^{2x+y}, \quad (x, y) \in \mathbb{R}^2, \quad 2x > -y.$$

Let

$$g(x, y) = \log(f(x, y)), \quad (x, y) \in \mathbb{R}^2, \quad 2x > -y.$$

Then

$$g(x, y) = \log((2x + y)^{2x+y})$$
$$= (2x + y) \log(2x + y), \quad (x, y) \in \mathbb{R}^2, \quad 2x > -y,$$
$$g_x(x, y) = 2 \log(2x + y) + \frac{2(2x + y)}{2x + y}$$
$$= 2 + 2\log(2x + y)$$
$$= 2(1 + \log(2x + y)), \quad (x, y) \in \mathbb{R}^2, \quad 2x > -y,$$

and

$$g_y(x,y) = \log(2x+y) + \frac{2x+y}{2x+y}$$
$$= 1 + \log(2x+y), \quad (x,y) \in \mathbb{R}^2, \quad 2x > -y.$$

On the other hand,

$$g_x(x,y) = \frac{f_x(x,y)}{f(x,y)},$$
$$g_y(x,y) = \frac{f_y(x,y)}{f(x,y)}, \quad (x,y) \in \mathbb{R}^2, \quad 2x > -y,$$

whereupon

$$f_x(x,y) = f(x,y)g_x(x,y)$$
$$= 2(1+\log(2x+y))(2x+y)^{2x+y},$$
$$f_y(x,y) = f(x,y)g_y(x,y)$$
$$= (1+\log(2x+y))(2x+y)^{2x+y}, \quad (x,y) \in \mathbb{R}^2, \quad 2x > -y.$$

Hence

$$f_x(1,1) = 2(1+\log(2+1))(2+1)^{2+1}$$
$$= 54(1+\log 3)$$

and

$$f_y(1,1) = 27(1+\log 3).$$

Exercise 3.2. Find $f_x(1,1)$ and $f_y(1,1)$, where

$$f(x,y) = \frac{x}{y^2}, \quad (x,y) \in \mathbb{R}^2, \quad y \neq 0.$$

Example 3.5. Let us find

$$\sum_{j=1}^{4} f_{x_j}(x_1,x_2,x_3,x_4),$$

where

$$f(x_1,x_2,x_3,x_4) = \frac{x_1-x_2}{x_3-x_4} + \frac{x_4-x_1}{x_2-x_3}, \quad (x_1,x_2,x_3,x_4) \in \mathbb{R}^4, \quad x_2 \neq x_3, x_4.$$

We have

$$f_{x_1}(x_1, x_2, x_3, x_4) = \frac{1}{x_3 - x_4} - \frac{1}{x_2 - x_3}$$

$$= \frac{(x_2 - x_3) - (x_3 - x_4)}{(x_2 - x_3)(x_3 - x_4)},$$

$$f_{x_2}(x_1, x_2, x_3, x_4) = -\frac{1}{x_3 - x_4} - \frac{x_4 - x_1}{(x_2 - x_3)^2}$$

$$= -\frac{(x_2 - x_3)^2 + x_4 - x_1}{(x_3 - x_4)(x_2 - x_3)^2},$$

$$f_{x_3}(x_1, x_2, x_3, x_4) = -\frac{x_1 - x_2}{(x_3 - x_4)^2} + \frac{x_4 - x_1}{(x_2 - x_3)^2}$$

$$= \frac{(x_3 - x_4)(x_4 - x_1) - (x_1 - x_2)(x_2 - x_3)^2}{(x_3 - x_4)^2(x_2 - x_3)^2},$$

$$f_{x_4}(x_1, x_2, x_3, x_4) = \frac{x_1 - x_2}{(x_3 - x_4)^2} + \frac{1}{x_2 - x_3}$$

$$= \frac{(x_1 - x_2)(x_2 - x_3) + (x_3 - x_4)^2}{(x_2 - x_3)(x_3 - x_4)^2}, \quad (x_1, x_2, x_3, x_4) \in \mathbb{R}^4, \quad x_2 \neq x_3, x_4.$$

Hence

$$\sum_{j=1}^{4} f_{x_j}(x_1, x_2, x_3, x_4) = \frac{x_2 - x_3 - (x_3 - x_4)}{(x_2 - x_3)(x_3 - x_4)} - \frac{(x_2 - x_3)^2 + x_4 - x_1}{(x_3 - x_4)(x_2 - x_3)^2}$$

$$+ \frac{(x_3 - x_4)(x_4 - x_1) - (x_1 - x_2)(x_2 - x_3)^2}{(x_3 - x_4)^2(x_2 - x_3)^2} + \frac{(x_1 - x_2)(x_2 - x_3) + (x_3 - x_4)^2}{(x_2 - x_3)(x_3 - x_4)^2}$$

$$= \frac{1}{(x_3 - x_4)^2(x_2 - x_3)^2}\left((x_2 - x_3)^2(x_3 - x_4) - (x_2 - x_3)(x_3 - x_4)^2 - (x_2 - x_3)^2(x_3 - x_4)\right.$$

$$- (x_3 - x_4)(x_4 - x_1) + (x_3 - x_4)(x_4 - x_1) - (x_1 - x_2)(x_2 - x_3)^2$$

$$\left. + (x_1 - x_2)(x_2 - x_3)^2 + (x_2 - x_3)(x_3 - x_4)^2\right)$$

$$= 0, \quad (x_1, x_2, x_3, x_4) \in \mathbb{R}^4, \quad x_2 \neq x_3, x_4.$$

Exercise 3.3. Find

$$xf_x(x, y) + yf_y(x, y), \quad (x, y) \in X,$$

where
1.

$$f(x, y) = \frac{x}{\sqrt{x^2 + y^2}}, \quad (x, y) \in X, \quad X = \{(x, y) \in \mathbb{R}^2 : (x, y) \neq (0, 0)\}.$$

2.
$$f(x,y) = \log(x^2 + xy + y^2), \quad (x,y) \in X, \quad X = \{(x,y) \in \mathbb{R}^2 : (x,y) \neq (0,0)\}.$$

The term differential is used in calculus to refer to an infinitesimal change in some varying quantity. A precise definition is as follows.

Definition 3.2. Let $f : X \to \mathbb{R}$ and $x^0 \in X$. Suppose $f_{x_j}(x^0)$ exists for some $j \in \{1,\dots,n\}$. Then we define the jth partial differential of f at the point x^0 as follows:

$$d_{x_j}f(x^0) = f_{x_j}(x^0)dx_j,$$

where $dx_j = \Delta x_j$.

Example 3.6. Let f be as in Example 3.1. Then

$$d_x f(x,y) = (3x^2 - 3y)dx,$$
$$d_y f(x,y) = (3y^2 - 3x)dy, \quad (x,y) \in \mathbb{R}^2.$$

Example 3.7. Let f be as in Example 3.2. Then, for $(x,y,z) \in \mathbb{R}^3, x \neq 0, z \neq 0$, we have

$$d_x f(x,y,z) = \left(-\frac{z}{x^2} + \frac{1}{z}\right)dx,$$
$$d_y f(x,y,z) = dy,$$
$$d_z f(x,y,z) = \left(\frac{1}{x} - \frac{x}{z^2}\right)dz.$$

Example 3.8. Let f be the function in Example 3.3. Then

$$d_{x_j}f(x_1,\dots,x_n) = \sin(2x_j)dx_j, \quad j \in \{1,\dots,n\}, \quad (x_1,\dots,x_n) \in \mathbb{R}^n.$$

Exercise 3.4. Find the partial differentials of the functions in Exercise 3.1.

3.2 Differentiable functions

For a function of several variables, the differentiability of it is more complex than the existence of the partial derivatives of it. Consider the case $n = 2$. The idea is as follows. The differentiability of a function f at (x_0,y_0) means that f has a well-defined tangent plane at the point (x_0,y_0). All tangents to the surface at (x_0,y_0) lie in one plane (see Fig. 3.3). More formally, suppose that $x^0 \in \mathbb{R}^n$ and the function f is defined in a neighborhood $U(x^0,\delta)$ of the point x^0. For any $x \in U(x^0,\delta)$, define

$$\Delta x_j = x_j - x_j^0, \quad j \in \{1,\dots,n\},$$

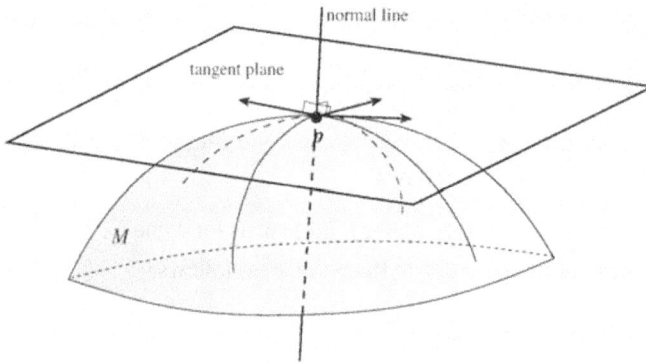

Figure 3.3: A tangent plane and a normal to a surface.

and

$$d = d(x, x^0)$$
$$= \sqrt{\sum_{j=1}^{n}(x_j - x_j^0)^2}$$
$$< \delta,$$

whereupon

$$|\Delta x_j| < \delta, \quad j \in \{1, \dots, n\},$$

and

$$d = \sqrt{\sum_{j=1}^{n} \Delta x_j^2}.$$

Set

$$\Delta f(x^0) = f(x_1^0 + \Delta x_1, \dots, x_n^0 + \Delta x_n) - f(x_1^0, \dots, x_n^0).$$

Definition 3.3. The quantity $\Delta f(x^0)$ is said to be the increment of the function f at the point x^0.

Definition 3.4. The function f is said to be differentiable at the point x^0 if there exist numbers $A_j, j \in \{1, \dots, n\}$, such that

$$\Delta f(x^0) = \sum_{j=1}^{n} A_j \Delta x_j + o(d) \quad \text{as} \quad d \to 0.$$

In this case, the linear function

$$\sum_{j=1}^{n} A_j \Delta x_j$$

is said to be the total differential, shortly differential, of the function f at the point x^0, denoted by $df(x^0)$. The term $o(d)$ is called the error.

Example 3.9. In the case $n = 2$, the equation of the tangent plane at (x_0, y_0) to a surface $z = f(x, y)$ is given by

$$z - z_0 = f_x(x_0, y_0)\Delta x + f_y(x_0, y_0)\Delta y$$

or

$$z = f(x_0, y_0) + f_x(x_0, y_0)\Delta x + f_y(x_0, y_0)\Delta y.$$

We rearrange this equation in the form of a plane and get

$$x f_x(x_0, y_0) + y f_y(x_0, y_0) - z = -f(x_0, y_0) + x_0 f_x(x_0, y_0) + y_0 f_y(x_0, y_0).$$

Hence the normal vector to the tangent plane is

$$n = (f_x(x_0, y_0), f_y(x_0, y_0), -1).$$

From this equation we find the equation of the normal line to the tangent plane at (x_0, y_0):

$$\frac{x - x_0}{f_x(x_0, y_0)} = \frac{y - y_0}{f_y(x_0, y_0)} = \frac{z - z_0}{-1}$$

or

$$(x, y, z) = (x_0, y_0, z_0) + t(f_x(x_0, y_0), f_y(x_0, y_0), -1).$$

Suppose that f is a differentiable function at the point x^0. Then there exists $\epsilon(\Delta x_1, \ldots, \Delta x_n)$ such that

$$\epsilon(\Delta x_1, \ldots, \Delta x_n) \to 0 \quad \text{as} \quad d \to 0$$

and

$$o(d) = \epsilon(\Delta x_1, \ldots, \Delta x_n)d.$$

Thus

$$\Delta f(x^0) = \sum_{j=1}^{n} A_j \Delta x_j + \epsilon(\Delta x_1, \ldots, \Delta x_n)d. \tag{3.1}$$

Since

$$\epsilon(\Delta x_1, \ldots, \Delta x_n)d = \Delta f(x^0) - \sum_{j=1}^{n} A_j \Delta x_j,$$

we get

$$\epsilon(0, \ldots, 0)d = 0,$$

whereupon

$$\epsilon(0, \ldots, 0) = 0.$$

Below we will give some criteria for differentiability of a function of several variables. For this aim, we need an equivalent representation of (3.1) given by the following result.

Theorem 3.1. *Condition (3.1) is equivalent to the condition*

$$\Delta f(x^0) = \sum_{j=1}^{n} A_j dx_j + \sum_{j=1}^{n} \epsilon_j(\Delta x_1, \ldots, \Delta x_n)\Delta x_j, \tag{3.2}$$

where $\epsilon_j \to 0$ as $d \to 0, j \in \{1, \ldots, n\}$.

Proof. 1.   Let (3.1) hold. Set

$$a = \epsilon(\Delta x_1, \ldots, \Delta x_n)d,$$

where $\epsilon(\Delta x_1, \ldots, \Delta x_n) \to 0$ as $d \to 0$. Hence, for $d \neq 0$, we have

$$a = \epsilon(\Delta x_1, \ldots, \Delta x_n)d$$

$$= \epsilon(\Delta x_1, \ldots, \Delta x_n)\sqrt{\sum_{j=1}^{n} \Delta x_j^2}$$

$$= \epsilon(\Delta x_1, \ldots, \Delta x_n)\frac{\sum_{j=1}^{n} \Delta x_j^2}{\sqrt{\sum_{j=1}^{n} \Delta x_j^2}}$$

$$= \epsilon(\Delta x_1, \ldots, \Delta x_n)\sum_{j=1}^{n} \frac{\Delta x_j}{\sqrt{\sum_{j=1}^{n} \Delta x_j^2}}\Delta x_j.$$

Set

$$\epsilon_j(\Delta x_1, \ldots, \Delta x_n) = \epsilon(\Delta x_1, \ldots, \Delta x_n) \frac{\Delta x_j}{\sqrt{\sum_{j=1}^n \Delta x_j^2}}, \quad j \in \{1, \ldots, n\}.$$

Then

$$a = \sum_{j=1}^n \epsilon_j(\Delta x_1, \ldots, \Delta x_n) \Delta x_j. \qquad (3.3)$$

Since

$$\left| \frac{\Delta x_j}{\sqrt{\sum_{j=1}^n \Delta x_j^2}} \right| \le 1, \quad j \in \{1, \ldots, n\}, \quad d \ne 0,$$

we get

$$\left| \epsilon_j(\Delta x_1, \ldots, \Delta x_n) \right| \le \left| \epsilon(\Delta x_1, \ldots, \Delta x_n) \right|, \quad d \ne 0. \qquad (3.4)$$

Then $\epsilon_j(\Delta x_1, \ldots, \Delta x_n) \to 0$ as $d \to 0, j \in \{1, \ldots, n\}$, and thus (3.2) holds.

2. Suppose (3.3) and (3.4) hold. Hence, for $d \ne 0$, we have

$$a = \left(\sum_{j=1}^n \epsilon_j(\Delta x_1, \ldots, \Delta x_n) \frac{\Delta x_j}{\sqrt{\sum_{j=1}^n \Delta x_j^2}} \right) \sqrt{\sum_{j=1}^n \Delta x_j^2}.$$

Let

$$\epsilon(\Delta x_1, \ldots, \Delta x_n) = \sum_{j=1}^n \epsilon_j(\Delta x_1, \ldots, \Delta x_n) \frac{\Delta x_j}{\sqrt{\sum_{j=1}^n \cdot \Delta x_j^2}}.$$

Then $\epsilon(\Delta x_1, \ldots, \Delta x_n) \to 0$ as $d \to 0$, and

$$a = \epsilon(\Delta x_1, \ldots, \Delta x_n) d.$$

Hence we get (3.1). This completes the proof. □

As in the case of functions of one variable, differentiability implies continuity.

Theorem 3.2. *If the function f is differentiable at x^0, then it is continuous at this point.*

Proof. Suppose that f is differentiable at x^0. Then (3.2) holds. Take $\epsilon > 0$ arbitrarily. Then, for $\delta \le \epsilon$, we have

$$|\Delta x_j| < \epsilon, \quad j \in \{1, \ldots, n\}.$$

Hence

$$\left| \Delta f(x^0) \right| = \left| \sum_{j=1}^{n} A_j \Delta x_j + \sum_{j=1}^{n} \epsilon_j(\Delta x_1, \dots, \Delta x_n) \Delta x_j \right|$$

$$\leq \left| \sum_{j=1}^{n} A_j \Delta x_j \right| + \left| \sum_{j=1}^{n} \epsilon_j(\Delta x_1, \dots, \Delta x_n) \Delta x_j \right|$$

$$\leq \sum_{j=1}^{n} |A_j| |\Delta x_j| + \sum_{j=1}^{n} |\epsilon_j(\Delta x_1, \dots, \Delta x_n)| |\Delta x_j|$$

$$< \sum_{j=1}^{n} \left(|A_j| + |\epsilon_j(\Delta x_1, \dots, \Delta x_n)| \right) \epsilon.$$

Because $\epsilon > 0$ was arbitrarily chosen, by the last chain of inequalities we get

$$\Delta f(x^0) \to 0 \quad \text{as} \quad d \to 0,$$

i. e., f is continuous at x^0. This completes the proof. $\qquad\square$

The converse of the preceding theorem is not true since the converse of the analogous theorem for functions of one variable is not true. An important property of differentiable functions of several variables is contained in the following result.

Theorem 3.3. *Let f be differentiable at x^0, and let*

$$df(x^0) = \sum_{j=1}^{n} A_j dx_j.$$

Then there exist $f_{x_j}(x^0), j \in \{1, \dots, n\}$, and

$$f_{x_j}(x^0) = A_j, \quad j \in \{1, \dots, n\}. \tag{3.5}$$

In this case,

$$df(x^0) = \sum_{j=1}^{n} f_{x_j}(x^0) dx_j. \tag{3.6}$$

Proof. Since f is differentiable at x^0, we have that

$$\Delta f(x^0) = \sum_{j=1}^{n} A_j dx_j + \sum_{j=1}^{n} \epsilon_j(\Delta x_1, \dots, \Delta x_n) \Delta x_j,$$

where $\epsilon_j(\Delta x_1, \dots, \Delta x_n) \to 0$ as $d \to 0, j \in \{1, \dots, n\}$. Note that by the definition of ϵ_j, $j \in \{1, \dots, n\}$ (see the proof of Theorem 3.1), it follows that

$$\lim_{\Delta x_j \to 0} \epsilon_j(\Delta x_1, \dots, \Delta x_n) = 0, \quad j \in \{1, \dots, n\}.$$

Take $j \in \{1, \ldots, n\}$ arbitrarily. Let

$$\Delta x_1 = \cdots = \Delta x_{j-1} = \Delta x_{j+1} = \cdots = \Delta x_n = 0.$$

Since $d \neq 0$ and $d = |\Delta x_j|$, we get that $\Delta x_j \neq 0$. Then

$$\Delta f(x^0) = f(x_1^0, \ldots, x_{j-1}^0, x_j^0 + \Delta x_j, x_{j+1}^0, \ldots, x_n^0) - f(x_1^0, \ldots, x_{j-1}^0, x_j^0, x_{j+1}^0, \ldots, x_n^0)$$

and

$$f(x_1^0, \ldots, x_{j-1}^0, x_j^0 + \Delta x_j, x_{j+1}^0, \ldots, x_n^0) - f(x_1^0, \ldots, x_{j-1}^0, x_j^0, x_{j+1}^0, \ldots, x_n^0)$$
$$= A_j \Delta x_j + \epsilon_j(\Delta x_1, \ldots, \Delta x_n) \Delta x_j,$$

whereupon

$$\frac{f(x_1^0, \ldots, x_{j-1}^0, x_j^0 + \Delta x_j, x_{j+1}^0, \ldots, x_n^0) - f(x_1^0, \ldots, x_{j-1}^0, x_j^0, x_{j+1}^0, \ldots, x_n^0)}{\Delta x_j}$$
$$= A_j + \epsilon_j(\Delta x_1, \ldots, \Delta x_n).$$

Hence

$$f_{x_j}(x^0) = \lim_{\Delta x_j \to 0} \frac{f(x_1^0, \ldots, x_{j-1}^0, x_j^0 + \Delta x_j, x_{j+1}^0, \ldots, x_n^0) - f(x_1^0, \ldots, x_{j-1}^0, x_j^0, x_{j+1}^0, \ldots, x_n^0)}{\Delta x_j}$$
$$= A_j + \lim_{\Delta x_j \to 0} \epsilon_j(\Delta x_1, \ldots, \Delta x_n)$$
$$= A_j,$$

and we obtain (3.6). This completes the proof. □

Remark 3.1. By (3.5), if the function f is differentiable at x^0, then its differential at x^0 is unique.

The natural question to ask is under what conditions can we conclude that a function of several variables is differentiable at a point. The answer is contained in the following theorem.

Theorem 3.4. *Suppose that the function f has partial derivatives f_{x_j}, $j \in \{1, \ldots, n\}$, in a neighborhood of the point x^0 and these derivatives are continuous at the point x^0. Then the function f is differentiable at x^0.*

Proof. Let $\delta > 0$ be such that f_{x_j}, $j \in \{1, \ldots, n\}$, exist in $U(x^0, \delta)$. Take Δx_j, $j \in \{1, \ldots, n\}$, such that

$$(x_1^0 + \Delta x_1, \ldots, x_n^0 + \Delta x_n) \in U(x^0, \delta),$$
$$(x_1^0, x_2^0 + \Delta x_2, \ldots, x_n^0 + \Delta x_n) \in U(x^0, \delta),$$

$$\vdots$$

$$(x_1^0, \ldots, x_{n-1}^0, x_n^0 + \Delta x_n) \in U(x^0, \delta).$$

Then

$$
\begin{aligned}
\Delta f(x^0) &= f(x_1^0 + \Delta x_1, \ldots, x_n^0 + \Delta x_n) - f(x_1^0, \ldots, x_n^0) \\
&= f(x_1^0 + \Delta x_1, x_2^0 + \Delta x_2, \ldots, x_n^0 + \Delta x_n) - f(x_1^0, x_2^0 + \Delta x_2, \ldots, x_n^0 + \Delta x_n) \\
&\quad + f(x_1^0, x_2^0 + \Delta x_2, x_3^0 + \Delta x_3, \ldots, x_n^0 + \Delta x_n) - f(x_1^0, x_2^0, x_3^0 + \Delta x_3, \ldots, x_n^0 + \Delta x_n)
\end{aligned}
$$

$$\vdots$$

$$+ f(x_1^0, \ldots, x_{n-1}^0, x_n^0 + \Delta x_n) - f(x_1^0, \ldots, x_{n-1}^0, x_n^0).$$

By the mean value theorem it follows that there is $\theta \in (0,1)$ such that

$$
\begin{aligned}
&f(x_1^0 + \Delta x_1, x_2^0 + \Delta x_2, \ldots, x_n^0 + \Delta x_n) - f(x_1^0, x_2^0 + \Delta x_2, \ldots, x_n^0 + \Delta x_n) \\
&= f_{x_1}(x_1^0 + \theta_1 \Delta x_1, x_2^0 + \Delta x_2, \ldots, x_n^0 + \Delta x_n) \Delta x_1.
\end{aligned}
$$

Again, by the mean value theorem it follows that there is $\theta_2 \in (0,1)$ such that

$$
\begin{aligned}
&f(x_1^0, x_2^0 + \Delta x_2, x_3^0 + \Delta x_3, \ldots, x_n^0 + \Delta x_n) - f(x_1^0, x_2^0, x_3^0 + \Delta x_3, \ldots, x_n^0 + \Delta x_n) \\
&= f_{x_2}(x_1^0, x_2^0 + \theta_2 \Delta x_2, x_3^0 + \Delta x_3, \ldots, x_n^0 + \Delta x_n) \Delta x_2,
\end{aligned}
$$

and so on. By the mean value theorem we get that there is $\theta_n \in (0,1)$ such that

$$f(x_1^0, \ldots, x_{n-1}^0, x_n^0 + \Delta x_n) - f(x_1^0, \ldots, x_{n-1}^0, x_n^0) = f_{x_n}(x_1^0, \ldots, x_{n-1}^0, x_n^0 + \theta_n \Delta x_n) \Delta x_n.$$

From this we get

$$
\begin{aligned}
\Delta f(x^0) &= f_{x_1}(x_1^0 + \theta_1 \Delta x_1, x_2^0 + \Delta x_2, \ldots, x_n^0 + \Delta x_n) \Delta x_1 \\
&\quad + f_{x_2}(x_1^0, x_2^0 + \theta_2 \Delta x_2, \ldots, x_3^0 + \Delta x_3, \ldots, x_n^0 + \Delta x_n) \Delta x_2
\end{aligned}
$$

$$\vdots$$

$$+ f_{x_n}(x_1^0, \ldots, x_{n-1}^0, x_n^0 + \theta_n \Delta x_n) \Delta x_n.$$

Let

$$
\begin{aligned}
\epsilon_1(\Delta x_1, \ldots, \Delta x_n) &= f_{x_1}(x_1^0 + \theta_1 \Delta x_1, x_2^0 + \Delta x_2, \ldots, x_n^0 + \Delta x_n) - f_{x_1}(x^0), \\
\epsilon_2(\Delta x_1, \ldots, \Delta x_n) &= f_{x_2}(x_1^0, x_2^0 + \theta_2 \Delta x_2, x_3^0 + \Delta x_3, \ldots, x_n^0 + \Delta x_n) - f_{x_2}(x^0),
\end{aligned}
$$

$$\vdots$$

$$\epsilon_n(\Delta x_1, \ldots, \Delta x_n) = f_{x_n}(x_1^0, \ldots, x_{n-1}^0, x_n^0 + \theta_n \Delta x_n) - f_{x_n}(x^0).$$

Since $f_{x_j}, j \in \{1, \ldots, n\}$, are continuous at x^0, we get

$$\lim_{d \to 0} \epsilon_j = 0, \quad j \in \{1, \ldots, n\}.$$

Therefore

$$\Delta f(x^0) = (f_{x_1}(x_1^0 + \theta_1 \Delta x_1, x_2^0 + \Delta x_2, \ldots, x_n^0 + \Delta x_n) - f_{x_1}(x^0))\Delta x_1$$
$$+ (f_{x_2}(x_1^0, x_2^0 + \theta_2 \Delta x_2, \ldots, x_3^0 + \Delta x_3, \ldots, x_n^0 + \Delta x_n) - f_{x_2}(x^0))\Delta x_2$$

$$\vdots$$

$$+ (f_{x_n}(x_1^0, \ldots, x_{n-1}^0, x_n^0 + \theta_n \Delta x_n) - f_{x_n}(x^0))\Delta x_n$$
$$+ f_{x_1}(x^0)\Delta x_1 + f_{x_2}(x^0)\Delta x_2 + \cdots + f_{x_n}(x^0)\Delta x_n$$

$$= \sum_{j=1}^{n} \epsilon_j \Delta x_j + \sum_{j=1}^{n} f_{x_j}(x^0)\Delta x_j.$$

Thus f is differentiable at x^0. This completes the proof. □

Example 3.10. We will prove that the function

$$f(x,y) = y \sin x, \quad (x,y) \in \mathbb{R}^2,$$

is a differentiable function at the point $(0,0)$. We have

$$f(0,0) = 0,$$
$$f(x,0) = 0,$$
$$f(0,y) = 0, \quad (x,y) \in \mathbb{R}^2.$$

In particular, we get

$$f_x(0,0) = 0,$$
$$f_y(0,0) = 0.$$

Next,

$$\Delta f(0,0) = f(\Delta x, \Delta y) - f(0,0)$$
$$= f(\Delta x, \Delta y)$$
$$= \Delta y \sin(\Delta x),$$

whereupon

$$|\Delta f(0,0)| = |\Delta y||\sin(\Delta x)|$$

$$\leq |\Delta x||\Delta y|$$
$$\to 0 \quad \text{as} \quad d \to 0.$$

Therefore

$$\Delta f(0,0) = o(d) \quad \text{as} \quad d \to 0,$$

i. e., f is differentiable at $(0,0)$.

Example 3.11. We will prove that the function

$$f(x,y) = x\left(\sqrt[3]{1 + \sqrt{|y|}} - 1\right), \quad (x,y) \in \mathbb{R}^2,$$

is a differentiable function at $(0,0)$. We have

$$f(x,0) = x(1-1)$$
$$= 0,$$
$$f(0,y) = 0, \quad (x,y) \in \mathbb{R}^2.$$

In particular, we get

$$f_x(0,0) = 0,$$
$$f_y(0,0) = 0,$$

and $f(0,0) = 0$. Next,

$$\Delta f(0,0) = f(\Delta x, \Delta y) - f(0,0)$$
$$= f(\Delta x, \Delta y)$$
$$= \Delta x\left(\sqrt[3]{1 - \sqrt{|\Delta y|}} - 1\right), \quad \Delta x, \Delta y \in \mathbb{R},$$

and

$$|\Delta f(0,0)| = \left|\Delta x\left(\sqrt[3]{1 - \sqrt{|\Delta y|}} - 1\right)\right|$$
$$= |\Delta x|\left|\sqrt[3]{1 - \sqrt{|\Delta y|}} - 1\right|$$
$$= |\Delta x|\frac{\left|\left(\sqrt[3]{1 - \sqrt{|\Delta y|}} - 1\right)\left(\sqrt[3]{(1 - \sqrt{|\Delta y|})^2} + \sqrt[3]{1 - \sqrt{|\Delta y|}} + 1\right)\right|}{\sqrt[3]{(1 - \sqrt{|\Delta y|})^2} + \sqrt[3]{1 - \sqrt{|\Delta y|}} + 1}$$
$$= \frac{|\Delta x|(1 + \sqrt{|\Delta y|} - 1)}{\sqrt[3]{(1 - \sqrt{|\Delta y|})^2} + \sqrt[3]{1 - \sqrt{|\Delta y|}} + 1}$$

$$= \frac{|\Delta x| \sqrt{|\Delta y|}}{\sqrt[3]{(1 - \sqrt{|\Delta y|})^2} + \sqrt[3]{1 - \sqrt{|\Delta y|}} + 1}, \quad \Delta x, \Delta y \in \mathbb{R},$$

whereupon

$$\frac{|\Delta f(0,0)|}{\sqrt{(\Delta x)^2 + (\Delta y)^2}} = \frac{|\Delta x| \sqrt{|\Delta y|}}{\sqrt{(\Delta x)^2 + (\Delta y)^2}(\sqrt[3]{(1 - \sqrt{|\Delta y|})^2} + \sqrt[3]{1 - \sqrt{|\Delta y|}} + 1)}$$

$$= \frac{|\sqrt{|\Delta x|} \sqrt{|\Delta x|} \sqrt{|\Delta y|}|}{\sqrt{(\Delta x)^2 + (\Delta y)^2}(\sqrt[3]{(1 - \sqrt{|\Delta y|})^2} + \sqrt[3]{1 - \sqrt{|\Delta y|}} + 1)}$$

$$\leq \frac{|\sqrt{|\Delta x|}|}{(\sqrt[3]{(1 - \sqrt{|\Delta y|})^2} + \sqrt[3]{1 - \sqrt{|\Delta y|}} + 1)}$$

$$\to 0 \quad \text{as} \quad d \to 0,$$

i. e.,

$$\Delta f(0,0) = o(d) \quad \text{as} \quad d \to 0.$$

Thus f is differentiable at $(0,0)$.

Example 3.12. We will prove that the function

$$f(x,y) = \sqrt[3]{xy}, \quad (x,y) \in \mathbb{R}^2,$$

is not differentiable at $(0,0)$. We have

$$f(x,0) = 0,$$
$$f(0,y) = 0, \quad (x,y) \in \mathbb{R}^2,$$

whereupon

$$f_x(x,0) = 0,$$
$$f_y(0,y) = 0, \quad (x,y) \in \mathbb{R}^2,$$

and in particular,

$$f_x(0,0) = 0,$$
$$f_y(0,0) = 0.$$

Next,

$$\Delta f(0,0) = \sqrt[3]{\Delta x \Delta y}, \quad \Delta x, \Delta y \in \mathbb{R}.$$

Assume that the function f is differentiable at $(0,0)$. Then

$$\sqrt[3]{\Delta x \Delta y} = o(d) \quad \text{as} \quad d \to 0.$$

For $\Delta x = \Delta y \in \mathbb{R}$, we get

$$\sqrt[3]{|\Delta x|^2} = o(|\Delta x|) \quad \text{as} \quad \Delta x \to 0,$$

which is impossible. Therefore f is not differentiable at $(0,0)$.

Exercise 3.5. Prove that the following functions are differentiable at the point $(0,0)$:

1.
$$f(x,y) = (\sin x + \sqrt[3]{xy})^2, \quad (x,y) \in \mathbb{R}^2.$$

2.
$$f(x,y) = \sqrt[5]{x^4}(\cos(\sqrt[5]{y}) - 1), \quad (x,y) \in \mathbb{R}^2.$$

3.
$$f(x,y) = \sqrt{|xy|}, \quad (x,y) \in \mathbb{R}^2.$$

4.
$$f(x,y) = \sqrt{x^2 + y^2}, \quad (x,y) \in \mathbb{R}^2.$$

5.
$$f(x,y) = \sqrt[3]{x^3 + y^3}, \quad (x,y) \in \mathbb{R}^2.$$

Example 3.13. We will find the values of $\alpha \in \mathbb{R}$ for which the function

$$f(x,y) = \begin{cases} |x + y|^{3\alpha} + |y|^{4-\alpha} & \text{if } x^2 + y^2 \neq 0, \\ 0 & \text{if } x^2 + y^2 = 0 \end{cases}$$

is differentiable at $(0,0)$. We have that

$$f(x,0) = |x|^{3\alpha},$$
$$f(0,y) = |y|^{3\alpha} + |y|^{4-\alpha}, \quad (x,y) \in \mathbb{R}^2.$$

Then, for the existence of $f_x(0,0)$ and $f_y(0,0)$, we must have

$$3\alpha > 1,$$
$$4 - \alpha > 1,$$

whereupon $\alpha \in (\frac{1}{3}, 3)$. Hence, for $\alpha \in (\frac{1}{3}, 3)$, we have

$$f_x(0,0) = 0,$$
$$f_y(0,0) = 0,$$

and

$$\Delta f(0,0) = |\Delta x + \Delta y|^{3a} + |\Delta y|^{4-a}, \quad |\Delta x|, |\Delta y| \in \mathbb{R}.$$

Now using the inequality

$$a + b \le 2\sqrt{a^2 + b^2}, \quad a, b \ge 0,$$

we get

$$\frac{|\Delta f(0,0)|}{\sqrt{\Delta x^2 + \Delta y^2}} = \frac{|\Delta x + \Delta y|^{3a} + |\Delta y|^{4-a}}{\sqrt{\Delta x^2 + \Delta y^2}}$$

$$\le \frac{(|\Delta x| + |\Delta y|)^{3a} + |\Delta y|^{4-a}}{\sqrt{\Delta x^2 + \Delta y^2}}$$

$$\le \frac{2^{3a}(\sqrt{\Delta x^2 + \Delta y^2})^{3a} + (\sqrt{\Delta x^2 + \Delta y^2})^{4-a}}{\sqrt{\Delta x^2 + \Delta y^2}}$$

$$= 2^{3a}(\sqrt{\Delta x^2 + \Delta y^2})^{3a-1} + (\sqrt{\Delta x^2 + \Delta y^2})^{3-a}$$

$$\to 0 \quad \text{as} \quad d \to 0.$$

Thus, for $a \in (\frac{1}{3}, 3)$, the considered function is differentiable at $(0,0)$.

Exercise 3.6. Find the values of $a \in \mathbb{R}$ such that the function

$$f(x,y) = \begin{cases} \frac{|x|^{3a} + |y|^{7-a}}{x^2 + y^2} & \text{if } x^2 + y^2 \neq 0, \\ 0 & \text{if } x^2 + y^2 = 0 \end{cases}$$

is differentiable at $(0,0)$.

Example 3.14. We will find $df(x,y)$, $(x,y) \in X$, where

$$f(x,y) = (y^3 + 2x^2y + 3)^4, \quad (x,y) \in \mathbb{R}^2.$$

We have

$$f_x(x,y) = 4(y^3 + 2x^2y + 3)^3(4xy)$$
$$= 16xy(y^3 + 2x^2y + 3)^3,$$
$$f_y(x,y) = 4(y^3 + 2x^2y + 3)^3(3y^2 + 2x^2)$$
$$= 4(3y^2 + 2x^2)(y^3 + 2x^2y + 3)^3, \quad (x,y) \in \mathbb{R}^2,$$

whereupon

$$df(x,y) = f_x(x,y)dx + f_y(x,y)dy$$
$$= 16xy(y^3 + 2x^2y + 3)^3 dx + 4(3y^2 + 2x^2)(y^3 + 2x^2 + 3)^3 dy$$
$$= 4(y^3 + 2x^2y + 3)^3(4xydx + (3y^2 + 2x^2)dy), \quad (x,y) \in \mathbb{R}^2.$$

Example 3.15. We will find $df(x,y,z)$, $(x,y,z) \in X$, where

$$f(x,y,z) = x^{\frac{y}{z}}, \quad (x,y,z) \in X,$$
$$X = \{(x,y,z) \in \mathbb{R}^3 : x > 0, \ z > 0\}.$$

Let

$$g(x,y,z) = \log f(x,y,z), \quad (x,y,z) \in X.$$

Then

$$g(x,y,z) = \log x^{\frac{y}{z}}$$
$$= \frac{y}{z} \log x, \quad (x,y,z) \in X,$$

and

$$g_x(x,y,z) = \frac{y}{xz}$$
$$= \frac{f_x(x,y,z)}{f(x,y,z)},$$
$$g_y(x,y,z) = \frac{1}{z} \log x$$
$$= \frac{f_y(x,y,z)}{f(x,y,z)},$$
$$g_z(x,y,z) = -\frac{y}{z^2} \log x$$
$$= \frac{f_z(x,y,z)}{f(x,y,z)}, \quad (x,y,z) \in X.$$

Hence

$$f_x(x,y,z) = \frac{y}{z} x^{\frac{y}{z}-1},$$
$$f_y(x,y,z) = \frac{1}{z} x^{\frac{y}{z}} \log x,$$
$$f_z(x,y,z) = -\frac{y}{z^2} x^{\frac{y}{z}} \log x, \quad (x,y,z) \in X,$$

and

$$df(x,y,z) = f_x(x,y,z)dx + f_y(x,y,z)dy + f_z(x,y,z)dz$$

$$= \frac{y}{z} x^{\frac{y}{z}-1} dx + \frac{1}{z} x^{\frac{y}{z}} \log x dy - \frac{y}{z^2} x^{\frac{y}{z}} \log x dz$$

$$= \frac{x^{\frac{y}{z}-1}}{z} \left(y dx + x \log x dy - \frac{xy \log x}{z} dz \right), \quad (x, y, z) \in X.$$

Example 3.16. We will find $df(x_1, \ldots, x_n)$, $(x_1, \ldots, x_n) \in X$, where

$$f(x_1, \ldots, x_n) = \sin \left(\sum_{j=1}^{n} x_j^2 \right), \quad (x_1, \ldots, x_n) \in X, \quad X = \mathbb{R}^n.$$

We have

$$f_{x_j}(x_1, \ldots, x_n) = 2 x_j \cos \left(\sum_{j=1}^{n} x_j^2 \right), \quad (x_1, \ldots, x_n) \in X, \quad j \in \{1, \ldots, n\}.$$

Hence

$$df(x_1, \ldots, x_n) = \sum_{j=1}^{n} f_{x_j}(x_1, \ldots, x_n) dx_j$$

$$= 2 \cos \left(\sum_{j=1}^{n} x_j^2 \right) \sum_{j=1}^{n} x_j dx_j, \quad (x_1, \ldots, x_n) \in X.$$

Exercise 3.7. Find $df(x_1, \ldots, x_n)$, $(x_1, \ldots, x_n) \in X$, where
1. $n = 2$,

$$f(x, y) = 2x^4 - 3x^2 y^2 + x^3 y, \quad (x, y) \in X, \quad X = \mathbb{R}^2.$$

2.

$$f(x, y, z) = \sqrt{x^2 + y^2 + z^2}, \quad (x, y, z) \in X,$$
$$X = \{(x, y, z) \in \mathbb{R}^3 : (x, y, z) \neq (0, 0, 0)\}.$$

3.

$$f(x_1, \ldots, x_n) = \sum_{j=1}^{n} x_j^3, \quad (x_1, \ldots, x_n) \in X, \quad X = \mathbb{R}^n.$$

3.3 Differentiation of a composition of functions

In this section, we investigate composition of functions for differentiability at a point.
The chain rule for functions of several variables reads as follows.

Theorem 3.5. *Let $t_0 \in \mathbb{R}$, let $y_j = y_j(t)$, $j \in \{1, \ldots, n\}$, be functions differentiable at t_0, and let*

$$y^0 = y(t_0) = (y_1(t_0), \ldots, y_n(t_0)).$$

Let also, the function $f = f(x)$, $x = (x_1, \ldots, x_n)$, be differentiable at y^0. Then the composition

$$g = g(t) = f(y(t)), \quad y = (y_1, \ldots, y_n),$$

is defined in a neighborhood of t_0 and differentiable at t_0, and

$$g'(t_0) = \sum_{j=1}^{n} \frac{\partial f}{\partial x_j}(y^0) \frac{dy_j}{dt}(t_0).$$

Proof. Since the functions y_j, $j \in \{1, \ldots, n\}$, are differentiable at t_0, they are continuous at t_0 and defined in some of its neighborhood. Because the function f is differentiable at y^0, it is continuous at y^0. Now applying Theorem 2.3 and Corollary 2.1, we get that the function g is defined in a neighborhood of t_0. Since f is differentiable at y^0, we have the following representation:

$$\Delta f(y^0) = \sum_{j=1}^{n} f_{y_j}(y^0)\Delta y_j + \epsilon d, \tag{3.7}$$

where $\epsilon = \epsilon(d) \to 0$ as $d = \sqrt{\sum_{j=1}^{n} \Delta y_j^2} \to 0$. Moreover, we have

$$\Delta y_j(t_0) = y_j(t_0 + \Delta t) - y_j(t_0), \quad \Delta t \in \mathbb{R}.$$

Dividing (3.7) by $\Delta t \neq 0$, we get

$$\frac{\Delta g(t_0)}{\Delta t} = \frac{\Delta f(y^0)}{\Delta t}$$

$$= \sum_{j=1}^{n} f_{y_j}(y^0) \frac{\Delta y_j}{\Delta t} + \epsilon \sqrt{\sum_{j=1}^{n} \left(\frac{\Delta y_j}{\Delta t}\right)^2}. \tag{3.8}$$

Since y_j, $j \in \{1, \ldots, n\}$, are continuous at t_0, we have that $\Delta y_j \to 0$, $j \in \{1, \ldots, n\}$, as $\Delta t \to 0$, whereupon

$$\lim_{d \to 0} d = 0,$$

and thus

$$\lim_{d \to 0} \epsilon(d) = 0.$$

Also,

$$\lim_{\Delta t \to 0} \sum_{j=1}^{n} \left(\frac{\Delta y_j}{\Delta t}\right)^2 = \sum_{j=1}^{n} (y_j'(t_0))^2,$$

$$\lim_{\Delta t \to 0} \frac{\Delta y_j}{\Delta t} = y_j'(t_0).$$

Hence by (3.8) we arrive at

$$g'(t_0) = \lim_{\Delta t \to 0} \frac{\Delta g(t_0)}{\Delta t}$$

$$= \sum_{j=1}^{n} f_{y_j}(y^0) \lim_{\Delta t \to 0} \frac{\Delta y_j}{\Delta t} + \epsilon \sqrt{\sum_{j=1}^{n} \left(\lim_{\Delta t \to 0} \frac{\Delta y_j}{\Delta t} \right)^2}$$

$$= \sum_{j=1}^{n} f_{y_j}(y^0) y_j'(t_0).$$

This completes the proof. ☐

Corollary 3.1. *Let $y = y(x)$, $x = (x_1, \ldots, x_n)$, be a function defined in a neighborhood of the point $x^0 = (x_1^0, \ldots, x_n^0)$, and let $x_j = x_j(t)$, $t = (t_1, \ldots, t_k)$, $j \in \{1, \ldots, n\}$, be functions defined in a neighborhood of the point $t^0 = (t_1^0, \ldots, t_k^0) \in \mathbb{R}^k$ such that $x_j(t^0) = x_j^0$, $j \in \{1, \ldots, n\}$. If the function y is differentiable at x^0 and if $\frac{\partial x_j}{\partial t_l}(t^0)$, $j \in \{1, \ldots, n\}$, $l \in \{1, \ldots, k\}$, exist, then the composition $y(x(t))$ has partial derivatives at t^0, and*

$$\frac{\partial y}{\partial t_l}(t^0) = \sum_{j=1}^{n} \frac{\partial y}{\partial x_j}(x(t^0)) \frac{\partial x_j}{\partial t_l}(t^0), \quad l \in \{1, \ldots, k\}.$$

Proof. Since y is differentiable at x^0, it is defined in a neighborhood of x^0. Because x_j, $j \in \{1, \ldots, n\}$, are differentiable at t^0, they are continuous in a neighborhood of t^0. Now by Theorem 2.3 and Corollary 2.1 it follows that the composition $y(x(t))$ is defined in a neighborhood of t^0. We fix $l \in \{1, \ldots, k\}$, and applying Theorem 3.5, we get the desired result. This completes the proof. ☐

Example 3.17. Let $f = f(u)$, $u \in \mathbb{R}$, be a differentiable function, and let

$$\phi(x, y) = yf(x^2 - y^2), \quad (x, y) \in \mathbb{R}^2.$$

We will prove that

$$y^2 \phi_x(x, y) + xy \phi_y(x, y) = x\phi(x, y), \quad (x, y) \in \mathbb{R}^2.$$

We have

$$\phi_x(x, y) = yf'(x^2 - y^2) \frac{\partial}{\partial x}(x^2 - y^2)$$
$$= yf'(x^2 - y^2)2x$$
$$= 2xyf'(x^2 - y^2),$$

$$\phi_y(x,y) = f(x^2 - y^2) + yf'(x^2 - y^2)\frac{\partial}{\partial y}(x^2 - y^2)$$
$$= f(x^2 - y^2) + yf'(x^2 - y^2)(-2y)$$
$$= f(x^2 - y^2) - 2y^2 f'(x^2 - y^2), \quad (x,y) \in \mathbb{R}^2.$$

Hence

$$y^2\phi_x(x,y) + xy\phi_y(x,y) = 2xy^3 f'(x^2 - y^2) + xyf(x^2 - y^2) - 2xy^3 f'(x^2 - y^2)$$
$$= xyf(x^2 - y^2)$$
$$= x\phi(x,y), \quad (x,y) \in \mathbb{R}^2.$$

Example 3.18. Let $f = f(u_1, u_2)$, $(u_1, u_2) \in \mathbb{R}^2$, be a differentiable function. We will prove that the function

$$\phi(x,y,z) = f\left(\frac{x}{y}, x^2 + y - z^2\right), \quad (x,y,z) \in \mathbb{R}^3,$$

satisfies the equation

$$2xz\phi_x(x,y,z) + 2yz\phi_y(x,y,z) + (2x^2 + y)\phi_z(x,y,z) = 0, \quad (x,y,z) \in \mathbb{R}^3.$$

We have

$$\phi_x(x,y,z) = f_{u_1}\left(\frac{x}{y}, x^2 + y - z^2\right)\frac{\partial}{\partial x}\left(\frac{x}{y}\right) + f_{u_2}\left(\frac{x}{y}, x^2 + y - x^2\right)\frac{\partial}{\partial y}(x^2 + y - z^2)$$
$$= \frac{1}{y}f_{u_1}\left(\frac{x}{y}, x^2 + y - z^2\right) + 2xf_{u_2}\left(\frac{x}{y}, x^2 + y - z^2\right),$$

$$\phi_y(x,y,z) = f_{u_1}\left(\frac{x}{y}, x^2 + y - z^2\right)\frac{\partial}{\partial y}\left(\frac{x}{y}\right) + f_{u_2}\left(\frac{x}{y}, x^2 + y - z^2\right)\frac{\partial}{\partial y}(x^2 + y - z^2)$$
$$= -\frac{x}{y^2}f_{u_1}\left(\frac{x}{y}, x^2 + y - z^2\right) + f_{u_2}\left(\frac{x}{y}, x^2 + y - z^2\right),$$

$$\phi_z(x,y,z) = f_{u_2}\left(\frac{x}{y}, x^2 + y - z^2\right)\frac{\partial}{\partial z}(x^2 + y - z^2)$$
$$= -2zf_{u_2}\left(\frac{x}{y}, x^2 + y - z^2\right), \quad (x,y,z) \in \mathbb{R}^3.$$

Hence

$$2xz\phi_x(x,y,z) + 2yz\phi_y(x,y,z) + (2x^2 + y)\phi_z(x,y,z)$$
$$= 2xz\left(\frac{1}{y}f_{u_1}\left(\frac{x}{y}, x^2 + y - z^2\right) + 2xf_{u_2}\left(\frac{x}{y}, x^2 + y - z^2\right)\right)$$
$$+ 2yz\left(-\frac{x}{y^2}f_{u_1}\left(\frac{x}{y}, x^2 + y - z^2\right) + f_{u_2}\left(\frac{x}{y}, x^2 + y - z^2\right)\right)$$

$$+ (2x^2 + y)\left(-2zf_{u_2}\left(\frac{x}{y}, x^2 + y - z^2\right)\right)$$

$$= 2\frac{xz}{y}f_{u_1}\left(\frac{x}{y}, x^2 + y - z^2\right) + 4x^2zf_{u_2}\left(\frac{x}{y}, x^2 + y - z^2\right)$$

$$- 2\frac{xz}{y}f_{u_1}\left(\frac{x}{y}, x^2 + y - z^2\right) + 2yzf_{u_2}\left(\frac{x}{y}, x^2 + y - z^2\right)$$

$$- 4x^2zf_{u_2}\left(\frac{x}{y}, x^2 + y - z^2\right) - 2yzf_{u_2}\left(\frac{x}{y}, x^2 + y - z^2\right)$$

$$= 0, \quad (x, y, z) \in \mathbb{R}^3.$$

Exercise 3.8. Let $f = f(u_1, u_2)$, $(u_1, u_2) \in \mathbb{R}^2$, be a differentiable function. Prove that the function

$$\phi(x, y, z) = f\left(\frac{x - y}{xy}, (x - y)e^{-\frac{z^2}{2}}\right), \quad (x, y, z) \in \mathbb{R}^3,$$

satisfies the equation

$$x^2 z \phi_x(x, y, z) + y^2 z \phi_y(x, y, z) + (x + y)\phi_z(x, y, z) = 0, \quad (x, y, z) \in \mathbb{R}^3.$$

Example 3.19. Suppose that y, x, t, t^0 satisfy the conditions of Corollary 3.1. Let $x^0 = x(t^0)$. Then

$$dy(x^0) = \sum_{j=1}^{n} \frac{\partial y}{\partial x_j}(x^0) dx_j(t^0)$$

$$= \sum_{j=1}^{n} \frac{\partial y}{\partial x_j}(x^0)\left(\sum_{l=1}^{k} \frac{\partial x_j}{\partial t_l}(t^0) dt_l\right)$$

$$= \sum_{l=1}^{k}\left(\sum_{j=1}^{n} \frac{\partial y}{\partial x_j}(x^0)\frac{\partial x_j}{\partial t_l}(t^0)\right) dt_l.$$

Exercise 3.9. Let $X \subset \mathbb{R}^n$, and let $f, g : X \to \mathbb{R}$ be differentiable at $x^0 \in X$. Prove that
1. $d(f \pm g)(x^0) = df(x^0) \pm dg(x^0)$.
2. $d(fg)(x^0) = g(x^0)df(x^0) + f(x^0)dg(x^0)$.
3. $d(\frac{f}{g})(x^0) = \frac{g(x^0)df(x^0) - f(x^0)dg(x^0)}{(g(x^0))^2}$, provided that $g(x^0) \neq 0$.

Definition 3.5. Let $X \subset \mathbb{R}^n$. The set of all functions $f : X \to \mathbb{R}$ that have continuous partial derivatives at each point of X is denoted by $C^1(X)$ and is called the set of continuously differentiable functions on X.

3.4 Gradient of a function

The gradient of a function is one of the fundamental pillars of mathematics, with far-reaching applications in various fields such as physics, engineering, machine learning, and optimization. The gradient of a function is a vector that describes how the function output changes as you move through its input space. It consists of partial derivatives with respect to each input variable and points in the direction of the steepest increase of the function at a particular point. In simple terms, the gradient provides information about the function's slope and direction of change, making it a fundamental concept in mathematics and machine learning.

The basic idea for a gradient is as follows. Consider the case $n = 2$ and a function of two variables $f(x, y)$, whose graph represents a surface in three dimensions. If x and y are themselves functions of another variable t, then $(x(t), y(t))$ is a curve $C = c(t)$ in the xy-plane. The function $f(x(t), y(t))$ will then represent a curve on the surface of $f(x, y)$ directly above the curve $C = c(t)$ in the xy-plane (see Fig. 3.4). The chain rule expresses $\frac{df}{dt}$ along the curve C as the dot product of the two vectors

$$v = \left(\frac{dx}{dt}, \frac{dy}{dt} \right) \quad \text{and} \quad \nabla f = (f_x, f_y).$$

Then

$$v \cdot \nabla f = f_x \frac{dx}{dt} + f_y \frac{dy}{dt}$$
$$= \frac{df}{dt}.$$

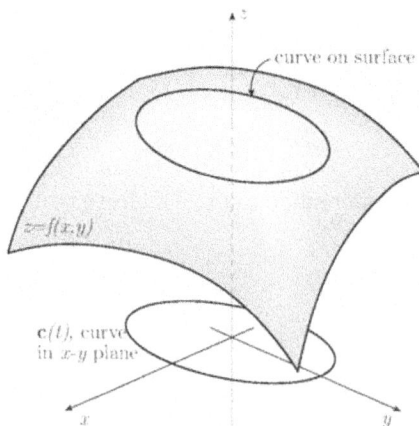

Figure 3.4: The curve on the surface of $z = f(x, y)$ directly above curve $C = c(t)$.

The vector v is the tangent vector to the curve $(x(t), y(t))$, and ∇f is called the gradient vector of f. These ideas can also be extended for functions of more than two variables. Let $X \subset \mathbb{R}^n$, let $f : X \to \mathbb{R}$ be a differentiable function at the point $x^0 \in X$, let Γ be a curve given by

$$\Gamma : x_j = x_j(t), \quad t \in [a, b], \quad j \in \{1, \ldots, n\}, \quad x_j \in C^1([a, b]),$$

and let $t_0 \in [a, b]$ be such that $x^0 = x(t_0)$. Let also,

$$f(x(t)) = 0, \quad t \in [a, b]. \tag{3.9}$$

Differentiating equation (3.9) with respect to t, we get

$$\sum_{j=1}^{n} f_{x_j}(x^0) x_j'(t_0) = 0.$$

Thus the vectors $(x_1'(t_0), \ldots, x_n'(t_0))$ and $(f_{x_1}(x^0), \ldots, f_{x_n}(x^0))$ are orthogonal. If $(x_1'(t_0), \ldots, x_n'(t_0)) \neq (0, \ldots, 0)$, then this vector is a tangent vector to Γ at the point x^0.

Definition 3.6. The vector

$$(f_{x_1}(x^0), \ldots, f_{x_n}(x^0))$$

is called the gradient of f at x^0. It is denoted by $\operatorname{grad} f(x^0)$.

Example 3.20. We will find $\operatorname{grad} f(-1, 1)$, where

$$f(x, y) = 1 + x^2 y^3, \quad (x, y) \in \mathbb{R}^2.$$

We have

$$f_x(x, y) = 2xy^3,$$
$$f_y(x, y) = 3x^2 y^2, \quad (x_1, x_2) \in \mathbb{R}^2,$$

and

$$f_x(-1, 1) = 2(-1)1^3$$
$$= -2,$$
$$f_y(-1, 1) = 3(-1)^2 1^2$$
$$= 3.$$

Therefore

$$\operatorname{grad} f(-1, 1) = (-2, 3).$$

Example 3.21. We will find $\operatorname{grad} f(1, 2, 3)$, where

$$f(x, y, z) = \frac{1}{\sqrt{x^2 + y^2 + z^2}}, \quad (x, y, z) \in \mathbb{R}^3 \backslash \{(0, 0, 0)\}.$$

We have

$$f_x(x, y, z) = -\frac{x}{(x^2 + y^2 + z^2)^{\frac{3}{2}}},$$

$$f_y(x, y, z) = -\frac{y}{(x^2 + y^2 + z^2)^{\frac{3}{2}}},$$

$$f_z(x, y, z) = -\frac{z}{(x^2 + y^2 + z^2)^{\frac{3}{2}}}, \quad (x, y, z) \in \mathbb{R}^3 \backslash \{(0, 0, 0)\}.$$

Hence

$$f_x(1, 2, 3) = -\frac{1}{(1^2 + 2^2 + 3^2)^{\frac{3}{2}}}$$

$$= -\frac{1}{14^{\frac{3}{2}}},$$

$$f_y(1, 2, 3) = -\frac{2}{14^{\frac{3}{2}}},$$

$$f_z(1, 2, 3) = -\frac{3}{14^{\frac{3}{2}}}.$$

Therefore

$$\operatorname{grad} f(1, 2, 3) = -\frac{1}{14^{\frac{3}{2}}}(1, 2, 3).$$

Exercise 3.10. Find $\operatorname{grad} f(M)$, where
1. $f(x, y) = yx^y$, $M(2, 1)$.
2. $f(x, y, z) = \arctan(\frac{xy}{z^2})$, $M(0, 1, 2)$.

3.5 Directional derivatives

A directional derivative is a concept in multivariable calculus that measures the rate at which a function changes in a particular direction at a given point. There are various applications of directional derivatives in mathematics, physics, engineering, and economics. For instance, applications of directional derivatives can be used in determining the rate of switching inputs in production functions, which can be very helpful in determining/forecasting switching costs for a given bundle of inputs.

For the case $n = 2$, the idea for directional derivatives is as follows. Let $P_0(x_0, y_0)$ be a fixed point in the xy-plane. Let also, y be a line in the xy-plane that passes through P_0.

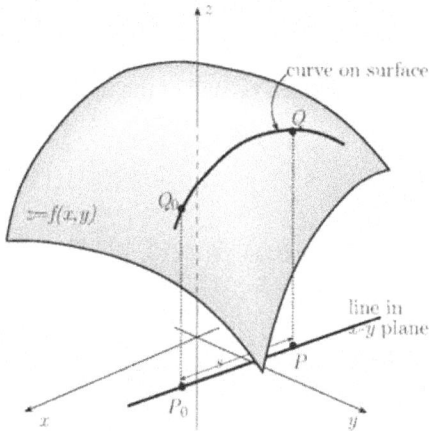

Figure 3.5: Curve on the surface $z = f(x,y)$ above the line y.

The point $P(x,y)$ moves along the line y. Directly above it, the point Q moves along the surface $z = f(x,y)$, tracing out a curve C (see Fig. 3.5). Let now u be the unit vector

$$u = u_1 e_1 + u_2 e_2$$

with initial point $P_0(x_0, y_0)$ and pointing in the direction of motion of $P(x,y)$. Then

$$\vec{P_0 P} = su,$$

where s is a parameter, and

$$(x - x_0)e_1 + (y - y_0)e_2 = su_1 e_1 + su_2 e_2.$$

Therefore

$$x = x_0 + su_1,$$
$$y = y_0 + su_2.$$

Hence

$$z = f(x,y)$$
$$= f(x_0 + su_1, y_0 + su_2).$$

By the chain rule we find

$$\frac{dz}{ds} = z_x \frac{dx}{ds} + z_y \frac{dy}{ds}$$
$$= f_x u_1 + f_y u_2$$

$$= \nabla f \cdot u.$$

Thus the rate of change of $f(x, y)$ at the point (x_0, y_0) in the direction of the unit vector u is given by the directional derivative

$$f_u(x_0, y_0) = \nabla f(x_0, y_0) \cdot u.$$

This can be extended to functions of n variables as follows. Let $x^0 = (x_1^0, \ldots, x_n^0) \in \mathbb{R}^n$, and let l be a nonzero vector in $\mathbb{R}^n$ such that

$$l = l_1 e_1 + \cdots + l_n e_n.$$

Let also, $\alpha_j = \angle(l, e_j)$. Then

$$\cos \alpha_j = \frac{l_j}{|l|}$$

$$= \frac{l_j}{\sqrt{l_1^2 + \cdots + l_n^2}}.$$

We have that

$$(\cos \alpha_1)^2 + \cdots + (\cos \alpha_n)^2 = 1.$$

Set $l_0 = \frac{l}{|l|}$. Then

$$l_0 = \cos \alpha_1 e_1 + \cdots + \cos \alpha_n e_n.$$

Take a line m through the point x^0 and direction vector l. Then

$$m: \quad \begin{matrix} x_1 = x_1^0 + t \cos \alpha_1, \\ \vdots \\ x_n = x_n^0 + t \cos \alpha_n, \end{matrix}$$

where $t > 0$ is the distance between the points $x = (x_1, \ldots, x_n) \in m$ and x^0. We have

$$d(x, x^0) = \sqrt{(t \cos \alpha_1)^2 + \cdots + (t \cos \alpha_n)^2}$$

$$= \sqrt{t^2((\cos \alpha_1)^2 + \cdots + (\cos \alpha_n)^2)}$$

$$= \sqrt{t^2}$$

$$= t.$$

Suppose that a function f is defined in a neighborhood of the point x^0.

Definition 3.7. The derivative of f at x^0 in the direction l is defined by

$$\frac{\partial f}{\partial l}(x^0) = \lim_{x \to x^0} \frac{f(x) - f(x^0)}{d(x, x^0)}.$$

Let f be differentiable at x^0. Then

$$\frac{\partial f}{\partial l}(x^0) = \lim_{t \to 0} \frac{f(x_1^0 + t \cos \alpha_1, \ldots, x_n^0 + t \cos \alpha_n) - f(x_1^0, \ldots, x_n^0)}{t}$$

$$= \frac{df}{dt}(x_1^0 + t \cos \alpha_1, \ldots, x_n^0 + t \cos \alpha_n)\big|_{t=0}$$

$$= f_{x_1}(x^0) \cos \alpha_1 + \cdots + f_{x_n}(x^0) \cos \alpha_n$$

$$= \langle \operatorname{grad} f(x^0), l_0 \rangle.$$

Example 3.22. We will find $\frac{\partial f}{\partial l}(M)$, where

$$f(x, y) = 3x^2 + 5y^2, \quad (x, y) \in \mathbb{R}^2,$$

$$l = \left(-\frac{1}{\sqrt{2}}, \frac{1}{\sqrt{2}} \right), \quad M(1, 1).$$

We have

$$|l| = \sqrt{\left(-\frac{1}{\sqrt{2}} \right)^2 + \left(\frac{1}{\sqrt{2}} \right)^2}$$

$$= \sqrt{\frac{1}{2} + \frac{1}{2}}$$

$$= 1.$$

Then

$$\cos \alpha_1 = -\frac{1}{\sqrt{2}},$$

$$\cos \alpha_2 = \frac{1}{\sqrt{2}}.$$

Next,

$$f_x(x, y) = 6x,$$

$$f_y(x, y) = 10y, \quad (x, y) \in \mathbb{R}^2.$$

Hence

$$f_x(M) = f_x(1, 1)$$

$$= 6,$$

$$f_y(M) = f_y(1,1)$$
$$= 10,$$

and

$$\frac{\partial f}{\partial l}(M) = f_x(M)\cos\alpha_1 + f_y(M)\cos\alpha_2$$

$$= 6\left(-\frac{1}{\sqrt{2}}\right) + 10\left(\frac{1}{\sqrt{2}}\right)$$

$$= 6\left(-\frac{\sqrt{2}}{2}\right) + 10\left(\frac{\sqrt{2}}{2}\right)$$

$$= -3\sqrt{2} + 5\sqrt{2}$$

$$= 2\sqrt{2}.$$

Example 3.23. We will find $\frac{\partial f}{\partial l}(M)$, where

$$f(x,y,z) = x^3 + 2xy^2 + 3yz^2, \quad (x,y,z) \in \mathbb{R}^3,$$
$$l = \left(\frac{2}{3}, \frac{2}{3}, \frac{1}{3}\right), \quad M(3,3,1).$$

We have

$$|l| = \sqrt{\left(\frac{2}{3}\right)^2 + \left(\frac{2}{3}\right)^2 + \left(\frac{1}{3}\right)^2}$$

$$= \sqrt{\frac{4}{9} + \frac{4}{9} + \frac{1}{9}}$$

$$= 1.$$

Thus

$$\cos\alpha_1 = \frac{2}{3},$$
$$\cos\alpha_2 = \frac{2}{3},$$
$$\cos\alpha_3 = \frac{1}{3}.$$

Next,

$$f_x(x,y,z) = 3x^2 + 2y^2,$$
$$f_y(x,y,z) = 4xy + 3z^2,$$
$$f_z(x,y,z) = 6yz, \quad (x,y,z) \in \mathbb{R}^3.$$

Hence

$$f_x(M) = f_x(3, 3, 1)$$
$$= 3 \cdot 3^2 + 2 \cdot 3^2$$
$$= 3 \cdot 9 + 2 \cdot 9$$
$$= 27 + 18$$
$$= 45,$$
$$f_y(M) = f_y(3, 3, 1)$$
$$= 4 \cdot 3 \cdot 3 + 3(1)^2$$
$$= 39,$$
$$f_z(M) = f_z(3, 3, 1)$$
$$= 6 \cdot 3 \cdot 1$$
$$= 18,$$

and

$$\frac{\partial f}{\partial l}(M) = f_x(M) \cos \alpha_1 + f_y(M) \cos \alpha_2 + f_z(M) \cos \alpha_3$$
$$= 45 \cdot \frac{2}{3} + 39 \cdot \frac{2}{3} + 18 \cdot \frac{1}{3}$$
$$= 30 + 26 + 6$$
$$= 62.$$

Example 3.24. We will find $\frac{\partial f}{\partial l}(M)$, where

$$f(x_1, \ldots, x_n) = \sum_{j=1}^{n} \arcsin x_j, \quad (x_1, \ldots, x_n) \in \mathbb{R}^n, \quad |(x_1, \ldots, x_n)| < 1,$$

$$l = \left(\frac{1}{\sqrt{n}}, \ldots, \frac{1}{\sqrt{n}} \right), \quad M\left(\frac{1}{4}, \ldots, \frac{1}{4} \right).$$

We have

$$|l| = \sqrt{\left(\frac{1}{\sqrt{n}} \right)^2 + \cdots + \left(\frac{1}{\sqrt{n}} \right)^2}$$
$$= \sqrt{\frac{1}{n} + \cdots + \frac{1}{n}}$$
$$= 1.$$

Thus

$$\cos \alpha_j = \frac{1}{\sqrt{n}}, \quad j \in \{1, \ldots, n\}.$$

Next,

$$f_{x_j}(x_1,\ldots,x_n) = \frac{1}{\sqrt{1-x_j^2}}, \quad (x_1,\ldots,x_n) \in \mathbb{R}^n, \quad |(x_1,\ldots,x_n)| < 1.$$

Therefore

$$f_{x_j}(M) = f_{x_j}\left(\frac{1}{4},\ldots,\frac{1}{4}\right)$$
$$= \frac{1}{\sqrt{1-\frac{1}{16}}}$$
$$= \frac{4}{\sqrt{15}}, \quad j \in \{1,\ldots,n\},$$

and

$$\frac{\partial f}{\partial l}(M) = \sum_{j=1}^{n} f_{x_j}(M)\cos\alpha_j$$
$$= \sum_{j=1}^{n} \frac{4}{\sqrt{15}} \cdot \frac{1}{\sqrt{n}}$$
$$= \frac{4n}{\sqrt{15n}}$$
$$= 4\sqrt{\frac{n}{15}}.$$

Exercise 3.11. Find $\frac{\partial f}{\partial l}(M)$, where
1.
$$f(x,y) = x\sin(x+y), \quad (x,y) \in \mathbb{R}^2,$$
$$l = (-1,0), \quad M\left(\frac{\pi}{4},\frac{\pi}{4}\right).$$

2.
$$f(x,y,z) = \log(x^2+y^2+z^2), \quad (x,y,z) \in \mathbb{R}^3, \quad x^2+y^2+z^2 \neq 0,$$
$$l = \left(-\frac{1}{3},\frac{2}{3},\frac{2}{3}\right), \quad M(1,2,1).$$

3.
$$f(x_1,x_2,x_3,x_4) = x_1^2 + x_2^2 - x_3^2 + x_4^2, \quad (x_1,x_2,x_3,x_4) \in \mathbb{R}^4,$$
$$l = \left(\frac{2}{3},\frac{1}{3},0,-\frac{2}{3}\right), \quad M(1,3,2,1).$$

The Euler[1] theorem states that every homogeneous function of given degree is the solution of a specific partial differential equation. The Euler identity has a wide range

---

[1] Leonhard Euler (15 April 1707–18 September 1783) was a Swiss mathematician, physicist, astronomer, geographer, logician, and engineer, who founded the studies of graph theory and topology and made pi-

of applications in the theory of ordinary and partial differential equations. It reads as follows.

Theorem 3.6 (The Euler identity). *Let $X \subset \mathbb{R}^n$, let $f : X \to \mathbb{R}$ be a homogeneous function of degree a, and suppose $f \in C^1(X)$. Then*

$$af(x_1,\ldots,x_n) = x_1 f_{x_1}(x_1,\ldots,x_n) + \cdots + x_n f_{x_n}(x_1,\ldots,x_n), \quad (x_1,\ldots,x_n) \in X. \qquad (3.10)$$

Definition 3.8. Identity (3.10) is called the Euler identity.

Proof. Since f is a homogeneous function of degree a on X, for any $t \in \mathbb{R}$ and $(x_1,\ldots,x_n) \in X$ such that $(tx_1,\ldots,tx_n) \in X$, we have

$$f(tx_1,\ldots,tx_n) = t^a f(x_1,\ldots,x_n), \quad (x_1,\ldots,x_n) \in X.$$

Differentiating with respect to t, we get

$$at^{a-1}f(x_1,\ldots,x_n) = x_1 f_{x_1}(tx_1,\ldots,tx_n) + \cdots + x_n f_{x_n}(tx_1,\ldots,tx_n).$$

Taking $t = 1$ in the last equation, we get (3.10). This completes the proof. $\qquad \square$

3.6 Advanced practical problems

Problem 3.1. Find the first partial derivatives of the following functions:

1.
$$f(x,y) = \sin x - x^2 y, \quad (x,y) \in \mathbb{R}^2.$$

2.
$$f(x,y) = \sin \frac{x}{y} \cos \frac{y}{x}, \quad (x,y) \in \mathbb{R}^2, \quad x \neq 0, \quad y \neq 0.$$

3.
$$f(x,y) = e^x(\cos y + x \sin y), \quad (x,y) \in \mathbb{R}^2.$$

4.
$$f(x,y) = \log \frac{\sqrt{x^2+y^2} - x}{\sqrt{x^2+y^2} + x}, \quad (x,y) \in \mathbb{R}^2, \quad x \neq 0, \quad y \neq 0.$$

5.
$$f(x,y) = \arcsin \sqrt{\frac{x^2-y^2}{x^2+y^2}}, \quad (x,y) \in \mathbb{R}^2, \quad (x,y) \neq (0,0).$$

---

oneering and influential discoveries in many other branches of mathematics such as analytic number theory, complex analysis, and infinitesimal calculus. He introduced much of modern mathematical terminology and notation, including the notion of a mathematical function. He is also known for his work in mechanics, fluid dynamics, optics, astronomy, and music theory.

6.
$$f(x,y) = (1 + (\sin x)^2)^{\log y}, \quad (x,y) \in \mathbb{R}^2, \quad y > 0.$$

7.
$$f(x,y,z) = \frac{1}{\sqrt{x^2 + y^2 + z^2}}, \quad (x,y,z) \in \mathbb{R}^3, \quad (x,y,z) \neq (0,0,0).$$

8.
$$f(x,y,z) = \frac{z}{x} + \frac{x}{z}, \quad (x,y,z) \in \mathbb{R}^3, \quad x \neq 0, \quad z \neq 0.$$

9.
$$f(x,y,z) = \frac{y}{z} + \arctan\frac{z}{x} + \arctan\frac{x}{z}, \quad (x,y,z) \in \mathbb{R}^3, \quad x \neq 0, \quad z \neq 0.$$

10.
$$f(x,y,z) = z^{xy}, \quad (x,y,z) \in \mathbb{R}^3, \quad z > 0.$$

11.
$$f(x,y,z) = \left(\frac{x}{y}\right)^z, \quad (x,y,z) \in \mathbb{R}^3, \quad x > 0, \quad y > 0.$$

12.
$$f(x_1,\ldots,x_n) = \sum_{j=1}^{n}(\cos x_j)^2, \quad (x_1,\ldots,x_n) \in \mathbb{R}^n.$$

Problem 3.2. Find $f_x(1,2)$ and $f_y(1,2)$, where
$$f(x,y) = \log\left(1 + \frac{x}{y}\right), \quad (x,y) \in \mathbb{R}^2, \quad y \neq 0, \quad 1 + \frac{x}{y} > 0.$$

Problem 3.3. Find $f_x(1,1)$ and $f_y(1,1)$, where
$$f(x,y) = xye^{\sin(\pi xy)}, \quad (x,y) \in \mathbb{R}^2.$$

Problem 3.4. Find
$$f_x(x,y,z) + f_y(x,y,z) + f_z(x,y,z), \quad (x,y,z) \in X,$$
where
1.
$$f(x,y,z) = (x-y)(y-z)(z-x), \quad (x,y,z) \in X, \quad X = \mathbb{R}^3.$$

2.
$$f(x,y,z) = x + \frac{x-y}{y-z}, \quad (x,y,z) \in X, \quad X = \{(x,y,z) \in \mathbb{R}^3 : y \neq z\}.$$

Problem 3.5. Find
$$f_x(1,1,1) + f_y(1,1,1) + f_z(1,1,1),$$
where
$$f(x,y,z) = \log(1 + x + y^2 + z^3), \quad (x,y,z) \in \mathbb{R}^3, \quad x > 0.$$

Problem 3.6. Check if the following functions are differentiable at $(0,0)$:

1.
$$f(x,y) = |y| \sin x, \quad (x,y) \in \mathbb{R}^2.$$

2.
$$f(x,y) = \log(2 - |x|^{\frac{7}{6}} + |y|^{\frac{5}{4}}), \quad (x,y) \in \mathbb{R}^2, \quad |x|, |y| < \frac{1}{100}.$$

3.
$$f(x,y) = y^{\frac{3}{5}} \arcsin(\sqrt{|x|}), \quad (x,y) \in \mathbb{R}^2, \quad |x| < \frac{1}{100}, \quad y \geq 0.$$

4.
$$f(x,y) = 1 + xy + \sin(\sqrt[5]{x^2 y^4}), \quad (x,y) \in \mathbb{R}^2.$$

5.
$$f(x,y) = \sqrt[5]{\sin x (1 - \cos(xy))}, \quad (x,y) \in \mathbb{R}^2.$$

6.
$$f(x,y) = \sin\left(\frac{\pi}{4} + \sqrt[3]{xy^2}\right), \quad (x,y) \in \mathbb{R}^2.$$

7.
$$f(x,y) = \log(3 + \sqrt[3]{x^2 y}), \quad (x,y) \in \mathbb{R}^2, \quad |x|, |y| < \frac{1}{100}.$$

8.
$$f(x,y) = \arctan(2x + \sqrt[3]{x^3 - 27y^3}), \quad (x,y) \in \mathbb{R}^2, \quad |x|, |y| < \frac{1}{100}.$$

9.
$$f(x,y) = \arcsin(y + \sqrt[3]{x^3 + 8y^3}), \quad (x,y) \in \mathbb{R}^2, \quad |x|, |y| < \frac{1}{100}.$$

10.
$$f(x,y) = \log(1 + x + \sqrt[3]{x^3 + 27y^3}), \quad (x,y) \in \mathbb{R}^2, \quad |x|, |y| < \frac{1}{100}.$$

11.
$$f(x,y) = y\sqrt[3]{1 + \sqrt[3]{x^2}}, \quad (x,y) \in \mathbb{R}^2.$$

12.
$$f(x,y) = \sqrt[3]{x^3 + y^4}, \quad (x,y) \in \mathbb{R}^2.$$

13.
$$f(x,y) = 2y + x \cos(\sqrt[3]{xy}), \quad (x,y) \in \mathbb{R}^2.$$

14.
$$f(x,y) = y + \cos(\sqrt[3]{x^2 + y^2}), \quad (x,y) \in \mathbb{R}^2.$$

15.
$$f(x,y) = \sqrt[3]{(\sin x)^4 + (\cos y)^4}, \quad (x,y) \in \mathbb{R}^2.$$

16.
$$f(x,y) = y + \log(3 + \sqrt[3]{x^2 y}), \quad (x,y) \in \mathbb{R}^2, \quad y \geq 0.$$

17.
$$f(x,y) = \arcsin(xy + \sqrt[3]{x^3 + y^3}), \quad (x,y) \in \mathbb{R}^2, \quad |x|, |y| < \frac{1}{100}.$$

18.
$$f(x,y) = \arctan(xy + y + \sqrt[3]{x^2 y}), \quad (x,y) \in \mathbb{R}^2, \quad |x|, |y| < \frac{1}{100}.$$

19.
$$f(x,y) = \begin{cases} \frac{y^3 - x^3}{x^2 + 2y^2} & \text{if } x^2 + y^2 \neq 0, \\ 0 & \text{if } x^2 + y^2 = 0. \end{cases}$$

20.
$$f(x,y) = \begin{cases} (x + y)\arctan\left(\left(\frac{x}{y}\right)^2\right) & \text{if } y \neq 0, \\ \frac{\pi x}{2} & \text{if } y = 0. \end{cases}$$

21.
$$f(x,y) = \begin{cases} e^{-\frac{1}{x^2 + y^2}} & \text{if } x^2 + y^2 \neq 0, \\ 0 & \text{if } x^2 + y^2 = 0. \end{cases}$$

22.
$$f(x,y) = \begin{cases} (x^2 + y^2)\sin\left(\frac{1}{x^2 + y^2}\right) & \text{if } x^2 + y^2 \neq 0, \\ & \text{if } x^2 + y^2 = 0. \end{cases}$$

23.
$$f(x,y) = \begin{cases} (x^2 + y^2)^3 & \text{if } x, y \in \mathbb{Q}, \\ 0 & \text{if } x \in \mathbb{Q}, \quad y \notin \mathbb{Q}. \end{cases}$$

24.
$$f(x,y) = \begin{cases} (x^2 + y^2)^{\frac{1}{4}} & \text{if } x, y \in \mathbb{Q}, \\ 0 & \text{if } x \in \mathbb{Q}, \quad y \notin \mathbb{Q}. \end{cases}$$

25.
$$f(x,y) = \begin{cases} (x + y)^{\frac{3}{2}} + |y|^{\frac{7}{2}} & \text{if } x^2 + y^2 \neq 0, \\ 0 & \text{if } x^2 + y^2 = 0. \end{cases}$$

Problem 3.7. Find the values of $\alpha \in \mathbb{R}$ such that the following functions are differentiable at $(0,0)$:

1.
$$f(x,y) = \begin{cases} \frac{|x|^\alpha |y|^{5-2\alpha}}{|x| + |y|} & \text{if } x^2 + y^2 \neq 0, \\ 0 & \text{if } x^2 + y^2 = 0. \end{cases}$$

2.
$$f(x,y) = \begin{cases} \frac{|xy|^\alpha}{1 - \log(x^2 + y^2)} & \text{if } x^2 + y^2 \neq 0, \\ 0 & \text{if } x^2 + y^2 = 0. \end{cases}$$

Problem 3.8. Find $df(x_1, \ldots, x_n)$, $(x_1, \ldots, x_n) \in X$, where

1. $n = 2$,
$$f(x,y) = \frac{y}{x} + \frac{x}{y}, \quad (x,y) \in X,$$
$$X = \{(x,y) \in \mathbb{R}^2 : x \neq 0, y \neq 0\}.$$

2. $n = 2$,

$$f(x,y) = \frac{x}{\sqrt{x^2 + y^2}}, \quad (x,y) \in X,$$

$$X = \{(x,y) \in \mathbb{R}^2 : (x,y) \neq (0,0)\}.$$

3. $n = 2$,

$$f(x,y) = \log(x + \sqrt{x^2 + y^2}), \quad (x,y) \in X,$$

$$X = \{(x,y) \in \mathbb{R}^2 : x > 0, \; y > 0\}.$$

4. $n = 2$,

$$f(x,y) = \log\left(\sin\left(\frac{x+1}{\sqrt{y}}\right)\right), \quad (x,y) \in X,$$

$$X = \left\{(x,y) \in \mathbb{R}^2 : y > 0, \; \frac{x+1}{\sqrt{y}} \in (0,\pi)\right\}.$$

5. $n = 2$,

$$f(x,y) = \arctan\frac{x}{y} + \arctan\frac{y}{x}, (x,y) \in X,$$

$$X = \left\{(x,y) \in \mathbb{R}^2 : x \neq 0, \; y \neq 0, \; \frac{x}{y}, \frac{y}{x} \in \left(-\frac{\pi}{2}, \frac{\pi}{2}\right)\right\}.$$

6. $n = 2$,

$$f(x,y) = \arctan\frac{x+y}{x-y}, \quad (x,y) \in X,$$

$$X = \left\{(x,y) \in \mathbb{R}^2 : x \neq y, \; \frac{x+y}{x-y} \in (0,\pi)\right\}.$$

7. $n = 3$,

$$f(x,y,z) = e^{xy\sin z}, \quad (x,y,z) \in X, \quad X = \mathbb{R}^3.$$

8. $n = 3$,

$$f(x,y,z) = (xy)^z, \quad (x,y,z) \in X,$$

$$X = \{(x,y,z) \in \mathbb{R}^3 : xy > 0\}.$$

9.

$$f(x_1,\ldots,x_n) = \cos\left(\sum_{j=1}^{n} x_j^2\right), \quad (x_1,\ldots,x_n) \in X, \quad X = \mathbb{R}^n.$$

Problem 3.9. Let $f = f(u_1, u_2)$, $(u_1, u_2) \in \mathbb{R}^2$, be a differentiable function. Prove that the function

$$\phi(x, y, z) = x^\alpha f(x^\beta y, x^\gamma z), \quad (x, y, z) \in \mathbb{R}^3,$$

where $\alpha, \beta, \gamma \in \mathbb{R}$, satisfies the equation

$$x\phi_x(x, y, z) - \beta y\phi_y(x, y, z) - \gamma z\phi_z(x, y, z) = \alpha\phi(x, y, z), \quad (x, y, z) \in \mathbb{R}^3.$$

Problem 3.10. Find $\operatorname{grad} f(M)$, where
1. $f(x, y, z) = e^{x+xy+xyz}$, $M(0, 1, 0)$.
2. $f(x, y, z) = \log(1 - x^2 - 2y^2 - 3z^2)$, $M(\frac{1}{2}, 0, 0)$.

Problem 3.11. Find $\frac{\partial f}{\partial l}(M)$, where
1.

$$f(x, y) = 5x + 10x^2y + y^5, \quad (x, y) \in \mathbb{R}^2,$$
$$l = (4, -3), \quad M(1, 2).$$

2.

$$f(x, y, z) = xy^2z^3, \quad (x, y, z) \in \mathbb{R}^3,$$
$$l = (4, 3, 0), \quad M(3, 2, 1).$$

3.

$$f(x, y, z) = \arcsin \frac{z}{\sqrt{x^2 + y^2}}, \quad (x, y, z) \in \mathbb{R}^3, \quad (x, y) \neq (0, 0),$$
$$l = (0, 4, 3), \quad M(1, 1, 1).$$

4.

$$f(x_1, x_2, x_3, x_4) = \frac{x_2}{x_1^2 + x_2^2 + x_3^2 + x_4^2}, \quad (x_1, x_2, x_3, x_4) \in \mathbb{R}^4, \quad (x_1, x_2, x_3, x_4) \neq (0, 0, 0, 0),$$
$$l = (3, 1, 0, 0), \quad M(0, 1, 1, 0).$$

Problem 3.12. Find $\max \frac{\partial f}{\partial l}(M)$, where
1.

$$f(x, y) = xy^2 - 3x^4y^5, \quad M(1, 1).$$

2.

$$f(x, y) = \frac{x + \sqrt{y}}{y}, \quad M(2, 1).$$

3.

$$f(x, y, z) = \log(xyz), \quad M(1, -2, -3).$$

4.

$$f(x, y, z) = \tan x - x + 3\sin y - (\sin y)^3 + 2z + \cot z, \quad M\left(\frac{\pi}{4}, \frac{\pi}{3}, \frac{\pi}{2}\right).$$

Problem 3.13. Find a unit vector l such that $\frac{\partial f}{\partial l}$ achieves its maximum at M, where

1.
$$f(x,y) = x^2 - xy + y^2, \quad M(-1,2).$$

2.
$$f(x,y) = x - 3y + \sqrt{3xy}, \quad M(3,1).$$

3.
$$f(x,y,z) = \arcsin(xy) + \arcsin(yz), \quad M(1,0,5,0).$$

4.
$$f(x,y,z) = xz^y, \quad M(-3,2,1).$$

Problem 3.14. Using the Euler identity, find

$$xf_x + yf_y + zf_z,$$

where

1.
$$f(x,y) = \frac{x}{x^2 + y^2}.$$

2.
$$f(x,y,z) = \frac{x+z}{\sqrt[3]{x^2 + z^2}}.$$

3.
$$f(x,y,z) = (x + 2y + 3z)^4.$$

4.
$$f(x,y,z) = (\log x - \log y)^{\frac{y}{z}}.$$

5.
$$f(x,y,z) = \frac{xy}{z} \log x + x\phi\left(\frac{y}{x}, \frac{z}{x}\right),$$

where ϕ is a differentiable function.

4 Higher-order partial derivatives and differentials

In this chapter, higher-order partial derivatives of functions of several variables are investigated. Minimum and maximum of functions of several variables are introduced and investigated. Implicit functions are defined and explored. The method of Lagrange multipliers is introduced. Vector and linear maps are investigated. The conception of functional dependence is given.

4.1 Higher-order partial derivatives

Let $X \subset \mathbb{R}^n$ and $f : X \to \mathbb{R}$. Notice that $\frac{\partial f}{\partial x_j}, j \in \{1, \ldots, n\}$, are functions of n variables. So they can also be partially differentiated. This motivates us to give the following definition.

Definition 4.1. Any partial derivative $f_{x_j}(x)$, $x \in \mathbb{R}, j \in \{1, \ldots, n\}$, is said to be a first-order partial derivative or a partial derivative of first order of the function f. The partial derivatives $\frac{\partial}{\partial x_j}(\frac{\partial f}{\partial x_j})(x)$, $x \in X, j \in \{1, \ldots, n\}$, are denoted by $f_{x_j x_j}(x)$ or $\frac{\partial^2 f}{\partial x_j^2}(x)$. The partial derivatives $\frac{\partial}{\partial x_l}(\frac{\partial f}{\partial x_j})(x)$, $j, l \in \{1, \ldots, n\}$, $x \in X$, are denoted by $f_{x_j x_l}(x)$ or $\frac{\partial^2 f}{\partial x_l \partial x_j}(x)$ and called the second-order partial derivatives or partial derivatives of second order of the function f. Similarly, we can define $f_{x_j x_l x_m}(x)$ or $\frac{\partial^3 f}{\partial x_l \partial x_m \partial x_j}(x), j, l, m \in \{1, \ldots, n\}$, which are called the third-order partial derivatives or partial derivatives of third order. Partial derivatives with respect to different variables are called mixed partial derivatives.

The application of higher-order derivatives plays a pivotal role in various fields such as physics, engineering, and economics, offering deeper insights into the rate of change beyond the immediate velocity or acceleration. These derivatives, beyond the first and second orders, help in identifying the curvature, inflection points, and optimizing functions within complex systems. Mastering higher-order derivatives equips students with the analytical tools needed to solve real-world problems, enhancing their understanding of the underlying mechanics of change.

Example 4.1. We will find $f_{xx}(x,y), f_{xy}(x,y), f_{yx}(x,y), f_{yy}(x,y), (x,y) \in X$, where

$$f(x,y) = xy(x^3 + y^3 - 3), \quad (x,y) \in X, \quad X = \mathbb{R}^2.$$

We have

$$f_x(x,y) = y(x^3 + y^3 - 3) + 3x^3 y$$
$$= y(x^3 + y^3 - 3 + 3x^3)$$
$$= y(4x^3 + y^3 - 3),$$
$$f_y(x,y) = x(x^3 + y^3 - 3) + 3xy^3$$

https://doi.org/10.1515/9783112218082-004

$$= x(x^3 + y^3 - 3 + 3y^3)$$
$$= x(x^3 + 4y^3 - 3), \quad (x,y) \in X.$$

Hence

$$f_{xx}(x,y) = 12x^2y,$$
$$f_{xy}(x,y) = 4x^3 + y^3 - 3 + 3y^3$$
$$= 4x^3 + 4y^3 - 3,$$
$$f_{yx}(x,y) = x^3 + 4y^3 - 3 + 3x^3$$
$$= 4x^3 + 4y^3 - 3,$$
$$f_{yy}(x,y) = 12xy^2, \quad (x,y) \in X.$$

We see that

$$f_{xy}(x,y) = f_{yx}(x,y), \quad (x,y) \in X.$$

Example 4.2. We will find $f_{xyz}(x,y,z)$, $(x,y,z) \in X$, where

$$f(x,y,z) = x^3 \sin y + y^3 \sin z + z^3 \sin x, \quad (x,y,z) \in X, \quad X = \mathbb{R}^3.$$

We have

$$f_x(x,y,z) = 3x^2 \sin y + z^3 \cos x,$$
$$f_{xy}(x,y,z) = 3x^2 \cos y,$$
$$f_{xyz}(x,y,z) = 0, \quad (x,y,z) \in \mathbb{R}^3.$$

Example 4.3. We will find $f_{xy}(0,0)$ and $f_{yx}(0,0)$, where

$$f(x,y) = \begin{cases} \frac{xy(x^2-y^2)}{x^2+y^2} & \text{if } x^2 + y^2 \neq 0, \\ 0 & \text{if } x^2 + y^2 = 0. \end{cases}$$

We have

$$f_x(x,y) = \begin{cases} \frac{x^4y+4x^2y^3-y^5}{(x^2+y^2)^2} & \text{if } x^2 + y^2 \neq 0, \\ 0 & \text{if } x^2 + y^2 = 0, \end{cases}$$

$$f_y(x,y) = \begin{cases} \frac{x^5-4x^3y^2-xy^4}{(x^2+y^2)^2} & \text{if } x^2 + y^2 \neq 0, \\ 0 & \text{if } x^2 + y^2 = 0. \end{cases}$$

Hence, for $h \neq 0$, we have

$$\frac{f_x(0,h) - f_x(0,0)}{h} = \frac{1}{h}\left(-\frac{h^5}{h^4}\right)$$
$$= -1,$$
$$\frac{f_y(h,0) - f_y(0,0)}{h} = \frac{1}{h}\left(\frac{h^5}{h^4}\right)$$
$$= 1.$$

Thus

$$f_{xy}(0,0) = \lim_{h \to 0} \frac{f_x(0,h) - f_x(0,0)}{h}$$
$$= -1,$$
$$f_{yx}(0,0) = \lim_{h \to 0} \frac{f_y(h,0) - f_y(0,0)}{h}$$
$$= 1.$$

Exercise 4.1. Find $f_{xx}(x,y), f_{xy}(x,y), f_{yy}(x,y), (x,y) \in X$, where

$$f(x,y) = e^{xy}, \quad (x,y) \in X.$$

Example 4.4. We will find

$$\frac{\partial^{m+k} f}{\partial y^k \partial x^m}(0,0),$$

where

$$f(x,y) = e^x \sin y, \quad (x,y) \in \mathbb{R}^2.$$

We have

$$\frac{\partial f}{\partial x}(x,y) = e^x \sin y,$$

$$\frac{\partial^2 f}{\partial x^2}(x,y) = e^x \sin y,$$

$$\vdots$$

$$\frac{\partial^m f}{\partial x^m}(x,y) = e^x \sin y,$$

$$\frac{\partial^{m+1} f}{\partial x^m \partial y}(x,y) = e^x \cos y,$$

$$\frac{\partial^{m+2} f}{\partial x^m \partial y^2}(x,y) = -e^x \sin y,$$

$$\frac{\partial^{m+3} f}{\partial x^m \partial y^3}(x,y) = -e^x \cos y.$$

Thus, for $k = 2l, l \in \mathbb{N}_0$, we have

$$\frac{\partial^{m+k} f}{\partial y^k \partial x^m}(0,0) = 0.$$

For $k = 2l + 1, l \in \mathbb{N}_0$, we have

$$\frac{\partial^{m+k} f}{\partial y^k \partial x^m}(0,0) = (-1)^l.$$

Exercise 4.2. Find

$$\frac{\partial^{m+k} f}{\partial y^k \partial x^m}(x,y), \quad (x,y) \in \mathbb{R}^2,$$

where

$$f(x,y) = (x-a)^m (y-b)^k, \quad (x,y) \in \mathbb{R}^2, \quad a, b \in \mathbb{R}.$$

By Examples 4.1 and 4.3 it follows that the mixed derivatives at a point do not coincide in the general case. In the next result, we will see a criterion for coincidence of the mixed derivatives at a point.

Theorem 4.1. Let $f_{x_j}, f_{x_l}, f_{x_j x_l},$ and $f_{x_l x_j}$ exist and be continuous in a neighborhood of the point x^0 for some $j, l \in \{1, \ldots, n\}, j \neq l$. Then

$$f_{x_j x_l}(x^0) = f_{x_l x_j}(x^0).$$

Proof. With out loss of generality, suppose that $j < l$. Note that

$$\Delta_{x_j} f(x^0) = f(x_1^0, \ldots, x_{j-1}^0, x_j^0 + \Delta x_j, x_{j+1}^0, \ldots, x_{l-1}^0, x_l^0, x_{l+1}^0, \ldots, x_n^0)$$
$$- f(x_1^0, \ldots, x_{j-1}^0, x_j^0, x_{j+1}^0, \ldots, x_{l-1}^0, x_l^0, x_{l+1}^0, \ldots, x_n^0),$$
$$\Delta_{x_l} f(x^0) = f(x_1^0, \ldots, x_{j-1}^0, x_j^0, x_{j+1}^0, \ldots, x_{l-1}^0, x_l^0 + \Delta x_l, x_{l+1}^0, \ldots, x_n^0)$$
$$- f(x_1^0, \ldots, x_{j-1}^0, x_j^0, x_{j+1}^0, \ldots, x_{l-1}^0, x_l^0, x_{l+1}^0, \ldots, x_n^0),$$
$$\Delta_{x_l}(\Delta_{x_j} f)(x^0) = f(x_1^0, \ldots, x_{j-1}^0, x_j^0 + \Delta x_j, x_{j+1}^0, \ldots, x_{l-1}^0, x_l^0 + \Delta x_l, x_{l+1}^0, \ldots, x_n^0)$$
$$- f(x_1^0, \ldots, x_{j-1}^0, x_j^0, x_{j+1}^0, \ldots, x_{l-1}^0, x_l^0 + \Delta x_l, x_{l+1}^0, \ldots, x_n^0)$$
$$- f(x_1^0, \ldots, x_{j-1}^0, x_j^0 + \Delta x_j, x_{j+1}^0, \ldots, x_{l-1}^0, x_l^0, x_{l+1}^0, \ldots, x_n^0)$$
$$+ f(x_1^0, \ldots, x_{j-1}^0, x_j^0, x_{j+1}^0, \ldots, x_{l-1}^0, x_l^0, x_{l+1}^0, \ldots, x_n^0),$$

and

$$\Delta_{x_j}(\Delta_{x_l} f)(x^0) = f(x_1^0, \ldots, x_{j-1}^0, x_j^0 + \Delta x_j, x_{j+1}^0, \ldots, x_{l-1}^0, x_l^0 + \Delta x_l, x_{l+1}^0, \ldots, x_n^0)$$

$$-f(x_1^0, \ldots, x_{j-1}^0, x_j^0 + \Delta x_j, x_{j+1}^0, \ldots, x_{l-1}^0, x_l^0, x_{l+1}^0, \ldots, x_n^0)$$
$$-f(x_1^0, \ldots, x_{j-1}^0, x_j^0, x_{j+1}^0, \ldots, x_{l-1}^0, x_l^0 + \Delta x_l, x_{l+1}^0, \ldots, x_n^0)$$
$$+f(x_1^0, \ldots, x_{j-1}^0, x_j^0, x_{j+1}^0, \ldots, x_{l-1}^0, x_l^0, x_{l+1}^0, \ldots, x_n^0).$$

Therefore

$$\Delta_{x_l}(\Delta_{x_j} f)(x^0) = \Delta_{x_j}(\Delta_{x_l} f)(x^0). \tag{4.1}$$

Set

$$\phi(x_l^0) = f(x_1^0, \ldots, x_{j-1}^0, x_j^0 + \Delta x_j, x_{j+1}^0, \ldots, x_{l-1}^0, x_l^0, x_{l+1}^0, \ldots, x_n^0)$$
$$-f(x_1^0, \ldots, x_{j-1}^0, x_j^0, x_{j+1}^0, \ldots, x_{l-1}^0, x_l^0, x_{l+1}^0, \ldots, x_n^0)$$

and

$$\psi(x_j^0) = f(x_1^0, \ldots, x_{j-1}^0, x_j^0, x_{j+1}^0, \ldots, x_{l-1}^0, x_l^0 + \Delta x_l, x_{l+1}^0, \ldots, x_n^0)$$
$$-f(x_1^0, \ldots, x_{j-1}^0, x_j^0, x_{j+1}^0, \ldots, x_{l-1}^0, x_l^0, x_{l+1}^0, \ldots, x_n^0).$$

Then

$$\Delta_{x_j} f(x^0) = \phi(x_l^0),$$
$$\Delta_{x_l} f(x^0) = \psi(x_j^0).$$

Hence

$$\Delta_{x_l}(\Delta_{x_j} f)(x^0) = \phi(x_l^0 + \Delta x_l) - \phi(x_l^0),$$
$$\Delta_{x_j}(\Delta_{x_l} f)(x^0) = \psi(x_j^0 + \Delta x_j) - \psi(x_j^0).$$

By the mean value theorem it follows that there are $\theta_1, \eta_1 \in (0, 1)$ such that

$$\phi(x_l^0 + \Delta x_l) - \phi(x_l^0) = \phi'(x_l^0 + \theta_1 \Delta x_l)\Delta x_l,$$
$$\psi(x_j^0 + \Delta x_j) - \psi(x_j^0) = \psi'(x_j^0 + \eta_1 \Delta x_j)\Delta x_j.$$

Hence, again applying the mean value theorem, it follows that there are $\theta_2, \eta_2 \in (0, 1)$ such that

$$\Delta_{x_l}(\Delta_{x_j} f)(x^0) = \phi'(x_l^0 + \theta_1 \Delta x_l)\Delta x_l$$
$$= (f_{x_l}(x_1^0, \ldots, x_{j-1}^0, x_j^0 + \Delta x_j, x_{j+1}^0, \ldots, x_{l-1}^0, x_l^0 + \theta_1 \Delta x_l, x_{l+1}^0, \ldots, x_n^0)$$
$$-f_{x_l}(x_1^0, \ldots, x_{j-1}^0, x_j^0, x_{j+1}^0, \ldots, x_{l-1}^0, x_l^0 + \theta_1 \Delta x_l, x_{l+1}^0, \ldots, x_n^0))\Delta x_l$$
$$= f_{x_l x_j}(x_1^0, \ldots, x_{j-1}^0, x_j^0 + \theta_2 \Delta x_j, x_{j+1}^0, \ldots, x_{l-1}^0, x_l^0 + \theta_1 \Delta x_l, x_{l+1}^0, \ldots, x_n^0)\Delta x_j \Delta x_l$$

and

$$\Delta_{x_j}(\Delta_{x_l}f)(x^0) = \psi'(x_j^0 + \eta_1\Delta x_j)\Delta x_j$$
$$= (f_{x_j}(x_1^0,\ldots,x_{j-1}^0,x_j^0 + \eta_1\Delta x_j, x_{j+1}^0,\ldots,x_{l-1}^0,x_l^0 + \Delta x_l, x_{l+1}^0,\ldots,x_n^0)$$
$$- f_{x_j}(x_1^0,\ldots,x_{j-1}^0,x_j^0 + \eta_1\Delta x_j, x_{j+1}^0,\ldots,x_{l-1}^0,x_l^0,x_{l+1}^0,\ldots,x_n^0))\Delta x_j$$
$$= f_{x_jx_l}(x_1^0,\ldots,x_{j-1}^0,x_j^0 + \eta_1\Delta x_j, x_{j+1}^0,\ldots,x_{l-1}^0,x_l^0 + \eta_2\Delta x_l, x_{l+1}^0,\ldots,x_n^0)\Delta x_j\Delta x_l.$$

Now using (4.1), we get

$$f_{x_lx_j}(x_1^0,\ldots,x_{j-1}^0,x_j^0 + \theta_2\Delta x_j, x_{j+1}^0,\ldots,x_{l-1}^0,x_l^0 + \theta_1\Delta x_l, x_{l+1}^0,\ldots,x_n^0)\Delta x_j\Delta x_l$$
$$= f_{x_jx_l}(x_1^0,\ldots,x_{j-1}^0,x_j^0 + \eta_1\Delta x_j, x_{j+1}^0,\ldots,x_{l-1}^0,x_l^0 + \eta_2\Delta x_l, x_{l+1}^0,\ldots,x_n^0)\Delta x_j\Delta x_l,$$

whereupon

$$f_{x_lx_j}(x_1^0,\ldots,x_{j-1}^0,x_j^0 + \theta_2\Delta x_j, x_{j+1}^0,\ldots,x_{l-1}^0,x_l^0 + \theta_1\Delta x_l, x_{l+1}^0,\ldots,x_n^0)$$
$$= f_{x_jx_l}(x_1^0,\ldots,x_{j-1}^0,x_j^0 + \eta_1\Delta x_j, x_{j+1}^0,\ldots,x_{l-1}^0,x_l^0 + \eta_2\Delta x_l, x_{l+1}^0,\ldots,x_n^0).$$

Since $f_{x_jx_l}$ and $f_{x_lx_j}$ are continuous at x^0, by the last equation we get

$$f_{x_lx_j}(x^0) = \lim_{\Delta x_j\to 0,\ \Delta x_l\to 0} f_{x_lx_j}(x_1^0,\ldots,x_{j-1}^0,x_j^0 + \theta_2\Delta x_j, x_{j+1}^0,\ldots,x_{l-1}^0,x_l^0 + \theta_1\Delta x_l, x_{l+1}^0,\ldots,x_n^0)$$
$$= f_{x_jx_l}(x^0)$$
$$= \lim_{\Delta x_j\to 0,\Delta x_l\to 0} f_{x_jx_l}(x_1^0,\ldots,x_{j-1}^0,x_j^0 + \eta_1\Delta x_j, x_{j+1}^0,\ldots,x_{l-1}^0,x_l^0 + \eta_2\Delta x_l, x_{l+1}^0,\ldots,x_n^0).$$

This completes the proof. □

Exercise 4.3. Let $X \subset \mathbb{R}^n$, and let $f : X \to \mathbb{R}$ have continuous partial derivatives up to order 3 on X. Prove that

$$f_{x_jx_lx_k} = f_{x_jx_kx_l} = f_{x_lx_jx_k} = f_{x_lx_kx_j}$$
$$= f_{x_kx_lx_j} = f_{x_kx_jx_l} \quad \text{on } X. \tag{4.2}$$

Exercise 4.4. Generalize Exercise 4.3.

4.2 Higher-order differentials

Consider a function f that is differentiable at some point $x^0 \in \mathbb{R}^n$. The first-order differential of the function f at the point x^0 is given as

$$df(x^0) = \sum_{j=1}^n f_{x_j}(x^0)dx_j.$$

We fix the increments $dx_j, j \in \{1,\dots,n\}$, i. e., we assume that $dx_j, j \in \{1,\dots,n\}$, are constants. Then df becomes a function only of the variables $x_j, j \in \{1,\dots,n\}$, for which we can also define the differential by taking the same differentials $dx_j, j \in \{1,\dots,n\}$, as the increments $\Delta x_j, j \in \{1,\dots,n\}$. As a result, we obtain the second-order differential or differential of the second order, which is denoted by $d^2f(x^0)$. This motivates us to give the following definition.

Definition 4.2. Suppose that $X \subset \mathbb{R}^n$ and $f : X \to \mathbb{R}$ has continuous partial derivatives up to order m on X. The differential of order m or mth-order differential of the function f is defined as

$$d^m f = \left(\sum_{k=1}^{n} \frac{\partial}{\partial x_k} dx_k \right)^m f \quad \text{on } X. \tag{4.3}$$

We can give the following alternative definition for mth-order differential of a function.

Definition 4.3. Suppose that $X \subset \mathbb{R}^n$ and $f : X \to \mathbb{R}$ has continuous partial derivatives up to order m on X. The differential of order m or mth-order differential of the function f is defined as

$$d^m f = d(d^{m-1}f). \tag{4.4}$$

In the next result, we show that the above two definitions are equivalent.

Theorem 4.2. *Suppose that $X \subset \mathbb{R}^n$ and $f : X \to \mathbb{R}$ has continuous partial derivatives up to order m on X. Then (4.3) and (4.4) are equivalent.*

Proof. Assume that (4.4) holds. Then we have

$$df = \sum_{j=1}^{n} f_{x_j} dx_j,$$

and using Theorem 4.1, we arrive at

$$d^2 f = d(df)$$

$$= d\left(\sum_{j=1}^{n} f_{x_j} dx_j \right)$$

$$= \sum_{l=1}^{n} \frac{\partial}{\partial x_l} \left(\sum_{j=1}^{n} f_{x_j} dx_j \right) dx_l$$

$$= \sum_{j,l=1}^{n} f_{x_j x_l} dx_j dx_l$$

$$= f_{x_1 x_1} dx_1^2 + f_{x_1 x_2} dx_1 dx_2 + \cdots + f_{x_1 x_n} dx_1 dx_n$$

$$+ f_{x_2x_1}dx_2dx_1 + f_{x_2x_2}dx_2^2 + \cdots + f_{x_2x_n}dx_2dx_n$$

$$+ \cdots$$

$$+ f_{x_nx_1}dx_ndx_1 + f_{x_nx_2}dx_ndx_2 + \cdots + f_{x_nx_n}dx_n^2$$

$$= \sum_{j=1}^{n} f_{x_jx_j}dx_j^2 + 2 \sum_{j,l=1,j<l}^{n} f_{x_jx_l}dx_jdx_l$$

$$= \left(\sum_{j=1}^{n} \frac{\partial}{\partial x_j}dx_j \right)^2 f.$$

Assume that (4.3) holds for some $k \in \{1,\ldots,m-1\}$, i. e., assume that

$$d^kf = \left(\sum_{l=1}^{n} \frac{\partial}{\partial x_l}dx_l \right)^k f.$$

We will prove that

$$d^{k+1}f = \left(\sum_{l=1}^{n} \frac{\partial}{\partial x_l}dx_l \right)^{k+1} f.$$

Indeed, we have

$$d^{k+1}f = d(d^kf)$$

$$= d\left(\left(\sum_{l=1}^{n} \frac{\partial}{\partial x_l}dx_l \right)^k f \right)$$

$$= \sum_{l=1}^{n} \frac{\partial}{\partial x_l}\left(\left(\sum_{l=1}^{n} \frac{\partial}{\partial x_l}dx_l \right)^k f \right)dx_l$$

$$= \sum_{l=1}^{n} \frac{\partial}{\partial x_l}\left(\left(\sum_{k_1+\cdots+k_n=k} \frac{k!}{k_1!\ldots k_n!} \frac{\partial^k}{\partial x_1^{k_1}\ldots\partial x_n^{k_n}}dx_1^{k_1}\ldots dx_n^{k_n} \right)f \right)dx_l$$

$$= \sum_{l=1}^{n} \frac{\partial}{\partial x_l}\left(\sum_{k_1+\cdots+k_n=k} \frac{k!}{k_1!\ldots k_n!} \frac{\partial^kf}{\partial x_1^{k_1}\ldots\partial x_n^{k_n}}dx_1^{k_1}\ldots dx_n^{k_n} \right)dx_l$$

$$= \sum_{l=1}^{n} \sum_{k_1+\cdots+k_n=k} \frac{k!}{k_1!\ldots k_n!} \frac{\partial^{k+1}f}{\partial x_1^{k_1}\ldots\partial x_{l-1}^{k_{l-1}}\partial x_l^{k_l}\partial x_{l+1}^{k_{l+1}}\ldots\partial x_n^{k_n}}dx_1^{k_1}\ldots dx_{l-1}^{k_{l-1}}dx_l^{k_l}dx_{l+1}^{k_{l+1}}\ldots dx_n^{k_n}$$

$$= \sum_{k_1+\cdots+k_n=k+1} \frac{(k+1)!}{k_1!\ldots k_n!} \frac{\partial^{k+1}}{\partial x_1^{k_1}\ldots\partial x_n^{k_n}}dx_1^{k_1}\ldots dx_n^{k_n}$$

$$= \left(\sum_{j=1}^{n} \frac{\partial}{\partial x_j}dx_j \right)^{k+1} f.$$

Thus (4.4) implies (4.3). The proof that (4.3) implies (4.4) is left to the reader as an exercise. This completes the proof. $\qquad\square$

Example 4.5. We will find $d^2f(x,y)$, $(x,y) \in \mathbb{R}^2$, where

$$f(x,y) = x(\sin y)^2, \quad (x,y) \in \mathbb{R}^2.$$

We have

$$f_x(x,y) = (\sin y)^2,$$
$$f_{xx}(x,y) = 0,$$
$$f_{xy}(x,y) = 2\sin y \cos y$$
$$= \sin(2y),$$
$$f_y(x,y) = 2x \sin y \cos y$$
$$= x \sin(2y),$$
$$f_{yy}(x,y) = 2x \cos(2y), \quad (x,y) \in \mathbb{R}^2.$$

Hence

$$d^2f(x,y) = f_{xx}(x,y)dx^2 + 2f_{xy}(x,y)dxdy + f_{yy}(x,y)dy^2$$
$$= 2\sin(2y)dxdy + 2x\cos(2y)dy^2, \quad (x,y) \in \mathbb{R}^2.$$

Example 4.6. We will find $d^3f(0,1,2)$, where

$$f(x,y,z) = x^4 + xy^2 + yz^2 + x^2z, \quad (x,y,z) \in \mathbb{R}^3.$$

We have

$$f_x(x,y,z) = 4x^3 + y^2 + 2xz,$$
$$f_{xx}(x,y,z) = 12x^2 + 2z,$$
$$f_{xxx}(x,y,z) = 24x,$$
$$f_{xxx}(0,1,2) = 0,$$
$$f_{xxy}(x,y,z) = 0,$$
$$f_{xxy}(0,1,2) = 0,$$
$$f_{xxz}(x,y,z) = 2,$$
$$f_{xxz}(0,1,2) = 2,$$
$$f_{xy}(x,y,z) = 2y,$$
$$f_{xyy}(x,y,z) = 2,$$

$$f_{xyy}(0,1,2) = 2,$$
$$f_{xz}(x,y,z) = 2x,$$
$$f_{xzz}(x,y,z) = 0,$$
$$f_{xzz}(0,1,2) = 0,$$
$$f_y(x,y,z) = 2xy + z^2,$$
$$f_{yz}(x,y,z) = 2z,$$
$$f_{yzz}(x,y,z) = 2,$$
$$f_{yzz}(0,1,2) = 2,$$
$$f_{yy}(x,y,z) = 2x,$$
$$f_{yyz}(x,y,z) = 0,$$
$$f_{yyz}(0,1,2) = 0,$$
$$f_{yyy}(x,y,z) = 0,$$
$$f_{yyy}(0,1,2) = 0,$$
$$f_z(x,y,z) = 2yz + x^2,$$
$$f_{zz}(x,y,z) = 2y,$$
$$f_{zzz}(x,y,z) = 0,$$
$$f_{zzz}(0,1,2) = 0,$$
$$f_{xyz}(x,y,z) = 0,$$
$$f_{xyz}(0,1,2) = 0, \quad (x,y,z) \in \mathbb{R}^3.$$

Hence

$$\begin{aligned}
d^3 f(0,1,2) &= f_{xxx}(0,1,2)dx^3 + f_{yyy}(0,1,2)dy^3 + f_{zzz}(0,1,2)dz^2 \\
&\quad + 3f_{xxy}(0,1,2)dx^2dy + 3f_{xxz}(0,1,2)dx^2dz + 3f_{xyy}(0,1,2)dxdy^2 \\
&\quad + 3f_{xzz}(0,1,2)dxdz^2 + 3f_{yyz}(0,1,2)dy^2dz + 3f_{yzz}(0,1,2)dydz^2 \\
&\quad + 6f_{xyz}(0,1,2)dxdydz \\
&= 6dx^2dz + 6dxdy^2 + 6dydz^2 \\
&= 6(dx^2dz + dxdy^2 + dydz^2).
\end{aligned}$$

Example 4.7. We will find $d^4 f(x,y)$, $(x,y) \in \mathbb{R}^2$, where

$$f(x,y) = \cos(x+y), \quad (x,y) \in \mathbb{R}^2.$$

We have

$$f_x(x,y) = -\sin(x+y),$$
$$f_{xx}(x,y) = -\cos(x+y),$$

$$f_{xxxx}(x,y) = \sin(x+y),$$
$$f_{xxxx}(x,y) = \cos(x+y),$$
$$f_{xxxy}(x,y) = \cos(x+y),$$
$$f_{xxy}(x,y) = \sin(x+y),$$
$$f_{xxyy}(x,y) = \cos(x+y),$$
$$f_{xy}(x,y) = -\cos(x+y),$$
$$f_{xyy}(x,y) = \sin(x+y),$$
$$f_{xyyy}(x,y) = \cos(x+y),$$
$$f_{y}(x,y) = -\sin(x+y),$$
$$f_{yy}(x,y) = -\cos(x+y),$$
$$f_{yyy}(x,y) = \sin(x+y),$$
$$f_{yyyy}(x,y) = \cos(x+y), \quad (x,y) \in \mathbb{R}^2.$$

Hence

$$d^4f(x,y) = f_{xxxx}(x,y)dx^4 + 4f_{xxxy}(x,y)dx^3dy$$
$$+ 6f_{xxyy}(x,y)dx^2dy^2 + 4f_{xyyy}(x,y)dxdy^3$$
$$+ f_{yyyy}(x,y)dy^4$$
$$= \cos(x+y)(dx^4 + 4dx^3dy + 6dx^2dy^2 + 4dxdy^3 + dy^4), \quad (x,y) \in \mathbb{R}^2.$$

Exercise 4.5. Find $d^2f(a,b)$, where

1.
$$f(x,y) = e^{xy}, \quad (x,y) \in \mathbb{R}^2, \quad (a,b) = (1,-1).$$

2.
$$f(x,y) = e^{\frac{x^2}{y}}, \quad (x,y) \in \mathbb{R}^2, \quad y \neq 0, \quad (a,b) = (1,1).$$

3.
$$f(x,y) = \frac{x}{y}e^{x^2}, \quad (x,y) \in \mathbb{R}^2, \quad y \neq 0, \quad (a,b) = (0,1).$$

4.
$$f(x,y) = x^2\cosh(3y) - y^2, \quad (x,y) \in \mathbb{R}^2, \quad (a,b) = (0,0).$$

5.
$$f(x,y) = x\cos(xy), \quad (x,y) \in \mathbb{R}^2, \quad (a,b) = \left(\frac{\pi}{2}, -1\right).$$

4.3 The Taylor formula

In calculus, the Taylor[1] formula gives an approximation of an m times differentiable function around a given point by a polynomial of degree m. This polynomial is called the mth-order Taylor polynomial. For a given smooth function, the Taylor polynomial is the truncation at the order m of the Taylor series of a function. The first-order Taylor polynomial is the linear approximation of the function, and the second-order Taylor polynomial is often called the quadratic approximation. It is fundamental in various areas of mathematics, numerical analysis, and mathematical physics. The Taylor formula is the starting point of the study of analytic functions.

The Taylor formula for multivariable functions reads as follows.

Theorem 4.3. *Suppose that a function f has continuous partial derivatives up to order $m + 1$ in a neighborhood of a point $x^0 \in \mathbb{R}^n$. Then*

$$\Delta f(x^0) = \sum_{k=1}^{n} \frac{1}{k!} \left(\sum_{j=1}^{n} \frac{\partial}{\partial x_j} \Delta x_j \right)^k f(x^0)$$

$$+ \frac{1}{(m+1)!} \left(\sum_{j=1}^{n} \frac{\partial}{\partial x_j} \Delta x_j \right)^{m+1} f(x_1^0 + \theta \Delta x_1, \dots, x_n^0 + \theta \Delta x_n) \qquad (4.5)$$

for some $\theta \in (0,1)$.

Definition 4.4. Formula (4.5) is called the Taylor formula for the function f in a neighborhood of the point x^0. If $x^0 = 0$, then formula (4.5) is called the Maclaurin[2] formula for the function f.

Proof. Denote

$$F(t) = f(x_1^0 + t\Delta x_1, \dots, x_n^0 + t\Delta x_n), \quad t \in [0,1].$$

Then

$$\Delta f(x^0) = f(x_1^0 + \Delta x_1, \dots, x_n^0 + \Delta x_n) - f(x_1^0, \dots, x_n^0)$$
$$= F(1) - F(0).$$

---

1 Brook Taylor (18 August 1685–29 December 1731) was an English mathematician and barrister, best known for several results in mathematical analysis. The Taylor most famous developments are the Taylor theorem and Taylor series, essential in the infinitesimal approach of functions in specific points.

2 Colin Maclaurin (February 1698–14 June 1746) was a Scottish mathematician, who made important contributions to geometry and algebra. He is also known for being a child prodigy and holding the record for being the youngest professor. The Maclaurin series, a particular case of the Taylor series, is named after him.

Since f has continuous partial derivatives up to order $m+1$, the function F has continuous derivatives up to order $m+1$ on the interval $[0,1]$. Now applying the Taylor formula for F, we get

$$F(1) = F(0) + F'(0) + \frac{1}{2!}F''(0) + \cdots + \frac{1}{m!}F^{(m)}(0) + \frac{1}{(m+1)!}F^{(m+1)}(\theta)$$

for some $\theta \in (0,1)$. We have

$$F(1) - F(0) = f(x_1^0 + \Delta x_1, \ldots, x_n^0 + \Delta x_n) - f(x_1^0, \ldots, x_n^0)$$
$$= \Delta f(x^0),$$

$$F'(t) = \sum_{j=1}^n f_{x_j}(x_1^0 + t\Delta x_1, \ldots, x_n^0 + t\Delta x_n)\Delta x_j,$$

$$F'(0) = \sum_{j=1}^n f_{x_j}(x_1^0, \ldots, x_n^0)\Delta x_j,$$

$$F''(t) = \sum_{j,l=1}^n f_{x_j x_l}(x_1^0 + t\Delta x_1, \ldots, x_n^0 + t\Delta x_n)\Delta x_j \Delta x_l,$$

$$F''(0) = \sum_{j,l=1}^n f_{x_j x_l}(x_1^0, \ldots, x_n^0)\Delta x_j \Delta x_l,$$

$$\vdots$$

$$F^{(m)}(0) = \left(\sum_{j=1}^n \frac{\partial}{\partial x_j}\Delta x_j\right)^m f(x^0),$$

$$F^{(m+1)}(\theta) = \left(\sum_{j=1}^n \frac{\partial}{\partial x_j}\Delta x_j\right)^{m+1} f(x_1^0 + \theta\Delta x_1, \ldots, x_n^0 + \theta\Delta x_n),$$

whereupon we get (4.5). This completes the proof. □

Remark 4.1. Suppose that all conditions of Theorem 4.3 hold. Define

$$r_m(\Delta x) = \frac{1}{(m+1)!}\left(\sum_{j=1}^n \frac{\partial}{\partial x_j}\Delta x_j\right)^{m+1} f(x_1^0 + \theta\Delta x_1, \ldots, x_n^0 + \theta\Delta x_n).$$

Then

$$r_m(\Delta x) = \sum_{m_1 + \cdots + m_n = m} \epsilon_{m_1 \ldots m_n}(\Delta x)\Delta x_1^{m_1} \ldots \Delta x_n^{m_n},$$

where

$$\lim_{d \to 0} \epsilon_{m_1 \dots m_n}(\Delta x) = 0, \quad d = \sqrt{\sum_{j=1}^{n} \Delta x_j^2},$$

and

$$r_m(\Delta x) = \epsilon(\Delta x) d^m, \quad \lim_{d \to 0} \epsilon(\Delta x) = 0,$$

i. e.,

$$r_m(\Delta x) = o(d^m) \quad \text{as} \quad d \to 0.$$

Any n-tuple

$$k = (k_1, \dots, k_n), \quad k_j \in \mathbb{N}_0, \quad j \in \{1, \dots, n\},$$

is said to be an n-dimensional multiindex. The set of all n-dimensional multiindices is denoted by $\mathbb{N}_0^n$. For $k \in \mathbb{N}_0^n$, $k = (k_1, \dots, k_n)$, $x = (x_1, \dots, x_n) \in \mathbb{R}^n$, we denote

$$k! = k_1! \dots k_n!,$$
$$\Delta x^k = \Delta x_1^{k_1} \dots \Delta x_n^{k_n},$$
$$\Delta x = (\Delta x_1, \dots, \Delta x_n),$$
$$x^k = x_1^{k_1} \dots x_n^{k_n},$$
$$f^{(k)}(x^0) = \frac{\partial^{k_1 + \dots + k_n} f}{\partial x_1^{k_1} \dots \partial x_n^{k_n}}(x^0).$$

Then the Taylor formula takes the form

$$\Delta f(x^0) = \sum_{|k|=1}^{m} \frac{1}{k!} f^{(k)}(x^0)(x - x^0)^k + o(d^{m+1}) \quad \text{as} \quad d \to 0,$$

or

$$f(x^0 + \Delta x) = \sum_{|k|=0}^{m} \frac{1}{k!} f^{(k)}(x^0)(x - x^0)^k + o(d^{m+1}) \quad \text{as} \quad d \to 0.$$

Now we will show that the Taylor formula is uniquely determined.

Theorem 4.4. *Suppose that all conditions of Theorem 4.3 hold. Then the representation (4.5) is unique.*

Proof. Assume that for the function f, we have two representations

$$f(x^0 + \Delta x) = \sum_{|k|=0}^{m} a_k \Delta x^k + o(d^m) \quad \text{as} \quad d \to 0$$

and

$$f(x^0 + \Delta x) = \sum_{|k|=0}^{m} b_k \Delta x^k + o(d^m) \quad \text{as} \quad d \to 0,$$

where $a_k, b_k \in \mathbb{R}$, $k \in \mathbb{N}_0^n$, $|k| \in \{0, \ldots, m\}$. Let

$$c_k = a_k - b_k, \quad k \in \mathbb{N}_0^n, \quad |k| \in \{0, \ldots, m\}.$$

Then

$$0 = \sum_{|k|=0}^{m} c_k \Delta x^k + o(d^m), \quad k \in \mathbb{N}_0^n, \quad |k| \in \{0, \ldots, m\} \tag{4.6}$$

We fix Δx such that $|\Delta x| < \delta$ for some $\delta > 0$. Then, for any $t \in [0,1]$, we have $|t\Delta x| < \delta$. Then (4.6) holds for $t\Delta x$, and we get

$$0 = \sum_{|k|=0}^{m} c_k t^{|k|} \Delta x^k + o(t^m d^m) \quad \text{as} \quad t \to 0.$$

Hence

$$0 = \sum_{l=0}^{m} t^l \sum_{|k|=l} c_k \Delta x^k + o(t^m) \quad \text{as} \quad t \to 0,$$

whereupon

$$\sum_{|k|=l} c_k \Delta x^k = 0, \quad l \in \{0, \ldots, m\},$$

and

$$c_k = 0, \quad k \in \mathbb{N}_0^n, \quad |k| \in \{0, \ldots, m\}.$$

This completes the proof. □

Example 4.8. We will find the Taylor formula for the function

$$f(x,y) = -x^2 + 2xy + 3y^2 - 6x - 2y - 4, \quad (x,y) \in \mathbb{R}^2,$$

around the point $(-2,1)$. We have

$$f(-2,1) = -(-2)^2 + 2 \cdot (-2) \cdot 1 + 3 \cdot 1^2 - 6 \cdot (-2) - 2 \cdot 1 - 4$$
$$= -4 - 4 + 3 + 12 - 2 - 4$$
$$= 1,$$

$$f_x(x, y) = -2x + 2y - 6,$$
$$f_x(-2, 1) = -2 \cdot (-2) + 2 \cdot 1 - 6$$
$$= 4 + 2 - 6$$
$$= 0,$$
$$f_y(x, y) = 2x + 6y - 2,$$
$$f_y(-2, 1) = 2 \cdot (-2) + 6 \cdot 1 - 2$$
$$= -4 + 6 - 2$$
$$= 0,$$
$$f_{xx}(x, y) = -2,$$
$$f_{xx}(-2, 1) = -2,$$
$$f_{xy}(x, y) = 2,$$
$$f_{xy}(-2, 1) = 2,$$
$$f_{yy}(x, y) = 6,$$
$$f_{yy}(-2, 1) = 6, \quad (x, y) \in \mathbb{R}^2.$$

Hence

$$f(x, y) = f(-2, 1) + f_x(-2, 1)(x + 2) + f_y(-2, 1)(y - 1) + \frac{1}{2} f_{xx}(-2, 1)(x + 2)^2$$
$$+ \frac{1}{2} \cdot 2 f_{xy}(-2, 1)(x + 2)(y - 1) + \frac{1}{2} f_{yy}(-2, 1)(y - 1)^2$$
$$= 1 - (x + 2)^2 + 2(x + 2)(y - 1) + 3(y - 1)^2, \quad (x, y) \in \mathbb{R}^2.$$

Example 4.9. We will find the Taylor formula for the function

$$f(x, y, z) = (x + y + z)^2, \quad (x, y, z) \in \mathbb{R}^3,$$

in a neighborhood of the point $(1, 1, -2)$. We have

$$f(1, 1, -2) = (1 + 1 - 2)^2$$
$$= 0,$$
$$f_x(x, y, z) = 2(x + y + z),$$
$$f_y(x, y, z) = 2(x + y + z),$$
$$f_z(x, y, z) = 2(x + y + z), \quad (x, y, z) \in \mathbb{R}^3,$$
$$f_x(1, 1, -2) = 2(1 + 1 - 2)$$
$$= 0,$$
$$f_y(1, 1, -2) = 0,$$
$$f_z(1, 1, -2) = 0,$$

and

$$f_{xx}(x,y,z) = 2,$$
$$f_{xy}(x,y,z) = 2,$$
$$f_{xz}(x,y,z) = 2,$$
$$f_{yy}(x,y,z) = 2,$$
$$f_{yz}(x,y,z) = 2,$$
$$f_{zz}(x,y,z) = 2, \quad (x,y,z) \in \mathbb{R}^3.$$

Hence

$$f(x,y,z) = f(1,1,-2) + f_x(1,1,-2)(x-1) + f_y(1,1,-2)(y-1) + f_z(1,1,-2)(z+2)$$
$$+ \frac{1}{2}f_{xx}(1,1,-2)(x-1)^2 + f_{xy}(1,1,-2)(x-1)(y-1) + f_{xz}(1,1,-2)(x-1)(z+2)$$
$$+ \frac{1}{2}f_{yy}(1,1,-2)(y-1)^2 + f_{yz}(1,1,-2)(y-1)(z+2) + \frac{1}{2}f_{zz}(1,1,-2)(z+2)^2$$
$$= (x-1)^2 + (y-1)^2 + (z+2)^2 + 2(x-1)(y-1) + 2(x-1)(z+2) + 2(y-1)(z+2),$$

$(x,y,z) \in \mathbb{R}^3$.

Example 4.10. We will find the Taylor formula for the function

$$f(x,y) = \sin x \sin y, \quad (x,y) \in \mathbb{R}^2,$$

in a neighborhood of the point $(\frac{\pi}{4}, \frac{\pi}{4})$. We have

$$f\left(\frac{\pi}{4}, \frac{\pi}{4}\right) = \sin\frac{\pi}{4} \sin\frac{\pi}{4}$$
$$= \frac{\sqrt{2}}{2} \cdot \frac{\sqrt{2}}{2}$$
$$= \frac{1}{2}$$

and

$$f_x(x,y) = \cos x \sin y,$$
$$f_y(x,y) = \sin x \cos y,$$
$$f_{xx}(x,y) = -\sin x \sin y,$$
$$f_{xy}(x,y) = \cos x \cos y,$$
$$f_{yy}(x,y) = -\sin x \sin y, \quad (x,y) \in \mathbb{R}^2.$$

Hence

$$f_x\left(\frac{\pi}{4},\frac{\pi}{4}\right) = f_y\left(\frac{\pi}{4},\frac{\pi}{4}\right)$$

$$= f_{xy}\left(\frac{\pi}{4},\frac{\pi}{4}\right)$$

$$= \frac{1}{2},$$

$$f_{xx}\left(\frac{\pi}{4},\frac{\pi}{4}\right) = f_{yy}\left(\frac{\pi}{4},\frac{\pi}{4}\right)$$

$$= -\frac{1}{2}$$

and

$$f(x,y) = f\left(\frac{\pi}{4},\frac{\pi}{4}\right) + f_x\left(\frac{\pi}{4},\frac{\pi}{4}\right)\left(x-\frac{\pi}{4}\right) + f_y\left(\frac{\pi}{4},\frac{\pi}{4}\right)\left(y-\frac{\pi}{4}\right)$$

$$+ \frac{1}{2}f_{xx}\left(\frac{\pi}{4},\frac{\pi}{4}\right)\left(x-\frac{\pi}{4}\right)^2 + f_{xy}\left(\frac{\pi}{4},\frac{\pi}{4}\right)\left(x-\frac{\pi}{4}\right)\left(y-\frac{\pi}{4}\right)$$

$$+ \frac{1}{2}f_{yy}\left(\frac{\pi}{4},\frac{\pi}{4}\right)\left(y-\frac{\pi}{4}\right)^2 + o(d^2)$$

$$= \frac{1}{2} + \frac{1}{2}\left(x-\frac{\pi}{4}\right) + \frac{1}{2}\left(y-\frac{\pi}{4}\right)$$

$$- \frac{1}{4}\left(x-\frac{\pi}{4}\right)^2 + \frac{1}{2}\left(x-\frac{\pi}{4}\right)\left(y-\frac{\pi}{4}\right) - \frac{1}{4}\left(y-\frac{\pi}{4}\right)^2 + o(d^2),$$

$(x,y) \in \mathbb{R}^2$.

Exercise 4.6. Find the Taylor formula for the function $f = f(x,\ldots,x_n)$, $(x,\ldots,x_n) \in \mathbb{R}^n$, in a neighborhood of the point $(a_1,\ldots,a_n)$, where
1.

$$f(x,y) = 2x^2 - xy - y^2 - 6x - 3y, \quad (x,y) \in \mathbb{R}^2, \quad (a_1,a_2) = (1,-2).$$

2.

$$f(x,y,z) = x^2 + 3z^2 - 2yz - 3z, \quad (x,y,z) \in \mathbb{R}^3, \quad (a_1,a_2,a_3) = (0,1,2).$$

4.4 Extremum of a function

In one variable calculus, the derivative is used to find the maximum and minimum values (extrema) of differentiable functions. Recall the following useful facts.
-   (Extreme value theorem) If $f : [a,\beta] \to \mathbb{R}$ is continuous, then it assumes a maximum and a minimum on the interval $[a,\beta]$.
-   (Critical point theorem) Suppose that $f : [a,\beta]$ is differentiable on (a,β) and that f assumes a maximum or minimum at interior point a of $[a,\beta]$. Then $f'(a) = 0$.

– (Second derivative test) Suppose that $f : [a, \beta] \rightarrow \mathbb{R}$ is a C^2 function on $R(a, \beta)$ and that $f'(a) = 0$ at an interior point a of $[a, \beta]$. If $f''(a) > 0$, then $f(a)$ is a local minimum of f, and if $f''(a) < 0$, then $f(a)$ is a local maximum of f.

Geometrically, the idea is that just as the affine function

$$A(a + h) = f(a) + f'(a)h$$

specifies the tangent line to the graph of f at $(a, f(a))$, the quadratic function

$$B(a + h) = f(a) + f'(a)h + \frac{1}{2}f''(a)h^2$$

determines the best fitting parabola. When $f'(a) = 0$, the tangent line is horizontal, and the sign of $f''(a)$ specifies whether the parabola opens upward or downward. When $f'(a) = 0$ and $f''(a) = 0$, the parabola degenerates to the horizontal tangent line, and the second derivative provides no information (see Fig. 4.1). In this section, we generalize these facts to functions of n variables. Let $X \subset \mathbb{R}^n$, $x^0 \in X$, and $f : X \rightarrow \mathbb{R}$.

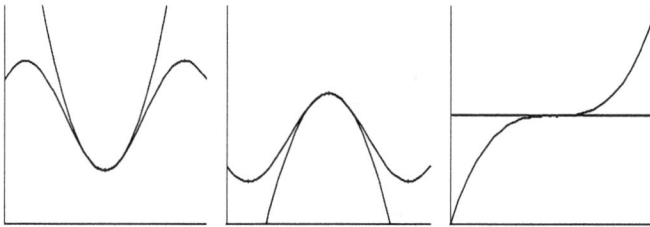

Figure 4.1: Approximating parabolas.

Definition 4.5. The point x^0 is said to be a point of strong local maximum (minimum) for the function f if there is a neighborhood $U(x^0)$ of the point x^0 such that for all $x \in U(x^0) \cap X$, we have

$$f(x) < (>)f(x^0).$$

If for all $x \in U(x^0) \cap X$, we have

$$f(x) \leq (\geq)f(x^0),$$

then the point x^0 is said to be a point of local maximum (minimum) for the function f. If the point x^0 is a point of local maximum (minimum) for the function f, we have

$$\Delta f = f(x) - f(x^0) \leq (\geq)0$$

for all $x \in U(x^0) \cap X$.

Definition 4.6. The points of (strong) local maximum and (strong) local minimum of the function f are called extreme points for the function f.

The critical point theorem has the following extension to functions of several variables, a test for local extrema at interior points.

Theorem 4.5. *Let f have continuous partial derivatives in a neighborhood of x^0. If the point x^0 is an extreme point of f, then*

$$f_{x_j}(x^0) = 0, \quad j \in \{1, \ldots, n\}. \tag{4.7}$$

Proof. Let $U(x^0)$ be a neighborhood of x^0 in which f has continuous partial derivatives:

$$g_j(x_j) = f(x_1^0, \ldots, x_{j-1}^0, x_j, x_{j+1}^0, \ldots, x_n^0), \quad (x_1^0, \ldots, x_{j-1}^0, x_j, x_{j+1}^0, \ldots, x_n^0) \in U(x^0),$$

$j \in \{1, \ldots, n\}$. Then $g_j, j \in \{1, \ldots, n\}$, has an extreme at x_j^0. Hence

$$g_j'(x_j^0) = 0, \quad j \in \{1, \ldots, n\},$$

whereupon we get (4.7). This completes the proof. □

Corollary 4.1. *Let f be differentiable at its extreme point x^0. Then*

$$df(x^0) = 0.$$

Definition 4.7. Let f be differentiable at x^0. If

$$df(x^0) = 0,$$

then x^0 is called a stationary point of f.

The next result is the second derivative test for extremum of a function of several variables.

Theorem 4.6. *Let f have continuous partial derivatives up to order 2 in a neighborhood of a point x^0. Let also, x^0 be a stationary point of the function f. If the quadratic form*

$$A(dx_1, \ldots, dx_n) = \sum_{i,j=1}^{n} f_{x_i x_j}(x^0) dx_i dx_j \tag{4.8}$$

is positive (negative) definite, then x^0 is a point of strong local minimum (maximum). If the quadratic form (4.8) is indefinite, then x^0 is not an extreme point of the function f. In this case, the point x^0 is called a saddle point.

Figure 4.2 shows a typical saddle point.

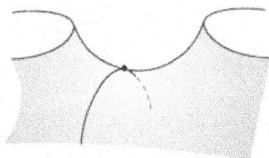

Figure 4.2: A typical saddle point.

Proof. Let $U(x^0, \delta_0)$ be a neighborhood of the point x^0 in which f has continuous partial derivatives up to order 2. Let also,

$$x^0 + dx = (x_1^0 + dx_1, \ldots, x_n^0 + dx_n) \in U(x^0, \delta_0).$$

Then applying the Taylor formula, we get

$$\Delta f(x^0) = f(x^0 + dx) - f(x^0)$$
$$= \frac{1}{2} \sum_{i,j=1}^{n} f_{x_i x_j}(x^0) dx_i dx_j + \epsilon(dx) d^2,$$

where

$$dx = (dx_1, \ldots, dx_n), \quad d^2 = dx_1^2 + \cdots + dx_n^2,$$

and

$$\lim_{d \to 0} \epsilon(dx) = 0.$$

Thus

$$\Delta f(x^0) = \frac{d^2}{2} \left(\sum_{i,j=1}^{n} f_{x_i x_j}(x^0) \frac{dx_i}{d} \frac{dx_j}{d} + 2\epsilon(dx) \right)$$
$$= \frac{d^2}{2} \left(A\left(\frac{dx_1}{d}, \ldots, \frac{dx_n}{d} \right) + 2\epsilon(dx) \right), \quad d \neq 0.$$

1. Suppose that A is positive definite. Then

$$\inf_{S} |A| = r > 0,$$

where S is the unit sphere. Take $\delta \in (0, \delta_0)$ such that

$$2|\epsilon(dx)| < r.$$

Hence $x^0 + dx \in U(x^0, \delta)$, $dx \neq 0$, and

$$\operatorname{sign} \Delta f(x^0) = \operatorname{sign} A\left(\frac{dx_1}{d},\dots,\frac{dx_n}{d}\right).$$

Therefore

$$\Delta f(x^0) > 0 \quad \text{in } U(x^0,\delta),$$

and x^0 is a point of strong local minimum of f.

2. The case where A is negative definite is similar to the previous case and is left to the reader as an exercise.

3. Let (4.8) be an indefinite quadratic form. Then there are dx' and dx'' such that

$$A(dx'_1,\dots,dx'_n) > 0,$$
$$A(dx''_1,\dots,dx''_n) < 0.$$

Let

$$d' = ((dx'_1)^2 + \cdots + (dx'_n)^2)^{\frac{1}{2}},$$
$$d'' = ((dx''_1)^2 + \cdots + (dx''_n)^2)^{\frac{1}{2}}.$$

Let l' denote the semi-line through the points x^0 and $x^0 + dx'$. For any $x \in l'$, we set

$$dx_j = x_j - x_j^0,$$

$$d = \sqrt{\sum_{j=1}^{n} dx_j^2},$$

$$\frac{dx_j}{d} = \cos\alpha_j, \quad j \in \{1,\dots,n\}.$$

Note that the point

$$\left(\frac{dx_1}{d},\dots,\frac{dx_n}{d}\right) \tag{4.9}$$

lies on the unit sphere with center x^0, and it will be the same for any point lying on l'. Thus the point (4.9) does not depend on the distance between the points x^0 and x. Thus

$$\left(\frac{dx_1}{d},\dots,\frac{dx_n}{d}\right) = \left(\frac{dx'_1}{d},\dots,\frac{dx'_n}{d}\right),$$

and then

$$A\left(\frac{dx_1}{d},\dots,\frac{dx_n}{d}\right) = A\left(\frac{dx'_1}{d},\dots,\frac{dx'_n}{d}\right)$$

$$= \frac{1}{d'^2} A(dx'_1, \ldots, dx'_n)$$
$$> 0.$$

Let

$$r' = A\left(\frac{dx_1}{d}, \ldots, \frac{dx_n}{d}\right).$$

Then $r' > 0$. We choose $d_0 > 0$ such that $d < d_0$ implies

$$2|\epsilon(dx)| < r'$$

and

$$\epsilon(dx) \to 0 \quad \text{as} \quad d \to 0.$$

Then, for any point $x^0 + dx \in l'$ such that

$$0 < d$$
$$= \sqrt{\sum_{j=1}^{n} dx_j^2}$$
$$< d_0,$$

we have

$$\Delta f(x^0 + dx) > 0.$$

As above, considering $(dx''_1, \ldots, dx''_n)$, in any neighborhood of x^0, there exists a point y^0 such that $\Delta f(y^0) < 0$. Therefore x^0 is not a local extreme point for f. This completes the proof. □

Example 4.11. Let $n = 2$. Then

$$A(dx, dy) = f_{xx}(x^0)dx^2 + 2f_{xy}(x^0)dxdy + f_{yy}(x^0)dy^2.$$

Hence A is positive definite if

$$f_{xx}(x^0) > 0,$$
$$f_{xx}(x^0)f_{yy}(x^0) - (f_{xy}(x^0))^2 > 0 \tag{4.10}$$

and negative definite if

$$f_{xx}(x^0) < 0,$$
$$f_{xx}(x^0)f_{yy}(x^0) - (f_{xy}(x^0))^2 > 0. \tag{4.11}$$

Thus f has strong local minimum at x^0 if (4.10) holds and strong local maximum at x^0 if (4.11) holds.

Example 4.12. We will find the local extreme points of the function

$$f(x,y) = x^3 + 3xy^2 - 39x - 36y + 26, \quad (x,y) \in \mathbb{R}^2.$$

Firstly, we will find the stationary points of the function f. For this aim, we will find the first partial derivatives of f at arbitrary point $(x,y) \in \mathbb{R}^2$. We have

$$f_x(x,y) = 3x^2 + 3y^2 - 39,$$
$$f_y(x,y) = 6xy - 36, \quad (x,y) \in \mathbb{R}^2.$$

Then the stationary points of the function f satisfy the system

$$3x^2 + 3y^2 - 39 = 0,$$
$$6xy - 36 = 0,$$

or

$$x^2 + y^2 = 13,$$
$$xy = 6,$$

whereupon

$$y = \frac{6}{x},$$
$$x^2 + \frac{36}{x^2} = 13,$$

and

$$x^4 - 13x^2 + 36 = 0.$$

Hence

$$(x^2)_{1,2} = \frac{13 \pm \sqrt{169 - 144}}{2}$$
$$= \frac{13 \pm 5}{2},$$

or

$$x^2 = 9 \quad \text{or} \quad x^2 = 4.$$

Thus the stationary points of the function f are as follows:

$$x = 3, \quad x = -3, \quad x = 2, \quad x = -2,$$
$$y = 2, \quad y = -2, \quad y = 3, \quad y = -3.$$

To determine if the above points are local extreme points, we have to determine if the quadratic form d^2f is positive definite or negative definite at each of them. For this aim, we will find the second partial derivatives of f at arbitrary point $(x,y) \in \mathbb{R}^2$. We have

$$f_{xx}(x,y) = 6x,$$
$$f_{xy}(x,y) = 6y,$$
$$f_{yy}(x,y) = 6x, \quad (x,y) \in \mathbb{R}^2.$$

Then

$$d^2f(x,y) = 6(xdx^2 + 2ydxdy + xdy^2), \quad (x,y) \in \mathbb{R}^2.$$

Let

$$A(x,y) = \begin{pmatrix} x & y \\ y & x \end{pmatrix},$$
$$\Delta_1(x,y) = x,$$
$$\Delta_2(x,y) = x^2 - y^2, \quad (x,y) \in \mathbb{R}^2.$$

Thus

$$\Delta_1(3,2) = 3$$
$$> 0,$$
$$\Delta_2(3,2) = 3^2 - 2^2$$
$$= 5$$
$$> 0,$$
$$\Delta_1(-3,-2) = -3$$
$$< 0,$$
$$\Delta_2(-3,-2) = (-3)^2 - (-2)^2$$
$$= 9 - 4$$
$$= 5$$
$$> 0,$$
$$\Delta_1(2,3) = 2$$
$$> 0,$$
$$\Delta_2(2,3) = 2^2 - 3^2$$

$$= 4 - 9$$

$$= -5$$

$$< 0,$$

$$\Delta_1(-2, -3) = -2$$

$$< 0,$$

$$\Delta_2(-2, -3) = (-2)^2 - (-3)^2$$

$$= 4 - 9$$

$$= -5$$

$$< 0.$$

From this we conclude that the points $(2, 3)$ and $(-2, -3)$ are not extreme points for the function f. The point $(3, 2)$ is a local minimum point for f, and

$$f(3, 2) = 3^3 + 3 \cdot 3 \cdot 2^2 - 39 \cdot 3 - 36 \cdot 2 + 26$$

$$= 27 + 36 - 117 - 72 + 26$$

$$= 89 - 189$$

$$= -100,$$

whereas the point $(-3, -2)$ is a local maximum point of f, and

$$f(-3, -2) = (-3)^3 + 3 \cdot (-3) \cdot (-2)^2 - 39 \cdot (-3) - 36 \cdot (-2) + 26$$

$$= -27 - 36 + 117 + 72 + 26$$

$$= -37 + 189$$

$$= 152.$$

Example 4.13. We will investigate for local extreme points the function

$$f(x, y, z) = 3x^3 + y^2 + z^2 + 6xy - 2z + 1, \quad (x, y, z) \in \mathbb{R}^3.$$

Firstly, we will find the stationary points of the function f. For this aim, we will find the first partial derivatives of f at arbitrary point $(x, y, z) \in \mathbb{R}^3$. We have

$$f_x(x, y, z) = 9x^2 + 6y,$$

$$f_y(x, y, z) = 2y + 6x,$$

$$f_z(x, y, z) = 2z - 2, \quad (x, y, z) \in \mathbb{R}^3.$$

Then the stationary points of f satisfy the system

$$9x^2 + 6y = 0,$$

$$2y + 6x = 0,$$
$$2z - 2 = 0,$$

whereupon

$$z = 1,$$
$$y = -3x,$$
$$9x^2 - 18x = 0,$$

and thus $x = 2, y = -6, z = 1$ and $x = 0, y = 0, z = 1$ are the stationary points of f. Now we will find $d^2f(x, y, z), (x, y, z) \in \mathbb{R}^3$. For this aim, we find the second partial derivatives of f at arbitrary point $(x, y, z) \in \mathbb{R}^3$. We have

$$f_{xx}(x, y, z) = 18x,$$
$$f_{xy}(x, y, z) = 6,$$
$$f_{xz}(x, y, z) = 0,$$
$$f_{yy}(x, y, z) = 2,$$
$$f_{yz}(x, y, z) = 0,$$
$$f_{zz}(x, y, z) = 2, \quad (x, y, z) \in \mathbb{R}^3.$$

Hence

$$d^2f(x, y, z) = 18x dx^2 + 12 dx dy + 2 dy^2 + 2 dz^2$$
$$= 2(9x dx^2 + 6 dx dy + dy^2 + dz^2), \quad (x, y, z) \in \mathbb{R}^3.$$

Let

$$A(x, y, z) = \begin{pmatrix} 9x & 3 & 0 \\ 3 & 1 & 0 \\ 0 & 0 & 1 \end{pmatrix},$$

$$\Delta_1(x, y, z) = 9x,$$
$$\Delta_2(x, y, z) = 9x - 9$$
$$= 9(x - 1),$$
$$\Delta_3(x, y, z) = 9(x - 1), \quad (x, y, z) \in \mathbb{R}^3.$$

Hence

$$\Delta_1(2, -6, 1) = 18$$
$$> 0,$$
$$\Delta_2(2, -6, 1) = 9$$

$$> 0,$$
$$\Delta_3(2, -6, 1) = 9$$
$$> 0,$$
$$\Delta_1(0, 0, 1) = 0,$$
$$\Delta_2(0, 0, 1) = -9,$$
$$\Delta_3(0, 0, 1) = -9.$$

Thus $(2, -6, 1)$ is a local minimum point for f, and

$$f(2, -6, 1) = 3 \cdot 2^3 + (-6)^2 + 1^2 + 6 \cdot 2 \cdot (-6) - 2 \cdot 1 + 1$$
$$= 24 + 36 + 1 - 72 - 2 + 1$$
$$= 62 - 74$$
$$= -12.$$

Take arbitrary $\epsilon \in \mathbb{R}$. Then

$$f(\epsilon, 0, 1) = 3\epsilon^3 + 1 - 2 + 1$$
$$= 3\epsilon^3.$$

Thus, for $\epsilon > 0$, we have $f(\epsilon, 0, 1) > 0$, and for $\epsilon < 0$, we have $f(\epsilon, 0, 1) < 0$. Consequently, $(0, 0, 1)$ is not a local extreme point for f.

Exercise 4.7. Find the local extreme points of the following functions:

1.
$$f(x, y) = x^2 + xy + y^2 - 12x - 3y, \quad (x, y) \in \mathbb{R}^2.$$

2.
$$f(x, y) = 3 + 2x - y - x^2 + xy - y^2, \quad (x, y) \in \mathbb{R}^2.$$

3.
$$f(x, y) = 3x + 6y - x^2 - xy + y^2, \quad (x, y) \in \mathbb{R}^2.$$

4.
$$f(x, y) = 4x^2 - 4xy + y^2 + 4x - 2y + 1, \quad (x, y) \in \mathbb{R}^2.$$

5.
$$f(x, y, z) = x^2 + y^2 + (z + 1)^2 - xy + x, \quad (x, y, z) \in \mathbb{R}^3.$$

6.
$$f(x, y, z) = 8 - 6x + 4y - 2z - x^2 - y^2 - z^2, \quad (x, y, z) \in \mathbb{R}^3.$$

7.
$$f(x, y, z) = x^2 + y^2 - z^2 - 4x + 6y - 2z, \quad (x, y, z) \in \mathbb{R}^3.$$

8.
$$f(x, y, z) = x^3 + y^2 + z^2 + 6xy - 4z, \quad (x, y, z) \in \mathbb{R}^3.$$

9.
$$f(x, y, z) = xyz(16 - x - y - 2z), \quad (x, y, z) \in \mathbb{R}^3.$$

10.
$$f(x, y, z) = \frac{xy + xz^2 + y^2z}{xyz} + x + 1, \quad (x, y, z) \in \mathbb{R}^3.$$

4.5 Implicit functions

An implicit function, as its name implies, is a function that is defined implicitly rather than explicitly. Unlike an explicit function, where the dependent variable is clearly defined as a function of the independent variable, an implicit function embeds the relationships between the variables within an equation. This fundamental concept, while seemingly abstract, underpins many areas of mathematics and engineering. An example of an implicit function is an equation

$$y^2 + xy = 0.$$

Also, a function

$$f(x, y, z) = 0$$

such that one variable is dependent on the other two variables, is an implicit function.

Implicit functions possess some unique characteristics. Here is more in-depth information about these:

–   Implicit functions can often represent complex relationships that are difficult to express explicitly.
–   Implicit functions can describe multi-valued functions, where one input can produce multiple outputs.
–   Implicit functions can express relationships between variables that do not have a clear cause-effect relationship.

Implicit functions carry significant importance in the field of engineering mathematics. They are used to represent real-world phenomena that are difficult to express in an explicit form. Implicit functions are essential for solving differential equations, a critical task in engineering. As they provide a way of dealing with multi-valued functions, they find significant use in disciplines like fluid dynamics, structural analysis, etc. In electrical engineering studies, implicit functions are used to analyze passive networks and to study RLC circuits.

In this section, we investigate the system

$$F_1(x_1,\ldots,x_n,y_1,\ldots,y_m) = 0,$$
$$F_2(x_1,\ldots,x_n,y_1,\ldots,y_m) = 0,$$
$$\vdots$$
$$F_m(x_1,\ldots,x_n,y_1,\ldots,y_m) = 0$$

(4.12)

for solvability with respect to $y_1,\ldots,y_m$ in a neighborhood of the point

$$(x_1^0,\ldots,x_n^0,y_1^0,\ldots,y_m^0)$$

so that

$$F_j(x_1^0,\ldots,x_n^0,y_1^0,\ldots,y_m^0) = 0, \quad j \in \{1,\ldots,m\}.$$

Definition 4.8. Suppose that the functions $u_j = u_j(t_1,\ldots,t_n), j \in \{1,\ldots,m\}, j \in \{1,\ldots,n\},$ have first partial derivatives at the point $t^0 = (t_1^0,\ldots,t_n^0)$. The matrix

$$\begin{pmatrix} \frac{\partial u_1}{\partial t_1} & \frac{\partial u_1}{\partial t_2} & \cdots & \frac{\partial u_1}{\partial t_n} \\ \frac{\partial u_2}{\partial t_1} & \frac{\partial u_2}{\partial t_2} & \cdots & \frac{\partial u_2}{\partial t_n} \\ \vdots & \vdots & \vdots & \vdots \\ \frac{\partial u_m}{\partial t_1} & \frac{\partial u_m}{\partial t_2} & \cdots & \frac{\partial u_m}{\partial t_n} \end{pmatrix}$$

or shortly $(\frac{\partial u_j}{\partial t_l}), j \in \{1,\ldots,m\}, l \in \{1,\ldots,n\},$ is said to be the Jacobian matrix of the system $u_j, j \in \{1,\ldots,m\}.$ If $n = m$, then the determinant

$$\det \begin{pmatrix} \frac{\partial u_1}{\partial t_1} & \frac{\partial u_1}{\partial t_2} & \cdots & \frac{\partial u_1}{\partial t_n} \\ \frac{\partial u_2}{\partial t_1} & \frac{\partial u_2}{\partial t_2} & \cdots & \frac{\partial u_2}{\partial t_n} \\ \vdots & \vdots & \vdots & \vdots \\ \frac{\partial u_n}{\partial t_1} & \frac{\partial u_n}{\partial t_2} & \cdots & \frac{\partial u_n}{\partial t_n} \end{pmatrix}$$

is called the Jacobian determinant or Jacobian of the system $u_j, j \in \{1,\ldots,n\}.$ It is denoted by $\frac{\partial(u_1,\ldots,u_n)}{\partial(t_1,\ldots,t_n)}.$

Firstly, we will investigate the case $m = 1$, i. e., we will consider the equation

$$F(x_1,\ldots,x_n,y) = 0.$$

(4.13)

In multivariable calculus, the implicit function theorem is a tool that allows relations to be converted to functions of several real variables. It does so by representing the relation as the graph of a function. There may not be a single function whose graph can represent the entire relation, but there may be such a function on a restriction of the domain of the relation. The implicit function theorem gives a sufficient condition to ensure that there is such a function. The implicit function theorem for the equation (4.13) reads as follows.

Theorem 4.7. *Let $F = F(x_1,\ldots,x_n,y)$ be a continuous function in a rectangular neighborhood*

$$U(x^0,y_0) = U(x_1^0,\ldots,x_n^0,y_0)$$
$$= \{(x_1,\ldots,x_n,y) \in \mathbb{R}^{n+1} : |x - x^0| < \xi,\ |y - y_0| < \eta,\ x = (x_1,\ldots,x_n)\}$$

of a point $(x^0,y_0) = (x_1^0,\ldots,x_n^0,y_0)$ such that for any fixed $x \in \mathbb{R}^n$ for which $|x - x^0| < \xi$, the function $F = (x_1,\ldots,x_n,\cdot)$ is strongly monotonic. If

$$F(x_1^0,\ldots,x_n^0,y_0) = 0, \tag{4.14}$$

then there exists a neighborhood

$$U(x^0) = \{x \in \mathbb{R}^n : |x - x^0| < \delta\}$$

of the point x^0 and a neighborhood

$$U(y_0) = \{y \in \mathbb{R} : |y - y_0| < \epsilon\}$$

of the point y_0 such that for any $x \in U(x^0)$, there exists a unique solution $y \in U(y_0)$ of equation (4.13). This solution is a function of $x_1,\ldots,x_n$ and is denoted by

$$y = f(x_1,\ldots,x_n),$$

which is continuous at the point x^0, and

$$y_0 = f(x^0).$$

Proof. By the conditions of the theorem we have that for any $x \in U(x^0), x = (x_1,\ldots,x_n)$, the function $F(x_1,\ldots,x_n,\cdot)$ is strongly monotonic on $(y_0 - \eta, y_0 + \eta)$. In particular, the function $F(x_1^0,\ldots,x_n^0,\cdot)$ is strongly monotonic on $(y_0-\eta,y_0+\eta)$. Without loss of generality, suppose that $F(x_1^0,\ldots,x_n^0,\cdot)$ is strongly increasing on $(y_0 - \eta, y_0 + \eta)$. Take arbitrary $\epsilon \in (0,\eta)$. Since $F(x_1^0,\ldots,x_n^0,\cdot)$ is strongly increasing on $(y_0 - \epsilon, y_0 + \epsilon)$ and (4.14) holds, we get that

$$F(x_1^0,\ldots,x_n^0,y_0 - \epsilon) < 0 \quad \text{and} \quad F(x_1^0,\ldots,x_n^0,y_0 + \epsilon) > 0.$$

Because F is continuous on (x^0,y_0), there are a δ-neighborhood of $(x^0,y_0 - \epsilon)$ in which $F < 0$ and a δ-neighborhood of $(x^0,y_0 + \epsilon)$ in which $F > 0$. In particular, for any fixed $x \in \mathbb{R}^n$ for which $|x - x^0| < \delta$, we have

$$F(x, y_0 - \epsilon) < 0, \quad F(x, y_0 + \epsilon) > 0. \tag{4.15}$$

We set

$$U(x^0) = \{x \in \mathbb{R}^n : |x - x^0| < \delta\},$$
$$U(y_0) = \{y \in \mathbb{R} : |y - y_0| < \delta\}.$$

Since for any fixed $x \in U(x^0)$, the function $F(x, \cdot)$ is continuous and strongly increasing on $(y_0 - \epsilon, y_0 + \epsilon)$ and (4.15) holds, there is a unique $y^* \in U(y_0)$ such that

$$F(x, y^*) = 0.$$

Thus we get a map $x \to y^*$, $x \in U(x^0)$, $y^* \in U(y_0)$, which we denote by $y^* = f(x)$. By the definition of this map we get: for any $x \in U(x^0)$ and a unique $y^* = f(x) \in U(y_0)$, we have

$$F(x, y^*) = 0.$$

By (4.14), using that $x^0 \in U(x^0)$, $y_0 \in U(y_0)$, and f is unique, we have

$$y_0 = f(x^0).$$

Since $\epsilon \in (0, \eta)$ was arbitrarily chosen and we have found $\delta > 0$ such that the inequality $|x - x^0| < \delta$ implies $f(x) \in U(y_0)$, we obtain

$$|f(x) - f(x^0)| < \epsilon,$$

i. e., f is continuous at x^0. This completes the proof. $\qquad\square$

In the next theorem, we give a rule for differentiating an implicit function given by equation (4.13).

Theorem 4.8. *Let $F = F(x_1, \ldots, x_n, y)$ be a continuous function in a neighborhood of the points $(x^0, y_0) = (x_1^0, \ldots, x_n^0, y_0)$ having a continuous partial derivative $F_y(x_1, \ldots, x_n, y)$ at the point (x^0, y_0). If*

$$F(x_1^0, \ldots, x_n^0, y_0) = 0, \quad F_y(x_1^0, \ldots, x_n^0, y_0) \neq 0,$$

then there are neighborhoods $U(x^0) = U(x_1^0, \ldots, x_n^0)$ and $U(y_0)$ of the points x^0 and y_0, respectively, such that for any $x \in U(x^0)$, there exists a unique solution

$$y = f(x) = f(x_1, \ldots, x_n) \in U(y_0)$$

of equation (4.13). This solution is continuous in $U(x^0)$, and $y_0 = f(x^0)$. If in addition, the function F has continuous partial derivatives $F_{x_j}(x, y)$ in a neighborhood of (x^0, y_0), then f has continuous partial derivatives at x^0, and

$$f_{x_j}(x^0) = -\frac{F_{x_j}(x^0, y_0)}{F_y(x^0, y_0)}.$$

Proof. Since F is continuous in a neighborhood of (x^0, y_0) and there exists $F_y(x, y)$, which is continuous at (x^0, y_0), there is a rectangular neighborhood

$$U(x^0, y_0) = \{(x, y) \in \mathbb{R}^{n+1} : |x - x^0| < \xi, \ |y - y_0| < \eta\}$$

such that F is continuous in $U(x^0, y_0)$ and $F_y(x, y)$ has the same sign as the sign of $F_y(x^0, y_0)$ for any $(x^0, y_0) \in U(x^0, y_0)$. Thus, for any $x \in \mathbb{R}^n$ such that $|x - x^0| < \xi$, we have that the function

$$\phi(y) = F(x, y)$$

is differentiable in $(y_0 - \eta, y_0 + \eta)$ and $\phi'(y) = F_y(x, y)$ has the same sign in $(y_0 - \eta, y_0 + \eta)$. Therefore F satisfies all conditions of Theorem 4.7 in the neighborhood $U(x^0, y_0)$. Then there exist neighborhoods

$$U(x^0) = \{x \in \mathbb{R}^n : |x - x^0| < \delta\}$$

and

$$U(y_0) = \{y \in \mathbb{R} : |y - y_0| < \epsilon\}$$

and a unique function $y = f(x)$, defined in $U(x^0)$, such that for any $x \in U(x^0)$, we have $f(x) \in U(x^0)$,

$$F(x, y(x)) = 0,$$

and f is continuous at x^0. Hence, in the neighborhood

$$U_0(x^0, y_0) = \{(x, y) \in \mathbb{R}^{n+1} : |x - x^0| < \delta, \ |y - y_0| < \epsilon\},$$

all conditions of Theorem 4.7 hold. Since F is differentiable at (x^0, y_0), we have

$$F(x^0 + \Delta x, y_0 + \Delta y) - F(x^0, y_0) = \sum_{j=1}^{n} f_{x_j}(x^0, y_0)\Delta x_j + f_y(x^0, y_0)\Delta y + \sum_{j=1}^{n} \epsilon_j \Delta x_j + \epsilon_{n+1}\Delta y,$$

where

$$\lim_{d \to 0} \epsilon_j = 0, \quad j \in \{1, \ldots, n+1\}, \quad d^2 = \sum_{j=1}^{n} \Delta x_j^2 + \Delta y^2,$$

$x^0 + \Delta x \in U(x^0)$, $y_0 + \Delta y \in U(y_0)$. Note that

$$F(x^0 + \Delta x, y_0 + \Delta y) = F(x^0 + \Delta x, f(x^0 + \Delta x))$$
$$= 0$$

and

$$F(x^0, y_0) = 0.$$

Therefore

$$\sum_{j=1}^{n} f_{x_j}(x^0, y_0)\Delta x_j + f_y(x^0, y_0)\Delta y + \sum_{j=1}^{n} \epsilon_j \Delta x_j + \epsilon_{n+1}\Delta y = 0. \tag{4.16}$$

Take arbitrary $l \in \{1, \dots, n\}$ and $\Delta x_j = 0, j \in \{1, \dots, n\}, j \neq l$, and Δx_l such that

$$(x_1^0, \dots, x_{l-1}^0, x_l^0 + \Delta x_l, x_{l+1}^0, \dots, x_n^0) \in U(x^0).$$

Take Δy such that $y_0 + \Delta y \in U(y^0)$. Then by (4.16) we find

$$f_{x_l}(x^0, y_0)\Delta x_l + \epsilon_l \Delta x_l + \epsilon_{n+1}\Delta y + f_y(x^0, y_0)\Delta y = 0,$$

whereupon

$$(\epsilon_{n+1} + f_y(x^0, y_0))\Delta y = -(f_{x_l}(x^0, y_0) + \epsilon_l)\Delta x_l,$$

and thus

$$\frac{\Delta y}{\Delta x_l} = -\frac{f_{x_l}(x^0, y_0) + \epsilon_l}{f_y(x^0, y_0) + \epsilon_{n+1}}.$$

Hence

$$y_{x_l}(x^0) = \lim_{d \to 0} \frac{\Delta y}{\Delta x_l}$$
$$= -\lim_{d \to 0} \frac{f_{x_l}(x^0, y_0) + \epsilon_l}{f_y(x^0, y_0) + \epsilon_{n+1}}$$
$$= -\frac{f_{x_l}(x^0, y_0)}{f_y(x^0, y_0)}.$$

This completes the proof. □

Example 4.14. Find $u_x(0, 1)$ and $u_y(0, 1)$, where

$$u^3 + 3xyu + 1 = 0, \quad (x, y) \in \mathbb{R}^2.$$

We have

$$\left(u(0,1)\right)^3 + 1 = 0,$$

whereupon $u(0,1) = -1$. Let

$$F(x,y,u) = u^3 + 3xyu + 1, \quad (x,y,u) \in \mathbb{R}^3.$$

We have

$$F_x(x,y,u) = 3yu,$$
$$F_y(x,y,u) = 3xu,$$
$$F_u(x,y,u) = 3u^2 + 3xy, \quad (x,y,u) \in \mathbb{R}^3.$$

Hence

$$u_x(x,y) = -\frac{F_x(x,y,u(x,y))}{F_u(x,y,u(x,y))}$$
$$= -\frac{3yu(x,y)}{3(u(x,y))^2 + 3xy}$$
$$= -\frac{yu(x,y)}{(u(x,y))^2 + xy},$$
$$u_y(x,y) = -\frac{F_y(x,y,u(x,y))}{F_u(x,y,u(x,y))}$$
$$= -\frac{3xu(x,y)}{3(u(x,y))^2 + 3xy}$$
$$= -\frac{xu(x,y)}{(u(x,y))^2 + xy}, \quad (x,y) \in \mathbb{R}^2,$$

and

$$u_x(0,1) = -\frac{u(0,1)}{(u(0,1))^2}$$
$$= -\frac{1}{u(0,1)}$$
$$= 1,$$
$$u_y(0,1) = 0.$$

Example 4.15. We will find $u_x(1,1)$ and $u_y(1,1)$, where

$$x^2 - 2y^2 + 3u^2 - yu + y = 0, \quad (x,y) \in \mathbb{R}^2.$$

We have

$$1 - 2 + 3\left(u(1,1)\right)^2 - u(1,1) + 1 = 0,$$

whereupon

$$3\big(u(1,1)\big)^2 - u(1,1) = 0.$$

Then

$$u(1,1) = 0 \quad \text{or} \quad u(1,1) = \frac{1}{3}.$$

Let

$$F(x,y,u) = x^2 - 2y^2 + 3u^2 - yu + y, \quad (x,y,u) \in \mathbb{R}^3.$$

Then

$$F_x(x,y,u) = 2x,$$
$$F_y(x,y,u) = -4y - u + 1,$$
$$F_u(x,y,u) = 6u - y, \quad (x,y,u) \in \mathbb{R}^3.$$

Hence

$$u_x(x,y) = -\frac{F_x(x,y,u(x,y))}{F_u(x,y,u(x,y))}$$
$$= -\frac{2x}{6u(x,y) - y},$$
$$u_y(x,y) = -\frac{F_y(x,y,u(x,y))}{F_u(x,y,u(x,y))}$$
$$= -\frac{-4y - u(x,y) + 1}{6u(x,y) - y}, \quad (x,y) \in \mathbb{R}^2.$$

1. Let $u(1,1) = 0$. Then

$$u_x(1,1) = -\frac{2}{-1}$$
$$= 2,$$
$$u_y(1,1) = -\frac{-4+1}{-1}$$
$$= -3.$$

2. $u(1,1) = \frac{1}{3}$. Then

$$u_x(1,1) = -\frac{2}{2-1}$$
$$= -2,$$

$$u_y(1,1) = -\frac{-4 - \frac{1}{3} + 1}{2 - 1}$$

$$= \frac{10}{3}.$$

Example 4.16. We will find $u_x(1,0)$ and $u_y(1,0)$, where

$$e^u - xyu - 2 = 0.$$

We have

$$e^{u(1,0)} - 2 = 0,$$

or

$$e^{u(1,0)} = 2.$$

Let

$$F(x,y,u) = e^u - xyu - 2, \quad (x,y,u) \in \mathbb{R}^3.$$

Then

$$F_x(x,y,u) = -yu,$$
$$F_y(x,y,u) = -xu,$$
$$F_u(x,y,u) = e^u - xy, \quad (x,y,u) \in \mathbb{R}^3.$$

Hence

$$u_x(1,0) = -\frac{F_x(1,0,u(1,0))}{F_u(1,0,u(1,0))}$$
$$= 0,$$
$$u_y(1,0) = -\frac{F_y(1,0,u(1,0))}{F_u(1,0,u(1,0))}$$
$$= -\frac{-u(1,0)}{e^{u(1,0)}}$$
$$= \frac{\log 2}{2}.$$

Exercise 4.8. Find $u_x(a,b)$ and $u_y(a,b)$, where
1.

$$u + \log(x + y + u) = 0, \quad (a,b) = (1,-1).$$

2.

$$\frac{u}{\sqrt{x^2 - y^2}} - \arctan \frac{u}{\sqrt{x^2 - y^2}} - 1 = 0, \quad (a,b) = (5,4).$$

3.
$$x \cos y + y \cos u + u \cos x = 1, \quad (a,b) = (0,1), \quad u(0,1) = 0.$$

Now we are ready to formulate and prove the implicit function theorem for system (4.12).

Theorem 4.9. *Let the functions* $F_j = F_j(x_1,\dots,x_n,y_1,\dots,y_m), j \in \{1,\dots,m\}$, *be continuously differentiable in a neighborhood of a point*

$$(x^0,y^0) = (x_1^0,\dots,x_n^0,y_1^0,\dots,y_m^0)$$

where

$$F_j(x^0,y^0) = 0, \quad j \in \{1,\dots,m\},$$

and

$$\frac{\partial(F_1,\dots,F_m)}{\partial(y_1,\dots,y_m)}(x^0,y^0) \neq 0.$$

Then there exist neighborhoods U_{x^0} *and* U_{y^0} *of the points* $x^0 = (x_1^0,\dots,x_n^0)$ *and* $y^0 = (y_1^0,\dots,y_m^0)$, *respectively, such that for any* $x \in U_{x^0}$, *equation (4.12) has a unique solution* $y_j = y_j(x_1,\dots,x_n), j \in \{1,\dots,m\}$, *and* $(y_1,\dots,y_m) \in U_{y^0}$. *If*

$$y = f(x) = (y_1 = f_1(x_1,\dots,x_n),\dots,y_n = f_n(x_1,\dots,x_n)), \quad x \in U_{x^0},$$

is this solution, then $f_j \in C^1(U_{x^0})$, *and* $y^0 = f(x^0)$.

Proof. For the proof of this theorem, we will use the principle of mathematical induction.
1. For $m = 1$, the statement follows by Theorem 4.8.
2. Assume that the statement holds for $m - 1$.
3. We will prove the statement for m. Consider the equation

$$F_m(x_1,\dots,x_n,y_1,\dots,y_m) = 0.$$

Let U be a neighborhood of the point (x^0,y^0) such that $F_j \in C^1(U), j \in \{1,\dots,m\}$. Set

$$\tilde{y} = (y_1,\dots,y_{m-1}), \quad \tilde{y}^0 = (y_1^0,\dots,y_m^0).$$

Since

$$\frac{\partial(F_1,\dots,F_m)}{\partial(y_1,\dots,y_m)} = \det \begin{pmatrix} \frac{\partial F_1}{\partial y_1} & \cdots & \frac{\partial F_1}{\partial y_m} \\ \vdots & \vdots & \vdots \\ \frac{\partial F_m}{\partial y_1} & \cdots & \frac{\partial F_m}{\partial y_m} \end{pmatrix}(x^0,y^0)$$

$$\neq 0,$$

there exists $j \in \{1, \ldots, m\}$ such that

$$\frac{\partial F_m}{\partial y_j}(x^0, y^0) \neq 0.$$

Without loss of generality, suppose that

$$\frac{\partial F_m}{\partial y_m}(x^0, y^0) \neq 0.$$

Hence, using that $F_m(x^0, y^0) = 0$ and applying Theorem 4.8, we conclude that there are a neighborhood U^{m+n-1} of the point $(x^0, \tilde{y}^0)$ and a neighborhood U^1 of the point y_m^0 such that

$$U^{m+n-1} \times U^1 \subset U$$

and in U^{m+n-1}, there is a unique function

$$y_m = \phi(x, \tilde{y})$$

such that if $(x, \tilde{y}) \in U^{m+n-1}$, then $\phi(x, \tilde{y}) \in U^1$, and $\phi(x^0, \tilde{y}^0) = y_m^0$. Thus, for $(x, \tilde{y}) \in U^{m+n-1}$ and $y_m \in U^1$, system (4.12) is equivalent to the system

$$F_j(x, y) = 0, \quad j \in \{1, \ldots, m\},$$
$$y_m = \phi(x, \tilde{y}). \tag{4.17}$$

Let

$$\Phi_j(x, \tilde{y}) = F_j(x, \tilde{y}, \phi(x, \tilde{y})), \quad j \in \{1, \ldots, m\}.$$

Then by (4.17) we find

$$\Phi_j(x, \tilde{y}) = 0, \quad j \in \{1, \ldots, m-1\},$$
$$y_m = \phi(x, \tilde{y}). \tag{4.18}$$

We will show that

$$\Phi_j(x, \tilde{y}) = 0, \quad j \in \{1, \ldots, m-1\},$$

satisfy all conditions of our statement in the case $m - 1$. We have $\Phi_j \in C^1(U^{m+n-1})$, $j \in \{1, \ldots, m-1\}$. Since

$$F_j(x^0, y^0) = 0, \quad j \in \{1, \ldots, m\},$$

we get

$$\Phi_j(x^0,\tilde{y}^0) = 0, \quad j \in \{1,\dots,m-1\}.$$

Note that

$$\frac{\partial \Phi_j}{\partial y_l} = \frac{\partial F_j}{\partial y_l} + \frac{\partial F_j}{\partial y_m}\frac{\partial \phi}{\partial y_l}, \quad j,l \in \{1,\dots,m-1\},$$

and

$$\frac{\partial \Phi_m}{\partial y_l} = \frac{\partial F_m}{\partial y_l} + \frac{\partial F_m}{\partial y_m}\frac{\partial \phi}{\partial y_l}$$
$$= 0.$$

Therefore

$$\frac{\partial(F_1,\dots,F_m)}{\partial(y_1,\dots,y_m)}(x^0,y^0) = \det\begin{pmatrix} \frac{\partial F_1}{\partial y_1}+\frac{\partial F_1}{\partial y_m}\frac{\partial \phi}{\partial y_1} & \cdots & \frac{\partial F_1}{\partial y_{m-1}}+\frac{\partial F_1}{\partial y_m}\frac{\partial \phi}{\partial y_{m-1}} & \frac{\partial F_1}{\partial y_m} \\ \vdots & \vdots & \vdots & \vdots \\ \frac{\partial F_m}{\partial y_1}+\frac{\partial F_m}{\partial y_m}\frac{\partial \phi}{\partial y_1} & \cdots & \frac{\partial F_m}{\partial y_{m-1}}+\frac{\partial F_m}{\partial y_m}\frac{\partial \phi}{\partial y_{m-1}} & \frac{\partial F_m}{\partial y_m} \end{pmatrix}(x^0,y^0)$$

$$= \det\begin{pmatrix} \frac{\partial \Phi_1}{\partial y_1} & \cdots & \frac{\partial \Phi_1}{\partial y_{m-1}} & \frac{\partial F_1}{\partial y_m} \\ \vdots & \vdots & \vdots & \vdots \\ \frac{\partial \Phi_m}{\partial y_1} & \cdots & \frac{\partial \Phi_m}{\partial y_{m-1}} & \frac{\partial F_m}{\partial y_m} \end{pmatrix}$$

$$= \det\begin{pmatrix} \frac{\partial \Phi_1}{\partial y_1} & \cdots & \frac{\partial \Phi_1}{\partial y_{m-1}} & \frac{\partial F_1}{\partial y_m} \\ \vdots & \vdots & \vdots & \vdots \\ \frac{\partial \Phi_{m-1}}{\partial y_1} & \cdots & \frac{\partial \Phi_{m-1}}{\partial y_{m-1}} & \frac{\partial F_{m-1}}{\partial y_m} \\ 0 & \cdots & 0 & \frac{\partial F_m}{\partial y_m} \end{pmatrix}(x^0,y^0)$$

$$= \frac{\partial F_m}{\partial y_m}(x^0,y^0)\frac{\partial(\Phi_1,\dots,\Phi_{m-1})}{\partial(y_1,\dots,y_{m-1})}(x^0,y^0)$$
$$\neq 0.$$

Let $U^{m+n-1} = U_x^2 \times U_{\tilde{y}}^2$, where U_x^2 is a neighborhood of the point x^0, and $U_{\tilde{y}}^2$ is a neighborhood of $\tilde{y}^0$. By the induction assumption it follows that there are a neighborhood $U_x \subset U_{x^0}^2$ of the point x^0, a neighborhood $U_y \subset U_{\tilde{y}^0}$ of the point $\tilde{y}^0$, and a unique system of functions

$$y_1 = f_1(x_1,\dots,x_m),$$
$$\vdots$$
$$y_{m-1} = f_{m-1}(x_1,\dots,x_m),$$

defined in U_x and satisfying the conditions: if $x \in U_x$, then

$$(f_1(x),\ldots,f_{m-1}(x)) \in U_{\bar{y}},$$

$f_j \in C^1(U_x), j \in \{1,\ldots,m-1\}$, and

$$\Phi_j(x,f_1(x),\ldots,f_{m-1}(x)) = 0, \quad j \in \{1,\ldots,m-1\}.$$

Therefore the system

$$y_k = f_k(x), \quad j \in \{1,\ldots,m-1\},$$
$$y_m = \phi(x,\bar{y}), \quad x \in U_x, \quad \bar{y} \in U_y,$$

is equivalent to system (4.18). Let

$$f_m(x_1,\ldots,x_n) = \phi(x,f_1(x),\ldots,f_{m-1}(x)), \quad x \in U_x.$$

Thus we get the system

$$y_k = f_k(x), \quad k \in \{1,\ldots,m\}, \quad x \in U_x.$$

Set

$$U_y = U_{\bar{y}} \times U^1.$$

Then for $x \in U_x$, we have

$$f(x) = (f_1(x),\ldots,f_m(x)) \in U_y$$

and

$$F(x,f(x)) = 0.$$

Moreover, $f_j \in C^1(U_x)$. Since for any $x \in U_x$, we have $y = f(x) \in U_y$ and

$$F_j(x^0,y^0) = 0, \quad j \in \{1,\ldots,m\},$$

we conclude that $y^0 = f(x^0)$. This completes the proof. $\square$

4.6 Vector maps. Linear maps

A vector map, also referred to as a vector function, is a mathematical function of one or more variables whose range is a set of multidimensional vectors or infinite-dimensional vectors. The input of a vector-valued function may be a scalar or a vector. Vector-valued functions provide a useful method for studying various curves both in the plane and in

three-dimensional space. We can apply this concept to calculate the velocity, acceleration, arc length, and curvature of an object's trajectory.

Suppose that $X \subset \mathbb{R}^n$.

Definition 4.9. A map $f : X \to \mathbb{R}^m$ that maps any vector $x \in \mathbb{R}^n$ to a vector $y = f(x) \in \mathbb{R}^m$
is called a vector map. Let $e_1, \ldots, e_n$ be a basis in $\mathbb{R}^n$, and let $\epsilon_1, \ldots, \epsilon_m$ be a basis of $\mathbb{R}^m$.
Then all

$$x = (x_1, \ldots, x_n) \in X$$

can be represented in the form

$$x = x_1 e_1 + \cdots + x_n e_n, \quad x_l \in \mathbb{R}, \quad l \in \{1, \ldots, n\},$$

all vectors $y \in \mathbb{R}^m$ can be represented in the form

$$y = y_1 \epsilon_1 + \cdots + y_m \epsilon_m, \quad y_j \in \mathbb{R}, \quad j \in \{1, \ldots, m\},$$

and every coordinate $y_j, j \in \{1, \ldots, m\}$, is a function of $x \in X$, and therefore

$$y_j = f_j(x), \quad x \in X, \quad j \in \{1, \ldots, m\}. \tag{4.19}$$

The functions given in (4.19) are said to be coordinate functions of the vector map f, and
we write

$$f = (f_1, \ldots, f_m).$$

The interpretation of a point as a vector does not preclude the consideration of properties of maps such as continuity and uniform continuity. Therefore all properties of
continuous maps of point sets remain valid in a natural interpretation for vector maps.

Definition 4.10. A map $f : \mathbb{R}^n \to \mathbb{R}^m$ is said to be a linear map if for all $x^1, x^2 \in \mathbb{R}^n$ and
$a_1, a_2 \in \mathbb{R}$, we have the equality

$$f(a_1 x^1 + a_2 x^2) = a_1 f(x^1) + a_2 f(x^2).$$

By the definition of a linear map, using induction, it follows that any linear map
$f : \mathbb{R}^n \to \mathbb{R}^m$ maps any finite linear combination of vectors $x^j \in \mathbb{R}^n, j \in \{1, \ldots, k\}$, to a
finite linear combination of the vectors $f(x^j), j \in \{1, \ldots, k\}$, i. e., if $f : \mathbb{R}^n \to \mathbb{R}^m$ is a linear
map and $x^1, \ldots, x^k \in \mathbb{R}^n, a_1, \ldots, a_k \in \mathbb{R}$, then

$$f(a_1 x^1 + \cdots + a_k x^k) = a_1 f(x^1) + \cdots + a_k f(x^k).$$

By the definition of a linear map we get the following important result.

Theorem 4.10. *Let $f : \mathbb{R}^n \to \mathbb{R}^m$ and $g : \mathbb{R}^m \to \mathbb{R}^l$ be linear maps. Then $g \circ f : \mathbb{R}^n \to \mathbb{R}^l$ is a linear map.*

Proof. Let $x^1, x^2 \in \mathbb{R}^n$ and $a_1, a_2 \in \mathbb{R}$ be arbitrarily chosen. Then

$$f(a_1 x^1 + a_2 x^2) = a_1 f(x^1) + a_2 f(x^2),$$

and hence

$$g(f(a_1 x^1 + a_2 x^2)) = g(a_1 f(x^1) + a_2 f(x^2))$$
$$= a_1 g(f(x^1)) + a_2 g(f(x^2)).$$

Thus $g \circ f : \mathbb{R}^n \to \mathbb{R}^l$ is a linear map. This completes the proof. $\qquad\square$

Suppose that $f : \mathbb{R}^n \to \mathbb{R}^m$ is a linear map and $e_1, \ldots, e_n$ and $\epsilon_1, \ldots, \epsilon_m$ are bases in $\mathbb{R}^n$ and $\mathbb{R}^m$, respectively. In addition, suppose that

$$f(e_j) = \sum_{k=1}^{m} a_{kj} \epsilon_k, \quad j \in \{1, \ldots, n\},$$

and

$$x = \sum_{i=1}^{n} x_i e_i,$$
$$y = f(x)$$
$$= \sum_{j=1}^{m} y_j \epsilon_j.$$

Since $f : \mathbb{R}^n \to \mathbb{R}^m$ is a linear map, we obtain

$$y = f(x)$$
$$= \sum_{j=1}^{m} y_j \epsilon_j$$
$$= \sum_{j=1}^{n} x_j f(e_j)$$
$$= \sum_{j=1}^{n} \sum_{k=1}^{m} x_j a_{kj} \epsilon_k$$
$$= \sum_{k=1}^{m} \left(\sum_{j=1}^{n} a_{kj} x_j \right),$$

whereupon

$$f(x) = \sum_{k=1}^{m}\left(\sum_{j=1}^{n} a_{kj}x_j\right) \tag{4.20}$$

and

$$y_k = \sum_{j=1}^{n} a_{kj}x_j, \quad k \in \{1,\ldots,m\},$$

or

$$y_1 = a_{11}x_1 + a_{12}x_2 + \cdots + a_{1n}x_n,$$
$$y_2 = a_{21}x_1 + a_{22}x_2 + \cdots + a_{2n}x_n,$$
$$\vdots$$
$$y_m = a_{m1}e_1 + a_{m2}e_2 + \cdots + a_{mn}e_n.$$

Definition 4.11. The matrix

$$\begin{pmatrix} a_{11} & a_{12} & \cdots & a_{1n} \\ a_{21} & a_{22} & \cdots & a_{2n} \\ \vdots & \vdots & \vdots & \vdots \\ a_{m1} & a_{m2} & \cdots & a_{mn} \end{pmatrix}$$

is called the matrix of the linear map $f : \mathbb{R}^n \to \mathbb{R}^m$.

Definition 4.12. If $m = 1$, then the linear map $f : \mathbb{R}^n \to \mathbb{R}$ is called a linear functional.

If $f : \mathbb{R}^n \to \mathbb{R}$ is a linear functional, then applying (4.20), we get

$$y = a_1 x_1 + \cdots + a_n x_n,$$

where $a_1,\ldots,a_n \in \mathbb{R}$.

Exercise 4.9. Let $f,g : \mathbb{R}^n \to \mathbb{R}^m$ be linear maps with matrices A and B, respectively. Prove that for $\alpha, \beta \in \mathbb{R}$, $\alpha f + \beta g : \mathbb{R}^n \to \mathbb{R}^m$ is a linear map with matrix $\alpha A + \beta B$.

Exercise 4.10. Let $f : \mathbb{R}^n \to \mathbb{R}^m$ and $g : \mathbb{R}^m \to \mathbb{R}^l$ be linear maps with matrices A and B, respectively. Prove that $g \circ f : \mathbb{R}^n \to \mathbb{R}^l$ is a linear map with matrix BA.

Definition 4.13. For a linear map $f : \mathbb{R}^n \to \mathbb{R}^m$, the number

$$\sup_{|x|\leq 1}|f(x)|$$

is called the norm of the map f and is denoted by $\|f\|$.

Let $f : \mathbb{R}^n \to \mathbb{R}^m$ be a linear map. Then $|f| : \mathbb{R}^n \to \mathbb{R}^m$ is a continuous map. Since the set $\{x \in \mathbb{R}^n : |x| \leq 1\}$ is a compact set, we conclude that the norm of f is finite. For $x \in \mathbb{R}^n, x \neq 0$, we set $\xi = \frac{x}{|x|}$. Then

$$|\xi| = \frac{|x|}{|x|}$$
$$= 1,$$

and since $f : \mathbb{R}^n \to \mathbb{R}^m$ is a linear map, we find

$$|f(x)| = \left| f\left(|x|\frac{x}{|x|} \right) \right|$$
$$= \left| |x| f\left(\frac{x}{|x|} \right) \right|$$
$$= |x| \left| f\left(\frac{x}{|x|} \right) \right|$$
$$= |x| |f(\xi)|$$
$$\leq |x| \sup_{\xi \in \mathbb{R}^n, |\xi| \leq 1} |f(\xi)|$$
$$= |x| \|f\|,$$

i.e.,

$$|f(x)| \leq |x| \|f\|, \quad x \in \mathbb{R}^n, \quad x \neq 0.$$

If $x \in \mathbb{R}^n$ and $|x| < 1$, then

$$|f(x)| < \|f\|. \tag{4.21}$$

Because $|f| : \mathbb{R}^n \to \mathbb{R}^m$ is a continuous map, we find

$$\sup_{x \in \mathbb{R}^n : |x| \leq 1} |f(x)| = \max_{x \in \mathbb{R}^n, |x| \leq 1} |f(x)|.$$

By the last equality and (4.21) we conclude that

$$\|f\| = \max_{|x|=1} |f(x)|.$$

Now we will give a way for finding a norm.

Theorem 4.11. *For any linear map $f : \mathbb{R}^n \to \mathbb{R}^m$, we have*

$$\|f\| = \sup_{x \in \mathbb{R}^n, x \neq 0} \frac{|f(x)|}{|x|}.$$

Proof. We have

$$\sup_{x\in\mathbb{R}^n, x\neq 0}\frac{|f(x)|}{|x|} = \sup_{x\in\mathbb{R}^n, x\neq 0}\frac{1}{|x|}|f(x)|$$

$$= \sup_{x\in\mathbb{R}^n, x\neq 0}\left|\frac{1}{|x|}f(x)\right|$$

$$= \sup_{x\in\mathbb{R}^n, x\neq 0}\left|f\left(\frac{x}{|x|}\right)\right|$$

$$= \sup_{\xi\in\mathbb{R}^n, |\xi|\leq 1}|f(\xi)|$$

$$= \|f\|.$$

This completes the proof. □

Now suppose that $f : \mathbb{R}^n \to \mathbb{R}^m$ is a linear map with matrix

$$\begin{pmatrix} a_{11} & a_{12} & \cdots & a_{1n} \\ a_{21} & a_{22} & \cdots & a_{2n} \\ \vdots & \vdots & \vdots & \vdots \\ a_{m1} & a_{m2} & \cdots & a_{mn} \end{pmatrix}.$$

Let $x \in \mathbb{R}^n$, $x \neq 0$. Then

$$f(x) = \sum_{k=1}^{m}\left(\sum_{j=1}^{n} a_{kj}x_j\right)\epsilon_k,$$

where $\epsilon_1,\ldots,\epsilon_m$ is a basis in $\mathbb{R}^m$. Applying the Cauchy–Schwarz inequality, we get

$$|f(x)|^2 = \sum_{k=1}^{m}\left(\sum_{j=1}^{n} a_{kj}x_j\right)^2$$

$$\leq \sum_{k=1}^{m}\left(\sum_{j=1}^{n} a_{kj}^2\right)\left(\sum_{j=1}^{n} x_j^2\right)$$

$$= \left(\sum_{k=1}^{m}\left(\sum_{j=1}^{n} a_{kj}^2\right)\right)|x|^2.$$

Therefore

$$|f(x)| \leq \left(\sum_{k=1}^{m}\sum_{j=1}^{n} a_{kj}^2\right)^{\frac{1}{2}}|x|$$

and

$$\frac{|f(x)|}{|x|} \le \left(\sum_{k=1}^{m} \sum_{j=1}^{n} a_{kj}^2 \right)^{\frac{1}{2}}.$$

By Theorem 4.11 we find

$$\|f\| = \sup_{x \in \mathbb{R}^n, x \ne 0} \frac{|f(x)|}{|x|}$$

$$\le \left(\sum_{k=1}^{m} \sum_{j=1}^{n} a_{kj}^2 \right)^{\frac{1}{2}}.$$

4.7 Differentiable maps

In this section, we give some criteria for differentiability of vector maps. The results in this section can be applied to functions of several variables.

Definition 4.14. We say that a map $\alpha : X \to \mathbb{R}^m$ is infinitely small as $x \to x_0$ with respect to the function $|x - x_0|^n$ and we will write

$$\alpha(x) = o(|x - x_0|^n), \quad x \to x_0,$$

if there exists a map $\epsilon : X \to \mathbb{R}^m$ such that

$$\alpha(x) = \epsilon(x)|x - x_0|^n, \quad x \in X,$$

and

$$\lim_{x \to x_0} \epsilon(x) = 0.$$

Note that the map a is determined at the point x_0. Therefore the map ϵ is determined at the point x_0. Hence, applying the definition for a limit of a function, we conclude that ϵ is continuous at x_0 and $\epsilon(x_0) = 0$.

Definition 4.15. A map $f : X \to \mathbb{R}^m$ is said to be differentiable at a point $x \in X$ if there exists a linear map $l(x) : \mathbb{R}^n \to \mathbb{R}^m$ such that

$$f(x + h) = f(x) + l(x)(h) + o(h) \quad \text{as} \quad h \to 0, \quad h \in \mathbb{R}^n. \tag{4.22}$$

The linear map $l(x) : \mathbb{R}^n \to \mathbb{R}^m$ is called the differential of f at x and is denoted by $Df(x)$. Then equation (4.22) can be rewritten in the form

$$f(x + h) = f(x) + Df(x)(h) + o(h) \quad \text{as} \quad h \to 0. \tag{4.23}$$

The matrix of the differential $Df(x)$ is said to be the derivative of the function f at x and is denoted by $f'(x)$. Note that by (4.23) it follows that

$$\lim_{h \to 0} f(x + h) = f(x).$$

The map f is said to be differentiable on X if it is differentiable at each point of X.

Theorem 4.12. *If the map $f : X \to \mathbb{R}^m$ is differentiable at the point $x \in X$, then its differential at this point is uniquely determined.*

Proof. Suppose that there are two linear maps $l, l_1 : \mathbb{R}^n \to \mathbb{R}^m$ such that equation (4.22) and the equation

$$f(x + h) = f(x) + l_1(x)(h) + o(h) \quad \text{as} \quad h \to 0$$

hold. Then

$$l(x)(h) - l_1(x)(h) = o(h) \quad \text{as} \quad h \to 0.$$

Therefore there exist a neighborhood V of the zero and a map $e : V \to \mathbb{R}^m$ such that

$$\lim_{h \to 0} e(h) = 0$$

and

$$l(x)(h) - l_1(x)(h) = e(h)h.$$

Hence

$$\left| l(x)(h) - l_1(x)(h) \right| = \left| e(h) \right| \|h\|, \quad h \in V.$$

Let $k \in \mathbb{R}^n$ be arbitrarily chosen. Then there is enough small $t \in \mathbb{R}$ such that $tk \in V$, and then

$$\left| l(x)(tk) - l_1(x)(tk) \right| = \left| e(tk) \right| \|tk\|.$$

Now, since l and l_1 are linear maps, we find

$$|t| \left| l(x)(k) - l_1(x)(k) \right| = \left| e(tk) \right| \|tk\|,$$

whereupon

$$\left| l(x)(k) - l_1(x)(k) \right| = \left| e(tk) \right| \|k\|. \tag{4.24}$$

Since

$$\lim_{t \to 0}(tk) = 0,$$

we have

$$\lim_{t \to 0} \epsilon(tk) = 0,$$

and by (4.24) we get

$$\left| l(x)(k) - l_1(x)(k) \right| = 0.$$

Because $k \in \mathbb{R}^n$ was arbitrarily chosen, we conclude that

$$l(x) \equiv l_1(x).$$

This completes the proof. □

Corollary 4.2. *The differential of a linear map coincides with the map.*

Proof. Let $X \to \mathbb{R}^m$ be a linear differentiable map at $x \in X$. Then by (4.22) we get

$$f(x + h) = f(x) + f(h)$$
$$= f(x) + l(x)(h) + o(h).$$

Therefore

$$f(h) = l(x)(h),$$
$$o(h) \equiv 0.$$

Thus

$$Df(x) = f(x).$$

This completes the proof. □

An elementary property of the differentiable maps reads as follows.

Theorem 4.13. *Let $f, g : X \to \mathbb{R}^m$ be maps differentiable at a point $x \in X$. Then, for any constants $a_1, a_2 \in \mathbb{R}$, the map $a_1 f + a_2 g : X \to \mathbb{R}^m$ is differentiable at x, and*

$$D(a_1 f + a_2 g)(x) = a_1 Df(x) + a_2 Dg(x). \tag{4.25}$$

Proof. Since $f, g : X \to \mathbb{R}^m$ are differentiable at x, we have

$$f(x + h) = f(x) + Df(x)(h) + o(h) \quad \text{as} \quad h \to 0,$$
$$g(x + h) = g(x) + Dg(x)(h) + o(h) \quad \text{as} \quad h \to 0,$$

whereupon

$$
\begin{aligned}
a_1 f(x+h) + a_2 g(x+h) &= a_1\big(f(x) + Df(x)(h) + o(h)\big) \\
&\quad + a_2\big(g(x) + Dg(x)(h) + o(h)\big) \\
&= \big(a_1 f(x) + a_2 g(x)\big) \\
&\quad + \big(a_1 Df(x) + a_2 Dg(x)\big)(h) + o(h) \quad \text{as} \quad h \to 0.
\end{aligned}
$$

Since $Df(x)$ and $Dg(x)$ are linear maps and a linear combination of linear maps is a linear map, we conclude that

$$
a_1 Df(x) + a_2 Dg(x)
$$

is a linear map. Now, applying the definition of a differential, we conclude that $a_1 f + a_2 g :$ $X \to \mathbb{R}^m$ is differentiable and (4.25) hold. This completes the proof. □

Now we give a criterion for differentiability of the composition of two maps.

Theorem 4.14. *Suppose $X \subset \mathbb{R}^n$, $Y \subset \mathbb{R}^m$, $f : X \to Y$, $g : Y \to \mathbb{R}^s$, f is differentiable at $x \in X$, and g is differentiable at $f(x)$. Then $g \circ f$ is differentiable at x, and*

$$
D(g \circ f)(x) = Dg(f(x)) \circ Df(x). \tag{4.26}
$$

Proof. Since $f : X \to Y$ is differentiable at x, we have

$$
f(x+h) = f(x) + Df(x)(h) + o(h) \quad \text{as} \quad h \to 0.
$$

Then

$$
g(f(x+h)) = g\big(f(x) + Df(x)(h) + o(h)\big) \quad \text{as} \quad h \to 0. \tag{4.27}
$$

Let

$$
\begin{aligned}
y &= f(x), \\
k &= Df(x)(h) + o(h) \quad \text{as} \quad h \to 0.
\end{aligned}
$$

Then

$$
\begin{aligned}
g(f(x+h)) &= g(y+k) \\
&= g(y) + Dg(y)(k) + o(k) \quad \text{as} \quad k \to 0.
\end{aligned} \tag{4.28}
$$

We have

$$
|Df(x)(h)| \le \|Df(x)\| |h|
$$

and

$$|k| = |Df(x)(h) + o(h)|$$
$$\leq |Df(x)(h)| + |o(h)|$$
$$\leq \|Df(x)\||h| + |o(h)|$$
$$\leq \|Df(x)\||h| + |h|$$
$$= (\|Df(x)\| + 1)|h|.$$

Note that by the definition of $o(k)$ it follows that there exists a function $\epsilon(h)$ such that

$$\lim_{k \to 0} \epsilon(k) = 0 \tag{4.29}$$

and

$$o(k) = \epsilon(k)|k|.$$

Then

$$|o(k)| = |\epsilon(k)||k|$$
$$\leq |\epsilon(k)|(\|Df(x)\| + 1)|h|.$$

Since

$$\lim_{h \to 0} k = 0,$$

applying (4.29), we get

$$\lim_{h \to 0} \epsilon(k) = 0.$$

Therefore

$$\epsilon(k)(\|Df(x)\| + 1)|h| = o(h) \quad \text{as} \quad h \to 0,$$

and then

$$o(k) = o(h) \quad \text{as} \quad h \to 0.$$

Consequently, (4.28) can be rewritten in the form

$$g(f(x + h)) = g(y) + Dg(y)(k) + o(h) \quad \text{as} \quad h \to 0.$$

Note that

$$Dg(y)(k) = Dg(y)(Df(x)(h) + o(h))$$
$$= Dg(y)(Df(x)(h)) + Dg(y)o(h) \quad \text{as} \quad h \to 0. \tag{4.30}$$

Because

$$\left|Dg(y)o(h)\right| \le \|Dg(y)\|\|o(h)\|,$$

we get

$$Dg(y)o(h) = o(h) \quad \text{as} \quad h \to 0.$$

Then (4.30) takes the form

$$Dg(y)(k) = Dg(y)(Df(x)(h)) + o(h)$$
$$= (Dg(y) \circ Df(x))(h) + o(h).$$

By the last expression and (4.27) we obtain

$$(g \circ f)(x + h) = g(f(x)) + (Dg(f(x)) \circ Df(x))(h) + o(h) \quad \text{as} \quad h \to 0.$$

Hence $Dg(f(x)) \circ Df(x)$ is the differential of $f \circ g$, and (4.26) holds. This completes the proof. $\square$

Next, we will give a necessary and sufficient condition for differentiability of a map.

Theorem 4.15. *The map $f = (f_1, \ldots, f_m) : X \to \mathbb{R}^m$ is differentiable at $x \in X$ if and only if the coordinate functions $f_j : X \to \mathbb{R}, j \in \{1, \ldots, m\}$, are differentiable at this point. In this case, the elements $a_{ij}, i \in \{1, \ldots, m\}, j \in \{1, \ldots, n\}$, of the matrix of the differential $Df(x)$ are the corresponding partial derivatives of the coordinate functions, i. e.,*

$$a_{ij} = \frac{\partial f_i}{\partial x_j}(x), \quad i \in \{1, \ldots, m\}, \quad j \in \{1, \ldots, n\},$$

and

$$f'(x) = Df(x)$$
$$= \left(\frac{\partial f_i}{\partial x_j}\right)$$
$$= \begin{pmatrix} \frac{\partial f_1}{\partial x_1}(x) & \cdots & \frac{\partial f_1}{\partial x_n}(x) \\ \vdots & \vdots & \vdots \\ \frac{\partial f_m}{\partial x_1}(x) & \cdots & \frac{\partial f_m}{\partial x_n}(x) \end{pmatrix},$$

which is called the Jacobian matrix of the map f.

Proof. 1. Let f be a differentiable function at $x \in X$. Let also, $\pi_j, j \in \{1, \ldots, m\}$, be the projector operator of the coordinate exes, i. e.,

$$\pi_j(y) = \pi_j(y_1, \ldots, y_m)$$
$$= y_j, \quad j \in \{1, \ldots, m\}.$$

Then

$$f_j = \pi_j \circ f, \quad j \in \{1, \ldots, m\}. \tag{4.31}$$

Therefore $f_j, j \in \{1, \ldots, m\}$, are compositions of two differentiable functions at x, and then $f_j, j \in \{1, \ldots, m\}$, are differentiable functions.

2. Let $f_j, j \in \{1, \ldots, m\}$, be differentiable functions at x. Then there exist constants

$$a_{ij} = \frac{\partial f_i}{\partial x_j}(x), \quad i \in \{1, \ldots, m\}, \quad j \in \{1, \ldots, n\},$$

such that

$$f_j(x + h) = f_j(x) + a_{j1}h_1 + \cdots + a_{jn}h_n + o_j(h) \quad \text{as} \quad h \to 0, \tag{4.32}$$

$j \in \{1, \ldots, m\}$. Let $l: \mathbb{R}^m \to \mathbb{R}^m$ be a linear operator with matrix (a_{ij}). Since

$$(o_1(h), \ldots, o_m(h)) = o(h),$$

equation (4.32) can be rewritten in the form

$$f(x + h) = f(x) + l(h) + o(h) \quad \text{as} \quad h \to 0.$$

Therefore f is differentiable at x. This completes the proof. □

Remark 4.2. Let $X \subset \mathbb{R}^n$ and $Y \subset \mathbb{R}^m$, and let $f: X \to Y$ and $g: Y \to \mathbb{R}^s$ be differentiable maps. Then by Theorem 4.14 and equality (4.26) the Jacobian matrix of the composition $g \circ f$ is the product of the Jacobian matrices of g and f. Indeed, let

$$z_k = g_k(y_1, \ldots, y_m), \quad k \in \{1, \ldots, s\},$$
$$y_j = f_j(x_1, \ldots, x_n), \quad j \in \{1, \ldots, m\}.$$

Then

$$\frac{\partial z_k}{\partial x_l} = \sum_{j=1}^m \frac{\partial z_k}{\partial y_j} \frac{\partial y_j}{\partial x_l}, \quad k \in \{1, \ldots, s\}, \quad l \in \{1, \ldots, n\}.$$

By the rule of multiplication of matrices we get

$$\left(\frac{\partial z_k}{\partial x_l} \right) = \left(\frac{\partial z_k}{\partial y_j} \right)\left(\frac{\partial y_j}{\partial x_l} \right).$$

An important determinant in differential and integral calculus is the Jacobian.

Definition 4.16. Let $x \in X$, and let a map $f : X \to \mathbb{R}^n$ be differentiable at x. The determinant

$$\det\left(\frac{\partial f_i}{\partial x_j}(x)\right)$$

is called the Jacobian of the map f and is denoted by

$$\frac{\partial(f_1,\ldots,f_n)}{\partial(x_1,\ldots,x_n)}, \quad \text{or} \quad \frac{D(f_1,\ldots,f_n)}{D(x_1,\ldots,x_n)}, \quad \text{or} \quad J_f.$$

Let now $Y \subset \mathbb{R}^n$, let $f : X \to Y$ and $g : Y \to \mathbb{R}^n$ be differentiable maps, and let

$$z_k = g_k(y_1,\ldots,y_n),$$
$$y_j = f_j(x_1,\ldots,x_n), \quad k,j \in \{1,\ldots,n\}.$$

Then

$$\frac{\partial(z_1,\ldots,z_n)}{\partial(x_1,\ldots,x_n)} = \det\left(\frac{\partial z_k}{\partial x_j}\right)$$

$$= \det\left(\left(\frac{\partial z_k}{\partial y_l}\right)\left(\frac{\partial y_l}{\partial x_j}\right)\right)$$

$$= \det\left(\frac{\partial z_k}{\partial y_l}\right)\det\left(\frac{\partial y_l}{\partial x_j}\right)$$

$$= \frac{\partial(z_1,\ldots,z_n)}{\partial(y_1,\ldots,y_n)}\frac{\partial(y_1,\ldots,y_n)}{\partial(x_1,\ldots,x_n)}.$$

If $h : X \to \mathbb{R}^n$ is the map

$$h_j(x) = x_j, \quad j \in \{1,\ldots,n\},$$

then

$$\frac{\partial h_j}{\partial x_l}(x) = \delta_{jl}, \quad j,l \in \{1,\ldots,n\},$$

and

$$\frac{\partial(h_1,\ldots,h_n)}{\partial(x_1,\ldots,x_n)} = \det\begin{pmatrix} 1 & 0 & \cdots & 0 \\ 0 & 1 & \cdots & 0 \\ \vdots & \vdots & \vdots & \vdots \\ 0 & 0 & \cdots & 1 \end{pmatrix}$$

$$= \det E$$
$$= 1,$$

where

$$E = \begin{pmatrix} 1 & 0 & \cdots & 0 \\ 0 & 1 & \cdots & 0 \\ \vdots & \vdots & \vdots & \vdots \\ 0 & 0 & \cdots & 1 \end{pmatrix}.$$

Suppose that $f : X \to Y$ is an injective differentiable map and $f^{-1} : Y \to \mathbb{R}^n$ is its inverse, which is differentiable. Then

$$f^{-1}(f(x)) = f(f^{-1}(x))$$
$$= x, \quad x \in X.$$

By Theorem 4.14 and equation (4.26) we get

$$(f^{-1})' f' = f' (f^{-1})'$$
$$= E$$

and

$$\det(f^{-1})' \det f' = 1.$$

In addition, if

$$y_j = f_j(x_1, \ldots, x_n), \quad x = (x_1, \ldots, x_n) \in X,$$

then

$$\frac{\partial(x_1, \ldots, x_n)}{\partial(y_1, \ldots, y_n)} \frac{\partial(y_1, \ldots, y_n)}{\partial(x_1, \ldots, x_n)} = 1,$$

and

$$\frac{\partial(x_1, \ldots, x_n)}{\partial(y_1, \ldots, y_n)} = \frac{1}{\frac{\partial(y_1, \ldots, y_n)}{\partial(x_1, \ldots, x_n)}}.$$

Example 4.17. Let $n = 2$. Consider the polar coordinates

$$x = r \cos \phi,$$
$$y = r \sin \phi, \quad r \geq 0, \quad \phi \in [0, 2\pi].$$

Then

$$\frac{\partial(x,y)}{\partial(r,\phi)} = \det \begin{pmatrix} \frac{\partial x}{\partial r} & \frac{\partial x}{\partial \phi} \\ \frac{\partial y}{\partial r} & \frac{\partial y}{\partial \phi} \end{pmatrix}$$

$$= \det \begin{pmatrix} \cos\phi & -r\sin\phi \\ \sin\phi & r\cos\phi \end{pmatrix}$$

$$= r(\cos\phi)^2 + r(\sin\phi)^2$$

$$= r, \quad r \geq 0, \quad \phi \in [0, 2\pi].$$

Example 4.18. Let $n = 3$. Consider the spherical coordinates

$$x = r\cos\phi \sin\theta,$$
$$y = r\sin\phi \sin\theta,$$
$$z = r\cos\theta, \quad r \geq 0, \quad \phi \in [0, 2\pi], \quad \theta \in [0, \pi].$$

Then

$$\frac{\partial(x,y,z)}{\partial(r,\phi,\theta)} = \det \begin{pmatrix} \frac{\partial x}{\partial r} & \frac{\partial x}{\partial \phi} & \frac{\partial x}{\partial \theta} \\ \frac{\partial y}{\partial r} & \frac{\partial y}{\partial \phi} & \frac{\partial y}{\partial \theta} \\ \frac{\partial z}{\partial r} & \frac{\partial z}{\partial \phi} & \frac{\partial z}{\partial \theta} \end{pmatrix}$$

$$= \det \begin{pmatrix} \cos\phi \sin\theta & -r\sin\phi \sin\theta & r\cos\phi \cos\theta \\ \sin\phi \sin\theta & r\cos\phi \sin\theta & r\sin\phi \cos\theta \\ \cos\theta & 0 & -r\sin\theta \end{pmatrix}$$

$$= -r^2(\cos\phi)^2(\sin\theta)^3 - r^2(\sin\phi)^2 \sin\theta(\cos\theta)^2$$
$$\quad - r^2(\cos\phi)^2 \sin\theta(\cos\theta)^2 - r^2(\sin\phi)^2(\sin\theta)^3$$
$$= -r^2(\cos\phi)^2 \sin\theta - r^2(\sin\phi)^2 \sin\theta$$
$$= -r^2 \sin\theta, \quad r \geq 0, \quad \phi \in [0, 2\pi], \quad \theta \in [0, \pi].$$

Example 4.19. Let $n = 3$. Consider the cylindrical coordinates

$$x = r\cos\phi,$$
$$y = r\sin\phi,$$
$$z = h, \quad r, h \geq 0, \quad \phi \in [0, 2\pi].$$

Then

$$\frac{\partial(x,y,z)}{\partial(r,\phi,h)} = \det \begin{pmatrix} \frac{\partial x}{\partial r} & \frac{\partial x}{\partial \phi} & \frac{\partial x}{\partial h} \\ \frac{\partial y}{\partial r} & \frac{\partial y}{\partial \phi} & \frac{\partial y}{\partial h} \\ \frac{\partial z}{\partial r} & \frac{\partial z}{\partial \phi} & \frac{\partial z}{\partial h} \end{pmatrix}$$

$$= \det \begin{pmatrix} \cos\phi & -r\sin\phi & 0 \\ \sin\phi & r\cos\phi & 0 \\ 0 & 0 & 1 \end{pmatrix}$$

$$= r(\cos\phi)^2 + r(\sin\phi)^2$$

$$= r, \quad r, h \geq 0, \quad \phi \in [0, 2\pi].$$

Definition 4.17. A function $f : X \to \mathbb{R}^m$ is said to be a homomorphism if it satisfies the following conditions:
1. $f \in C(X)$.
2. f^{-1} exists.
3. $f^{-1} \in C(f(X))$.

Definition 4.18. A function $f : X \to \mathbb{R}^m$ is said to be a diffeomorphism if it satisfies the following conditions:
1. $f \in C^1(X)$.
2. f^{-1} exists.
3. $f^{-1} \in C^1(f(X))$.

Definition 4.19. Let $x^0 \in X$. A map $f : X \to \mathbb{R}^m$ is said to be a local homomorphism (local diffeomorphism) at the point x^0 if there are neighborhoods U and V of x^0 and $f(x^0)$, respectively, such that $f(U) = V$ homomorphic (diffeomorphic).

Theorem 4.16. *Let $f : X \to \mathbb{R}^n$ be a continuously differentiable map such that $J_f(x^0) \neq 0$ for some $x^0 \in X$. Then f is a local diffeomorphism at the point x^0.*

Proof. For

$$x = (x_1, \ldots, x_n),$$
$$y = (f_1(x), \ldots, f_n(x))$$
$$= (y_1, \ldots, y_n) \in \mathbb{R}^n,$$

denote

$$F_j(x, y) = f_j(x_1, \ldots, x_n) - y_j, \quad j \in \{1, \ldots, n\},$$
$$y^0 = f(x^0),$$
$$F(x, y) = (F_1(x, y), \ldots, F_n(x, y)).$$

We have that $F : X \times \mathbb{R}^n \to \mathbb{R}^n$,

$$F_j(x^0, y^0) = 0, \quad j \in \{1, \ldots, n\},$$

and

$$\frac{\partial(F_1,\ldots,F_n)}{\partial(x_1,\ldots,x_n)}\bigg|_{(x^0,y^0)} = \frac{\partial(f_1,\ldots,f_n)}{\partial(x_1,\ldots,x_n)}\bigg|_{x^0}$$
$$\neq 0.$$

Consider the system

$$F_j(x,y) = 0, \quad j \in \{1,\ldots,n\},$$

or the system

$$f_j(x) - y_j = 0, \quad j \in \{1,\ldots,n\},$$

or the equation

$$y = f(x). \tag{4.33}$$

By Theorem 4.9 it follows that there are neighborhoods U and V of the points x^0 and y^0 such that

$$x^0 \in U \quad \text{and} \quad y^0 \in V$$

and for each $y \in V$, equation (4.33) is uniquely solvable, and its solution is denoted by $x = g(y)$. This solution is continuously differentiable in V, and

$$f(g(y)) = y.$$

Thus $V \subset f(U)$. Consider the restriction f_U of f on U. Let U_0 be such that

$$U_0 = f_U^{-1}(V).$$

Then

$$f_U(U_0) = V,$$

and $x^0 \in U_0$. Because V is an open set and f_U is continuous on U, we conclude that U_0 is an open set. Therefore U_0 is a neighborhood of x^0, and

$$f_{U_0}^{-1} = g \quad \text{on } V.$$

Thus f is a local diffeomorphism at the point x^0. This completes the proof. $\square$

Now we will investigate the equation

$$f(x,y) = 0, \tag{4.34}$$

where f is defined and is continuously differentiable in a neighborhood of the point $(x^0, y^0) \in \mathbb{R}^2$, and

$$f(x^0, y^0) = 0, \tag{4.35}$$

$$f_x(x^0, y^0) = 0,$$
$$f_y(x^0, y^0) = 0. \tag{4.36}$$

Definition 4.20. A point $(x^0, y^0) \in \mathbb{R}^2$ for which conditions (4.35) and (4.36) hold is said to be a singular point. If the point (x^0, y^0) is a singular point for which there exists a neighborhood in which it is unique, it is said to be an isolated singular point.

Theorem 4.17. *Let f be twice continuously differentiable in a neighborhood U of an isolated singular point (x^0, y^0), and let*

$$f_{xx}(x^0, y^0)f_{yy}(x^0, y^0) - (f_{xy}(x^0, y^0))^2 \neq 0.$$

If

$$f_{xx}(x^0, y^0)f_{yy}(x^0, y^0) - (f_{xy}(x^0, y^0))^2 > 0, \tag{4.37}$$

then equation (4.34) is uniquely solvable in a neighborhood of the point (x^0, y^0). If

$$f_{xx}(x^0, y^0)f_{yy}(x^0, y^0) - (f_{xy}(x^0, y^0))^2 < 0, \tag{4.38}$$

then equation (4.34) has at least two different solutions that are continuously differentiable in a neighborhood of the point (x^0, y^0).

Proof. Since conditions (4.35), (4.36), and (4.37) or (4.38) hold, we conclude that the point (x^0, y^0) is a point of local extremum for the function f in a neighborhood of the point (x^0, y^0). Then we have

$$f(x, y) > f(x^0, y^0)$$

or

$$f(x, y) < f(x^0, y^0)$$

for all $(x, y) \in U$, $(x, y) \neq (x^0, y^0)$. Thus the point (x^0, y^0) is an isolated singular point for the function f. By the Taylor formula, using (4.35) and (4.36), we find

$$f(x, y) = \frac{1}{2}(f_{xx}(x^0, y^0)(x - x^0)^2 + 2f_{xy}(x^0, y^0)(x - x^0)(y - y^0)$$
$$+ f_{yy}(x^0, y^0)(y - y^0)^2) + o(r^2) \quad \text{as} \quad r \to 0,$$

where

$$r^2 = (x - x^0)^2 + (y - y^0)^2, \quad (x,y) \in U.$$

Set

$$x - x^0 = r \cos \phi,$$
$$y - y^0 = r \sin \phi, \quad \phi \in [0, 2\pi], \quad r \geq 0, \quad (x,y) \in U.$$

Then

$$f(x,y) = \frac{1}{2}(f_{xx}(x^0,y^0)r^2(\cos \phi)^2 + 2f_{xy}(x^0,y^0)r^2 \sin \phi \cos \phi$$
$$+ f_{yy}(x^0,y^0)r^2(\sin \phi)^2) + o(r^2)$$
$$= \frac{r^2}{2}(f_{xx}(x^0,y^0)(\cos \phi)^2 + 2f_{xy}(x^0,y^0) \sin \phi \cos \phi$$
$$+ f_{yy}(x^0,y^0)(\sin \phi)^2) + o(r^2) \quad \text{as} \quad r \to 0,$$

$(x,y) \in U, r \geq 0, \phi \in [0, 2\pi]$. Set

$$g(\phi) = f_{xx}(x^0,y^0)(\cos \phi)^2 + 2f_{xy}(x^0,y^0) \sin \phi \cos \phi$$
$$+ f_{yy}(x^0,y^0)(\sin \phi)^2, \quad \phi \in [0, 2\pi].$$

Hence

$$f(x,y) = \frac{r^2}{2}g(\phi) + o(r^2) \quad \text{as} \quad r \to 0.$$

If $\phi = \pm\frac{\pi}{2}$, then

$$g(\phi) = f_{yy}(x^0,y^0), \quad \phi \in [0, 2\pi],$$

and

$$f(x,y) = \frac{r^2}{2}f_{xy}(x^0,y^0) + o(r^2) \quad \text{as} \quad r \to 0, \quad (x,y) \in U.$$

Thus, for $r \neq 0$ and $\phi = \pm\frac{\pi}{2}$, in a neighborhood of the point (x^0,y^0), we have

$$f(x,y) \neq 0.$$

Since we search for solutions of equation (4.34), we can suppose that there is an enough small neighborhood V_1 of the point (x^0, y^0) such that

$$\phi \neq \frac{\pi}{2}(2m + 1), \quad m \in \mathbb{Z}.$$

Then in V_1, we have

$$g(\phi) = (\cos\phi)^2(f_{xx}(x^0,y^0) + 2f_{xy}(x^0,y^0)\tan\phi + f_{yy}(x^0,y^0)(\tan\phi)^2).$$

Now consider the equation

$$f_{xx}(x^0,y^0) + 2f_{xy}(xc_1^0,y^0)k + f_{yy}(x^0,y^0)k^2 = 0. \qquad (4.39)$$

By condition (4.37) it follows that equation (4.39) has two solutions k_1 and k_2 such that

$$\phi_1 = \arctan k_1,$$
$$\phi_2 = \arctan k_2.$$

Then $\phi_{1,2} \neq \pm\frac{\pi}{2}$, and

$$g(\phi) = (\cos\phi)^2(\tan\phi - \tan\phi_1)(\tan\phi - \tan\phi_2).$$

By the last expression for the function g we get that

$$g(\phi) = 0$$

for $\phi \neq \frac{\pi}{2}(2m+1)$, $m \in \mathbb{Z}$, if and only if

$$\phi = \phi_1 + m\pi,$$
$$\phi = \phi_2 + m\pi, \quad m \in \mathbb{Z}.$$

Take arbitrary $\epsilon > 0$. Denote

$$U_1 = \{(r,\phi) : \phi_1 - \epsilon < \phi < \phi_1 + \epsilon\},$$
$$U_2 = \{(r,\phi) : \phi_2 - \epsilon < \phi < \phi_2 + \epsilon\}.$$

We choose $\epsilon > 0$ small enough such that

$$U_1 \cap U_2 = \{(0,0)\}$$

and

$$U_1 \cap \{x = 0\} = \{(0,0)\},$$
$$U_2 \cap \{x = 0\} = \{(0,0)\}.$$

Set

$$U_1^* = \{(r,\phi) : \phi_1 + \pi - \epsilon < \phi < \phi_1 + \pi + \epsilon\},$$
$$U_2^* = \{(r,\phi) : \phi_2 + \pi - \epsilon < \phi < \phi_2 + \pi + \epsilon\}.$$

Take $\epsilon > 0$ small enough such that

$$U_i \cap U_j^* = \{(0,0)\}, \quad U_l \cap U_m = \{(0,0)\}, \quad U_l^* \cap U_m^* = \{(0,0)\}, \quad i,j \in \{1,2\},$$

$l, m \in \{1,2\}$, $l \neq m$. Consider the function g on a circle C with center the point (x^0, y^0) and radius 1. For simplicity, any point $(1, \phi) \in C$ will be denoted by ϕ. Let B_1 be the union of the intervals in C with centers $\phi_1, \phi_2, \phi_1 + \pi, \phi_2 + \pi$ and length 2ϵ. By the choice of ϵ it follows that these intervals are disjoint. Let

$$B = C \backslash B_1.$$

Note that the set B is bounded and closed, and thus it is a compact set. We have that $g \in C(B)$ and $g \neq 0$ on B. Therefore

$$\inf_{\phi \in B} |g(\phi)| = \mu > 0.$$

Let

$$K_\rho = \{(r, \phi) : 0 \leq r \leq \rho\}$$

and

$$L_\rho = K_\rho \backslash (U_1 \cup U_1 \cup U_1^* \cup U_2^*).$$

Since the function g does not depend on r, we have

$$\inf_{(r,\phi) \in L_\rho} |g(\phi)| = \mu > 0.$$

Therefore

$$f(x, y) = \frac{r^2}{2}(g(\phi) + \alpha(r, \phi)),$$

where

$$\lim_{r \to 0} \alpha(r, \phi) = 0.$$

We choose $\rho > 0$ such that

$$|\alpha(r, \phi)| \leq \mu \quad \text{for all } r \leq \rho.$$

Therefore, for all $(r, \phi) \in L_\rho$, the functions $g(\phi)$ and $g(\phi) + \alpha(r, \phi)$ have the same sign. Now consider $U_1 = U_1(\epsilon)$. Without loss of generality, suppose that $\phi \in [0, \frac{\pi}{2})$. The intersection of $\overline{U_1}$ with the lines

$$x = x^*, \quad x^0 < x^* < x^0 + \rho \cos(\phi_1 + \epsilon)$$

are segments such that at their end points, the function $f(x^*, y)$ have different signs, and the function $f(x^*, \cdot)$ is a continuous function on these segments, and for any x^* such that $x^0 < x^* \leq x^0 + \rho \cos(\phi_1 + \epsilon)$, there exists a unique point y^* such that

$$f(x^*, y^*) = 0, \quad (x^*, y^*) \in U_1(\epsilon) \cap K_\rho. \tag{4.40}$$

Define the function f_1 so that

$$y = f_1(x),$$
$$f_1(x^*) = y^*, \quad x^0 < x^* < x^0 + \rho \cos(\phi_1 + \epsilon).$$

We will show that for any $\epsilon > 0$ and $\rho > 0$ small enough and for any given x^*, by condition (4.40) the point y^* is uniquely determined. Assume the contrary. Take the sequences $\{\epsilon_n\}_{n \in \mathbb{N}}$ and $\{\rho_n\}_{n \in \mathbb{N}}$ such that

$$\epsilon_n \to 0, \quad \rho_n \to 0 \quad \text{as} \quad n \to \infty.$$

Then there are sequences $\{x_n\}_{n \in \mathbb{N}}$, $\{y_n^1\}_{n \in \mathbb{N}}$, and $\{y_n^2\}_{n \in \mathbb{N}}$ such that

$$(x_n, y_n^1) \in U_1(\epsilon_n) \cap K_{\rho_n}, \quad f(x_n, y_n^1) = 0, \quad n \in \mathbb{N},$$

and

$$(x_n, y_n^2) \in U_1(\epsilon_n) \cap K_{\rho_n}, \quad f(x_n, y_n^2) = 0, \quad n \in \mathbb{N}.$$

Now by the Rolle theorem it follows that on the lines $x = x_n$, $n \in \mathbb{N}$, there are points $y_n \in (y_n^1, y_n^2)$ such that

$$f_y(x_n, y_n) = 0,$$

$(x_n, y_n) \in U_1(\epsilon_n) \cap K_{\rho_n}$, and

$$f_y(x^0, y^0) = 0.$$

Moreover,

$$f_y(x_n, y_n) - f_y(x^0, y^0) = f_{xy}(\xi_n, \eta_n)(x_n - x^0)$$
$$+ f_{yy}(\xi_n, \eta_n)(y_n - y^0)$$

for some $(\xi_n, \eta_n) \in U_1(\epsilon_n) \cap K_{\rho_n}$. Therefore

$$f_{xy}(\xi_n, \eta_n) + f_{yy}(\xi_n, \eta_n)\frac{y_n - y^0}{x_n - x^0} = 0, \quad n \in \mathbb{N}. \tag{4.41}$$

Let

$$(x_n, y_n) = (r_n, \psi_n), \quad n \in \mathbb{N}.$$

Then

$$|\psi_n - \phi_1| < \epsilon_n, \quad n \in \mathbb{N}.$$

Since $\epsilon_n \to 0$ as $n \to \infty$, we get

$$\psi_n \to \phi_1 \quad \text{as} \quad n \to \infty,$$

and since

$$\tan \psi_n = \frac{y_n - x_{20}}{x_n - x_{10}}, \quad n \in \mathbb{N},$$

we obtain

$$\frac{y_n - x_{20}}{x_n - x_{10}} \to \tan \phi_1 = k_1 \quad \text{as} \quad n \to \infty.$$

By the last equalities and (4.41) we arrive at

$$f_{xy}(x^0, y^0) + f_{yy}(x^0, y^0)k_1 = 0$$

or

$$k_1 = -\frac{f_{xy}(x^0, y^0)}{f_{yy}(x2^0, y^0)}.$$

Now by (4.39) we get

$$0 = f_{xx}(x^0, y^0) + 2f_{xy}(x^0, y^0)k_1 + f_{yy}(x^0, y^0)k_1^2$$
$$= f_{xx}(x^0, y^0) - 2\frac{(f_{xy}(x^0, y^0))^2}{f_{yy}(x^0, y^0)} + \frac{(f_{xy}(x^0, y^0))^2}{f_{yy}(x^0, y^0)},$$

whereupon

$$f_{xx}(x^0, y^0)f_{yy}(x^0, y^0) - (f_{xy}(x^0, y^0))^2 = 0.$$

This is contradiction. Thus the function

$$y = f_1(x)$$

is well defined for small enough $\epsilon > 0$ and $\rho > 0$. Now we set

$$y^0 = f_1(x^0).$$

Then

$$f(x, f_1(x)) = 0, \quad x^0 \le x \le x^0 + \rho \cos(\phi_1 + \epsilon).$$

Fix $\epsilon > 0$. Then there is $\rho = \rho(\epsilon) > 0$ such that

$$(x, f_1(x)) \in U_1(\epsilon) \cap K_\rho, \quad x^0 \le x \le x^0 + \rho \cos(\phi_1 + \epsilon). \tag{4.42}$$

Let

$$\delta = \rho \cos(\phi_1 + \epsilon)$$

and take x such that

$$0 < x - x^0 < \delta.$$

Then applying (4.42), we find

$$|\phi - \phi_1| < \epsilon.$$

Therefore

$$\lim_{x \to x^0+} \phi = \phi_1 \quad \text{and} \quad \lim_{x \to x^0+} \tan \phi = \tan \phi_1,$$

and

$$\lim_{x \to x^0} \frac{f_1(x) - f_1(x^0)}{x - x^0} = \lim_{x \to x^0+} \frac{y - y^0}{x - x^0}$$
$$= \tan \phi_1.$$

As above, we have that for some $\delta_1 > 0$ in the segment $[x^0 - \delta_1, x^0]$, there exists a function f_1 such that

$$f(x, f_1(x)) = 0, \quad (x, f_1(x)) \in U_1^*, \quad \text{and} \quad f_1(x^0) = y^0, \quad f_1'(x^0) = k_1.$$

If $\rho > 0$ is small enough such that in a circle with radius ρ, the point (x^0, y^0) is the unique singular point, then the function $f_1(x)$ is differentiable in this circle for all $x \ne x^0$. As above, there is a function $f_2(x)$ that satisfies equation (4.34) and the conditions of the theorem. If

$$f_{yy}(x^0, y^0) = 0 \quad \text{and} \quad f_{xx}(x^0, y^0) \ne 0,$$

then as above we obtain two solutions $f_1(y)$ and $f_2(y)$. If

$$f_{xx}(x^0, y^0) = f_{yy}(x^0, y^0) = 0,$$

then

$$f_{xy}(x^0, y^0) \neq 0.$$

In this case, using the change

$$x = \xi + \eta,$$
$$y = \xi - \eta,$$

we have

$$f_{\xi\xi}(x^0, y^0) = -f_{\eta\eta}(x^0, y^0)$$
$$= 2f_{xy}(x^0, y^0)$$
$$\neq 0$$

and

$$f_{\xi\eta}(x^0, y^0) = 0,$$

and this case is reduced to the previous cases. This completes the proof. ☐

4.8 The concept for functional dependence

A functional dependence is a constraint between two attribute sets in a relation. Functional dependences are an important part of designing databases in the relational model and in database normalization and denormalization. They have a wide range of applications in differential geometry and ordinary and partial differential equations.

Definition 4.21. Suppose that $X \subset \mathbb{R}^n$ is an open set and that on X, there are defined the functions

$$y_j = \phi_j(x), \quad x \in X, \quad j \in \{1, \ldots, m\}. \tag{4.43}$$

If there are an open set $Y \subset \mathbb{R}^{m-1}_{y_1, \ldots, y_{m-1}}$ and a continuously differentiable function Φ on Y such that $(\phi_1(x), \ldots, \phi_{m-1}(x)) \in Y$ for all $x \in X$ and

$$\phi_m(x) = \Phi(\phi_1(x), \ldots, \phi_{m-1}(x)), \quad x \in X,$$

then we say that the function ϕ_m is dependent on the functions $\phi_1, \ldots, \phi_{m-1}$ on the set X. If there is at least one function (4.43) that is dependent on the rest, we say that system (4.43) is dependent. If there is no function (4.43) that is dependent on the rest, then we say that system (4.43) is independent.

Theorem 4.18. *Suppose that $m \le n$ and system (4.43) is dependent. Then at each point of X the rank of the Jacobian matrix of system (4.43) is less than m.*

Proof. Since system (4.43) is dependent on X, then there is at least one function of system (4.43) that is dependent on the rest. Without loss of generality, suppose that ϕ_m is dependent on $\phi_1, \ldots, \phi_{m-1}$ on X. Hence there are an open set $Y \subset \mathbb{R}^{m-1}_{y_1, \ldots, y_{m-1}}$ and a continuously differentiable function Φ on Y such that

$$\phi_m(x) = \Phi(\phi_1(x), \ldots, \phi_{m-1}(x)), \quad x \in X.$$

Then

$$\frac{\partial \phi_m}{\partial x_j}(x) = \sum_{l=1}^{m-1} \frac{\partial \Phi}{\partial y_l}(\phi_1(x), \ldots, \phi_{m-1}(x)) \frac{\partial \phi_l}{\partial x_j}(x), \quad x \in X.$$

Thus the mth row of the Jacobian matrix for system (4.43) is dependent on its first $m - 1$ rows. Therefore the rank of the Jacobian matrix of system (4.43) is less than m. This completes the proof. $\quad\square$

Corollary 4.3. *Let $m = n$, and let system (4.43) be dependent. Then*

$$\frac{\partial(y_1, \ldots, y_n)}{\partial(x_1, \ldots, x_n)}(x) = 0, \quad x \in X.$$

Corollary 4.4. *Let $m \le n$, and let the rank of the Jacobian matrix for system (4.43) be equal to m. Then system (4.43) is independent.*

Now we give a criterion for the independence of a system of functions.

Theorem 4.19. *Let the rank of the Jacobian matrix for system (4.43) at each point of the open set $X \subset \mathbb{R}^n$ be less than r, where $r < m \le n$, and let at the point $x^0 \in X$ the rank of the Jacobian matrix for system (4.43) be equal to r, i. e., there exist points $x_{j_1}, \ldots, x_{j_r}$ and functions*

$$y_{j_k} = \phi_{j_k}(x), \quad k \in \{1, \ldots, r\}, \tag{4.44}$$

such that

$$\left.\frac{\partial(y_{j_1}, \ldots, y_{j_r})}{\partial(x_{j_1}, \ldots, x_{j_r})}\right|_{x=x^0} \ne 0.$$

Then all r functions (4.44) are independent on X, and there exists a neighborhood U of the point x^0 such that each function ϕ_l, $l \ne j_k$, $k \in \{1, \ldots, r\}$, depends on functions (4.44) at each point of U.

Proof. For convenience, suppose that

$$\left.\frac{\partial(y_1,\ldots,y_r)}{\partial(x_1,\ldots,x_r)}\right|_{x=x^0} \neq 0. \tag{4.45}$$

Then by Corollary 4.4 it follows that the system $\phi_1,\ldots,\phi_r$ is independent on X. Now we will prove that each function $\phi_l, l \in \{r+1,\ldots,n\}$, is dependent on the functions $\phi_1,\ldots,\phi_r$ in a neighborhood U of the point x^0. Let

$$y_j^0 = \phi_j(x^0), \quad j \in \{1,\ldots,m\}.$$

Consider the system

$$y_j = \phi_j(x_1,\ldots,x_n), \quad j \in \{1,\ldots,r\}. \tag{4.46}$$

Take $\eta_0 > 0$ small enough such that

$$|x_j - x_j^0| < \eta_0, \quad j \in \{1,\ldots,n\},$$

implies $x \in X$. Such η_0 exists because the set X is an open set and $x^0 \in X$. By condition (4.46) and the implicit function theorem we conclude that system (4.46) is solvable with respect to $x_1,\ldots,x_r$ in a neighborhood of the point (x^0, y^0) and

$$x_1 = f_1(y_1,\ldots,y_r, x_{r+1},\ldots,x_n),$$

$$\vdots \tag{4.47}$$

$$x_r = f_r(y_1,\ldots,y_r, x_{r+1},\ldots,x_n),$$

and the functions $f_j, j \in \{1,\ldots,r\}$, are continuously differentiable in a neighborhood of the point

$$(y_1^0,\ldots,y_r^0, x_{r+1}^0,\ldots,x_n^0).$$

Hence there exist $\delta > 0$ and $\eta > 0$ such that $\delta < \eta_0, \eta < \eta_0$, and for

$$U: \begin{array}{l} |y_j - y_j^0| < \delta, \quad j \in \{1,\ldots,r\}, \\ |x_j - x_j^0| < \delta, \quad j \in \{r+1,\ldots,n\}, \end{array}$$

we have $f_j \in C^1(U), j \in \{1,\ldots,r\}$, and for all $(y_1,\ldots,y_r, x_{r+1},\ldots,x_n) \in U$,

$$|f_k(y_1,\ldots,y_r, x_{r+1},\ldots,x_n) - x_k^0| < \eta, \quad k \in \{1,\ldots,r\}.$$

Consider functions (4.47) and

$$y_{r+1} = \phi_{r+1}(f_1,\ldots,f_r, x_{r+1},\ldots,x_n). \tag{4.48}$$

We have that $\phi_{r+1} \in C^1(U)$. Now we will show that the function ϕ_{r+1} does not depend on $x_{r+1}, \ldots, x_n$. For this aim, we will prove that

$$\frac{\partial y_{k+1}}{\partial x_j} = 0, \quad j \in \{r+1, \ldots, n\}. \tag{4.49}$$

Fix $j \in \{r+1, \ldots, n\}$ and let x_k^*, $k \in \{r+1, \ldots, j-1, j+1, \ldots, n\}$, be such that

$$|x_k^* - x_k^0| < \delta, \quad k \in \{r+1, \ldots, j-1, j+1, \ldots, n\}.$$

Let

$$f_k^* = f_k(y_1, \ldots, y_r, x_{r+1}^*, \ldots, x_{j-1}^*, x_j, x_{j+1}^*, \ldots, x_n^*),$$

$k \in \{r+1, \ldots, j-1, j+1, \ldots, n\}$. Consider the system

$$y_1 = y_1,$$
$$\vdots$$
$$y_r = y_r,$$
$$y_{r+1} = \phi_{r+1}(f_1^*, \ldots, f_r^*, x_{r+1}^*, \ldots, x_{j-1}^*, x_{j+1}^*, \ldots, x_n^*),$$

and the neighborhood U^j of the point $(y_1^0, \ldots, y_r^0, x_j^0)$ given by

$$|y_k - y_k^0| < \delta, \quad k \in \{1, \ldots, r\}, \quad |x_j - x_j^0| < \delta,$$

i. e., consider the map

$$(y_1, \ldots, y_r, x_j) \to (y_1, \ldots, y_r, y_{r+1}).$$

This map is continuously differentiable in U^j, and its Jacobian matrix is

$$\begin{pmatrix} 1 & 0 & 0 & \cdots & 0 \\ 0 & 1 & 0 & \cdots & 0 \\ 0 & 0 & 1 & \cdots & 0 \\ \vdots & \vdots & \vdots & \vdots & \vdots \\ \dfrac{\partial y_{r+1}}{\partial y_1} & \dfrac{\partial y_{r+1}}{\partial y_2} & \dfrac{\partial y_{r+1}}{\partial y_3} & \cdots & \dfrac{\partial y_{r+1}}{\partial x_j} \end{pmatrix},$$

and thus

$$\frac{\partial(y_1, \ldots, y_r, y_{r+1})}{\partial(y_1, \ldots, y_r, x_j)} = \frac{\partial y_{r+1}}{\partial x_j}.$$

Therefore on U^j the map $(y_1, \ldots, y_r, x_j) \to (y_1, \ldots, y_r, y_{r+1})$ can be represented as a composition of two continuously differentiable maps, one of which is the map

$$x_1 = f_1(y_1, \ldots, y_r, x_{r+1}^*, \ldots, x_{j-1}^*, x_j, x_{j+1}^*, \ldots, x_n^*),$$

$$\vdots$$

$$x_r = f_r(y_1, \ldots, y_r, x_{r+1}^*, \ldots, x_{j-1}^*, x_j, x_{j+1}^*, \ldots, x_n^*),$$

$$x_j = x_j,$$

which is continuously differentiable on U^j, and the second map is

$$y_1 = \phi_1(x_1, \ldots, x_r, x_{r+1}^*, \ldots, x_{j-1}^*, x_j, x_{j+1}^*, \ldots, x_n^*),$$

$$\vdots$$

$$y_r = \phi_r(x_1, \ldots, x_r, x_{r+1}^*, \ldots, x_{j-1}^*, x_j, x_{j+1}^*, \ldots, x_n^*),$$

$$y_{r+1} = \phi_{r+1}(x_1, \ldots, x_r, x_{r+1}^*, \ldots, x_{j-1}^*, x_j, x_{j+1}^*, \ldots, x_n^*),$$

which is continuously differentiable on

$$|x_l - x_l^0| < \eta, \quad l \in \{1, \ldots, r\}, \quad |x_j - x_j^0| < \delta.$$

Therefore

$$\frac{\partial y_{r+1}}{\partial x_j} = \frac{\partial(y_1, \ldots, y_r, y_{r+1})}{\partial(y_1, \ldots, y_r, x_j)}$$

$$= \frac{\partial(y_1, \ldots, y_r, y_{r+1})}{\partial(x_1, \ldots, x_r, x_j)} \frac{\partial(x_1, \ldots, x_r, x_j)}{\partial(y_1, \ldots, y_r, x_j)}.$$

By the conditions of the theorem we get

$$\frac{\partial(x_1, \ldots, x_r, x_j)}{\partial(y_1, \ldots, y_r, x_j)} = 0$$

on X. Consequently, $(y_1, \ldots, y_r, x_j) \in U^j$, and then

$$(y_1, \ldots, y_r, x_{r+1}^*, \ldots, x_{j-1}^*, x_j, x_{j+1}^*, \ldots, x_n^*) \in U,$$

and (4.49) holds. Then

$$\phi_{r+1}(f_1, \ldots, f_r, x_{r+1}, \ldots, x_n) = \phi_{r+1}(f_1, \ldots, f_r)$$

$$= \Phi(y_1, \ldots, y_r).$$

Take $\delta > 0$ such that $\delta_0 < \delta$ and $\delta_0 < \eta$ and the inequalities

$$|x_j - x_j^0| < \delta_0, \quad j \in \{1, \dots, n\},$$

imply the inequalities

$$|y_j - y_0^*| < \delta, \quad j \in \{1, \dots, r\}.$$

Then, for all x such that

$$|x_j - x_j^0| < \delta_0, \quad j \in \{1, \dots, r\},$$

we have

$$\phi_{r+1}(x) = \Phi(\phi_1(x), \dots, \phi_r(x)),$$

and the functions $\phi_1, \dots, \phi_r, \phi_{r+1}$ are dependent. As above, we can prove that each function $\phi_{r+2}, \dots, \phi_m$ depends on the functions $\phi_1, \dots, \phi_r$ on some neighborhood of the point x^0. This completes the proof. □

4.9 The method of Lagrange multipliers

The method of Lagrange multipliers is a strategy for finding the local maxima and minima of a function subject to equation constraints, i. e., subject to the condition that one or more equations have to be satisfied exactly by the chosen values of the variables. The Lagrange multipliers method is widely used to solve extreme value problems in science, economics, and engineering. An important application of the Lagrange multipliers method in power systems is the economic dispatch, which is at the crossroads of engineering and economics. In this problem, the objective function to minimize is the generating costs, and the variables are subjected to the power balance constraint.

Let $X \subset \mathbb{R}^n$ be an open set, and let $f, g_j : X \to \mathbb{R}, f, g_j \in C^1(X), j \in \{1, \dots, m\}, m < n$. Define

$$E = \{x \in X : g_j(x) = 0, j \in \{1, \dots, m\}\}.$$

Suppose that E is a nonempty set.

Definition 4.22. If $x_0 \in E$ and there is a neighborhood U of x^0 such that

$$f(x) \le (\ge)f(x^0)$$

for all $x \in U \cap E$, then we say that x^0 is a local maximum (minimum) point of f subject to the constraints

$$g_j(x) = 0, \quad j \in \{1, \dots, m\}, \quad x \in E. \tag{4.50}$$

One of the main results in this section is as follows.

Theorem 4.20. *Let $n > m$. If $x^0 \in X$ is a local extreme point of f subject to the constraints (4.50) and*

$$\frac{\partial(g_1, \ldots, g_m)}{\partial(x_{r_1}, \ldots, x_{r_m})}(x^0) \neq 0 \tag{4.51}$$

for some choice of $r_1 < r_2 < \cdots < r_m$ in $\{1, \ldots, n\}$. Then there are constants $\lambda_1, \ldots, \lambda_m \in \mathbb{R}$ such that

$$\frac{\partial f}{\partial x_j}(x^0) - \lambda_1 \frac{\partial g_1}{\partial x_j}(x^0) - \cdots - \lambda_m \frac{\partial g_m}{\partial x_j}(x^0) = 0, \quad j \in \{1, \ldots, m\}. \tag{4.52}$$

Proof. Without loss of generality, suppose that $r_j = j, j \in \{1, \ldots, m\}$. Let

$$\tilde{x} = (x_{m+1}, \ldots, x_n),$$
$$\tilde{x}^0 = (x_{m+1}^0, \ldots, x_n^0).$$

By condition (4.51) and Theorem 4.9 it follows that there are unique determined continuously differentiable functions $h_l = h_l(\tilde{x}), l \in \{1, \ldots, m\}$, defined in a neighborhood U of $\tilde{x}^0$ such that

$$(h_1(\tilde{x}), \ldots, h_m(\tilde{x}), \tilde{x}) \in X$$

for all $\tilde{x} \in U$,

$$(h_1(\tilde{x}^0), \ldots, h_m(\tilde{x}^0), \tilde{x}^0) = x^0,$$

and

$$g_j(h_1(\tilde{x}), \ldots, h_m(\tilde{x}), \tilde{x}) = 0, \quad j \in \{1, \ldots, m\}. \tag{4.53}$$

Moreover, by (4.51) it follows that the equation

$$\begin{pmatrix} \frac{\partial g_1}{\partial x_1}(x^0) & \frac{\partial g_1}{\partial x_2}(x^0) & \cdots & \frac{\partial g_1}{\partial x_m}(x^0) \\ \frac{\partial g_2}{\partial x_1}(x^0) & \frac{\partial g_2}{\partial x_2}(x^0) & \cdots & \frac{\partial g_2}{\partial x_m}(x^0) \\ \vdots & \vdots & \vdots & \vdots \\ \frac{\partial g_m}{\partial x_1}(x^0) & \frac{\partial g_m}{\partial x_2}(x^0) & \cdots & \frac{\partial g_m}{\partial x_m}(x^0) \end{pmatrix} \begin{pmatrix} \lambda_1 \\ \lambda_2 \\ \vdots \\ \lambda_m \end{pmatrix} = \begin{pmatrix} \frac{\partial f}{\partial x_1}(x^0) \\ \frac{\partial f}{\partial x_2}(x^0) \\ \vdots \\ \frac{\partial f}{\partial x_m}(x^0) \end{pmatrix} \tag{4.54}$$

has a unique solution and (4.52) holds for all $j \in \{1, \ldots, m\}$. Let $j \in \{m + 1, \ldots, n\}$ be arbitrarily chosen. By (4.53) we get

$$\frac{\partial g_l}{\partial x_j}(x^0) + \sum_{k=1}^{m} \frac{\partial g_l}{\partial x_k}(x^0)\frac{\partial h_l}{\partial x_k}(x^0) = 0, \quad l \in \{1, \ldots, m\}. \tag{4.55}$$

If x^0 is a local extreme point of f subject to constraints (4.50), then

$$\frac{\partial f}{\partial x_j}(x^0) + \sum_{k=1}^{m} \frac{\partial f}{\partial x_k}(x^0)\frac{\partial h_k}{\partial x_j}(x^0) = 0. \qquad (4.56)$$

By equations (4.55) and (4.56) we get

$$\det \begin{pmatrix} \frac{\partial f}{\partial x_j}(x^0) & \frac{\partial f}{\partial x_1}(x^0) & \cdots & \frac{\partial f}{\partial x_m}(x^0) \\ \frac{\partial g_1}{\partial x_j}(x^0) & \frac{\partial g_1}{\partial x_1}(x^0) & \cdots & \frac{\partial g_1}{\partial x_m}(x^0) \\ \vdots & \vdots & \vdots & \vdots \\ \frac{\partial g_m}{\partial x_j}(x^0) & \frac{\partial g_m}{\partial x_1}(x^0) & \cdots & \frac{\partial g_m}{\partial x_m}(x^0) \end{pmatrix} = 0,$$

whereupon

$$\det \begin{pmatrix} \frac{\partial f}{\partial x_j}(x^0) & \frac{\partial g_1}{\partial x_j}(x^0) & \cdots & \frac{\partial g_m}{\partial x_j}(x^0) \\ \frac{\partial f}{\partial x_1}(x^0) & \frac{\partial g_1}{\partial x_1}(x^0) & \cdots & \frac{\partial g_m}{\partial x_1}(x^0) \\ \vdots & \vdots & \vdots & \vdots \\ \frac{\partial f}{\partial x_m}(x^0) & \frac{\partial g_1}{\partial x_m}(x^0) & \cdots & \frac{\partial g_m}{\partial x_m}(x^0) \end{pmatrix} = 0.$$

Therefore there are constants $c_0, c_1, \ldots, c_m \in \mathbb{R}$, not all zero, such that

$$\begin{pmatrix} \frac{\partial f}{\partial x_j}(x^0) & \frac{\partial g_1}{\partial x_j}(x^0) & \cdots & \frac{\partial g_m}{\partial x_j}(x^0) \\ \frac{\partial f}{\partial x_1}(x^0) & \frac{\partial g_1}{\partial x_1}(x^0) & \cdots & \frac{\partial g_m}{\partial x_1}(x^0) \\ \vdots & \vdots & \vdots & \vdots \\ \frac{\partial f}{\partial x_m}(x^0) & \frac{\partial g_1}{\partial x_m}(x^0) & \cdots & \frac{\partial g_m}{\partial x_m}(x^0) \end{pmatrix} \begin{pmatrix} c_0 \\ c_1 \\ \vdots \\ c_m \end{pmatrix} = \begin{pmatrix} 0 \\ 0 \\ \vdots \\ 0 \end{pmatrix}.$$

If $c_0 = 0$, then by the last equation we get

$$\begin{pmatrix} \frac{\partial g_1}{\partial x_1}(x^0) & \cdots & \frac{\partial g_m}{\partial x_1}(x^0) \\ \vdots & \vdots & \vdots \\ \frac{\partial g_1}{\partial x_m}(x^0) & \cdots & \frac{\partial g_m}{\partial x_m}(x^0) \end{pmatrix} \begin{pmatrix} c_1 \\ \vdots \\ c_m \end{pmatrix} = \begin{pmatrix} 0 \\ \vdots \\ 0 \end{pmatrix}.$$

Hence by (4.51) we obtain

$$c_1 = \cdots = c_m = 0,$$

which is a contradiction. Thus $c_0 \neq 0$. Without loss of generality, suppose that $c_0 = 1$. Therefore

$$
\begin{pmatrix}
\frac{\partial f}{\partial x_j}(x^0) & \frac{\partial g_1}{\partial x_j}(x^0) & \cdots & \frac{\partial g_m}{\partial x_j}(x^0) \\
\frac{\partial f}{\partial x_1}(x^0) & \frac{\partial g_1}{\partial x_1}(x^0) & \cdots & \frac{\partial g_m}{\partial x_1}(x^0) \\
\vdots & \vdots & \vdots & \vdots \\
\frac{\partial f}{\partial x_m}(x^0) & \frac{\partial g_1}{\partial x_m}(x^0) & \cdots & \frac{\partial g_m}{\partial x_m}(x^0)
\end{pmatrix}
\begin{pmatrix} 1 \\ c_1 \\ \vdots \\ c_m \end{pmatrix}
=
\begin{pmatrix} 0 \\ 0 \\ \vdots \\ 0 \end{pmatrix},
$$

whereupon

$$
\begin{pmatrix}
\frac{\partial g_1}{\partial x_1}(x^0) & \frac{\partial g_1}{\partial x_2}(x^0) & \cdots & \frac{\partial g_1}{\partial x_m}(x^0) \\
\frac{\partial g_2}{\partial x_1}(x^0) & \frac{\partial g_2}{\partial x_2}(x^0) & \cdots & \frac{\partial g_2}{\partial x_m}(x^0) \\
\vdots & \vdots & \vdots & \vdots \\
\frac{\partial g_m}{\partial x_1}(x^0) & \frac{\partial g_m}{\partial x_1}(x^0) & \cdots & \frac{\partial g_m}{\partial x_m}(x^0)
\end{pmatrix}
\begin{pmatrix} -c_1 \\ -c_2 \\ \vdots \\ -c_m \end{pmatrix}
=
\begin{pmatrix} \frac{\partial f}{\partial x_1}(x^0) \\ \frac{\partial f}{\partial x_2}(x^0) \\ \vdots \\ \frac{\partial f}{\partial x_m}(x^0) \end{pmatrix}.
$$

Because (4.56) has a unique solution, we get

$$
c_j = -\lambda_j, \quad j \in \{1, \ldots, m\}.
$$

Hence by (4.56) we find

$$
\begin{pmatrix}
\frac{\partial f}{\partial x_j}(x^0) & \frac{\partial g_1}{\partial x_j}(x^0) & \cdots & \frac{\partial g_m}{\partial x_j}(x^0) \\
\frac{\partial f}{\partial x_1}(x^0) & \frac{\partial g_1}{\partial x_1}(x^0) & \cdots & \frac{\partial g_m}{\partial x_1}(x^0) \\
\vdots & \vdots & \vdots & \vdots \\
\frac{\partial f}{\partial x_m}(x^0) & \frac{\partial g_1}{\partial x_m}(x^0) & \cdots & \frac{\partial g_m}{\partial x_m}(x^0)
\end{pmatrix}
\begin{pmatrix} 1 \\ -\lambda_1 \\ \vdots \\ -\lambda_m \end{pmatrix}
=
\begin{pmatrix} 0 \\ 0 \\ \vdots \\ 0 \end{pmatrix},
$$

i. e., we obtain (4.52) for $j \in \{m + 1, \ldots, n\}$. This completes the proof. □

Definition 4.23. The function

$$
F(x) = f(x) + \sum_{j=1}^{m} \lambda_j g_j(x), \quad x \in X, \tag{4.57}
$$

where $\lambda_j, j \in \{1, \ldots, m\}$, satisfy (4.52), is said to be the Lagrange[3] function and the numbers $\lambda_j, j \in \{1, \ldots, m\}$, are said to be Lagrange multipliers.

The following implementation of the last theorem is the method of the Lagrange multipliers.

---

3 Joseph-Louis Lagrange (25 January 1736–10 April 1813), also reported as Giuseppe Luigi Lagrange or Lagrangia, was an Italian mathematician, physicist, and astronomer, later naturalized French. He made significant contributions to the fields of analysis, number theory, and both classical and celestial mechanics.

1. Find the stationary points of the function

$$f - \lambda_1 g_1 - \cdots - \lambda_m g_m,$$

treating $\lambda_1, \ldots, \lambda_m$ as unspecified constants.
2. Find $\lambda_1, \ldots, \lambda_m$ such that the stationary points obtained in step 1 satisfy constraints (4.50).
3. Determine which stationary points are constraint extreme points of f.

Remark 4.3. Note that the local extreme points of f subject to the constraints

$$g_1 \equiv g_2 \equiv \cdots \equiv g_m \equiv 0$$

are the same as the local extreme points of f.

Example 4.20. We will find the extreme points of the function

$$f(x,y) = 6 - 5x - 4y, \quad (x,y) \in \mathbb{R}^2,$$

under the constraint

$$g(x,y) = x^2 - y^2 - 9 = 0, \quad (x,y) \in \mathbb{R}^2.$$

Note that the set

$$\{(x,y) \in \mathbb{R}^2 : x^2 - y^2 - 9 = 0\}$$

is not empty. For instance, the point $(3,0)$ is its element. Next,

$$g_x(x,y) = 2x,$$
$$g_y(x,y) = -2y, \quad (x,y) \in \mathbb{R}^2.$$

Take $r_1 = 1$. Then

$$g_x(x,y) \neq 0 \quad \text{for } x \neq 0.$$

Consider the function

$$F(x,y) = f(x,y) - \lambda g(x,y)$$
$$= 6 - 5x - 4y - \lambda(x^2 - y^2 - 9), \quad (x,y) \in \mathbb{R}^2.$$

Here $\lambda \in \mathbb{R}$. We have

$$F_x(x,y) = -5 - 2\lambda x,$$

$$F_y(x,y) = -4 + 2\lambda y, \quad (x,y) \in \mathbb{R}^2.$$

Consider the system

$$F_x(x,y) = 0,$$
$$F_y(x,y) = 0,$$
$$g(x,y) = 0$$

or

$$-5 - 2\lambda x = 0,$$
$$-4 + 2\lambda y = 0,$$
$$x^2 - y^2 - 9 = 0.$$

Therefore

$$x = -\frac{5}{2\lambda},$$
$$y = \frac{2}{\lambda},$$

and

$$\frac{25}{4\lambda^2} - \frac{4}{\lambda^2} - 9 = 0,$$

or

$$\frac{9}{4\lambda^2} = 9,$$

or

$$\lambda^2 = \frac{1}{4},$$

or

$$\lambda_{1,2} = \pm\frac{1}{2}.$$

Thus we get

$$(x,y,\lambda) = \left(-5, 4, \frac{1}{2}\right),$$
$$(x,y,\lambda) = \left(5, -4, -\frac{1}{2}\right).$$

Now we will find $d^2F(x,y)$, $(x,y) \in \mathbb{R}^2$. We have

$$F_{xx}(x,y) = -2\lambda,$$
$$F_{xy}(x,y) = 0,$$
$$F_{yy}(x,y) = 2\lambda, \quad (x,y) \in \mathbb{R}^2.$$

Therefore

$$d^2F(x,y) = -2\lambda(dx^2 - dy^2), \quad (x,y) \in \mathbb{R}^2.$$

1. Let $\lambda = \frac{1}{2}$. Then

$$d^2F(x,y) = -dx^2 + dy^2, \quad (x,y) \in \mathbb{R}^2.$$

At the point $(5,-4)$, we have

$$dg(5,-4) = g_x(5,-4)dx + g_y(5,-4)dy$$
$$= 10dx - 8dy$$
$$= 0,$$

whereupon

$$dx = \frac{4}{5}dy.$$

Hence

$$d^2F(5,-4) = -\frac{16}{25}dy^2 + dy^2$$
$$= \frac{9}{25}dy^2$$
$$> 0,$$

i. e., $(5,-4)$ is a constrained minimum point for f, and

$$f(5,-4) = 6 - 25 + 16$$
$$= 22 - 25$$
$$= -3.$$

2. Let $\lambda = -\frac{1}{2}$. Then

$$d^2F(x,y) = dx^2 - dy^2, \quad (x,y) \in \mathbb{R}^2.$$

At the point $(-5,4)$, we have

$$dg(-5,4) = g_x(-5,4)dx + g_y(-5,4)dy$$

$$= -10dx + 8dy$$
$$= 0,$$

whereupon

$$dx = \frac{4}{5}dy,$$

and

$$d^2F(-5,4) = \frac{16}{25}dy^2 - dy^2$$
$$= -\frac{9}{25}dy^2$$
$$< 0.$$

Thus $(-5,4)$ is a constrained maximum point, and

$$f(-5,4) = 6 + 25 - 16$$
$$= 6 + 9$$
$$= 15.$$

Example 4.21. We will find the constrained extreme points of the function

$$f(x,y,z) = xyz, \quad (x,y,z) \in \mathbb{R}^3,$$

under the constraints

$$g_1(x,y,z) = x + y - z - 3 = 0,$$
$$g_2(x,y,z) = x - y - z - 8 = 0, \quad (x,y,z) \in \mathbb{R}^3.$$

We have

$$g_{1x}(x,y,z) = 1,$$
$$g_{1y}(x,y,z) = 1,$$
$$g_{1z}(x,y,z) = -1,$$
$$g_{2x}(x,y,z) = 1,$$
$$g_{2y}(x,y,z) = -1,$$
$$g_{2z}(x,y,z) = -1, \quad (x,y,z) \in \mathbb{R}^3.$$

Take

$$r_1 = 1, \quad r_2 = 2.$$

Then

$$\frac{\partial(g_1,g_2)}{\partial(x,y)}(x,y,z) = \det \begin{pmatrix} g_{1x}(x,y,z) & g_{1y}(x,y,z) \\ g_{2x}(x,y,z) & g_{2y}(x,y,z) \end{pmatrix}$$

$$= \det \begin{pmatrix} 1 & 1 \\ 1 & -1 \end{pmatrix}$$

$$= -1 - 1$$

$$= -2$$

$$\neq 0.$$

Thus we can apply the method of Lagrange multipliers. Consider the function

$$F(x,y,z) = f(x,y,z) - \lambda_1 g_1(x,y,z) - \lambda_2 g_2(x,y,z)$$
$$= xyz - \lambda_1(x+y-z-3) - \lambda_2(x-y-z-8), \quad (x,y,z) \in \mathbb{R}^3,$$

where $\lambda_1, \lambda_2 \in \mathbb{R}$ are constants. Then

$$F_x(x,y,z) = yz - \lambda_1 - \lambda_2,$$
$$F_y(x,y,z) = xz - \lambda_1 + \lambda_2,$$
$$F_z(x,y,z) = xy + \lambda_1 + \lambda_2,$$
$$F_{xx}(x,y,z) = 0,$$
$$F_{xy}(x,y,z) = z,$$
$$F_{xz}(x,y,z) = y,$$
$$F_{yy}(x,y,z) = 0,$$
$$F_{yz}(x,y,z) = x,$$
$$F_{zz}(x,y,z) = 0, \quad (x,y,z) \in \mathbb{R}^3.$$

Hence

$$d^2F(x,y,z) = 2(zdxdy + ydxdz + xdydz), \quad (x,y,z) \in \mathbb{R}^3.$$

Under the constraints, we have

$$dx + dy - dz = 0,$$
$$dx - dy - dz = 0,$$

whereupon

$$dx = dz,$$
$$dy = 0,$$

and

$$d^2F(x,y,z) = 2ydx^2.$$

Now we consider the system

$$F_x(x,y,z) = 0,$$
$$F_y(x,y,z) = 0,$$
$$F_z(x,y,z) = 0,$$
$$g_1(x,y,z) = 0,$$
$$g_2(x,y,z) = 0,$$

or

$$yz - \lambda_1 - \lambda_2 = 0,$$
$$xz - \lambda_1 + \lambda_2 = 0,$$
$$xy + \lambda_1 + \lambda_2 = 0,$$
$$x + y - z - 3 = 0,$$
$$x - y - z - 8 = 0,$$

or

$$\lambda_1 = \frac{(x+y)z}{2},$$
$$\lambda_2 = -\frac{x(y+z)}{2},$$

and

$$0 = xy + \lambda_1 + \lambda_2$$
$$= xy + \frac{(x+y)z}{2} - \frac{x(y+z)}{2}$$
$$= \frac{xy}{2} + \frac{yz}{2}$$
$$= \frac{(x+z)y}{2}.$$

1. Let $z = -x$. Then

$$x + y + x - 3 = 0,$$
$$x - y + x - 8 = 0,$$

   or

$$y = 3 - 2x,$$
$$0 = 2x + 2x - 3 - 8,$$

or

$$x = \frac{11}{4},$$
$$y = 3 - \frac{11}{2}$$
$$= -\frac{5}{2},$$
$$z = -\frac{11}{4},$$

and

$$d^2 F\left(\frac{11}{4}, -\frac{5}{2}, -\frac{11}{4}\right) = -5dx^2$$
$$< 0.$$

Thus $(\frac{11}{4}, -\frac{5}{2}, -\frac{11}{4})$ is a local constraint maximum point of f, and

$$f\left(\frac{11}{4}, -\frac{5}{2}, -\frac{11}{4}\right) = \frac{11}{4} \cdot \left(-\frac{5}{2}\right)\left(-\frac{11}{4}\right)$$
$$= \frac{605}{32}.$$

2. Let $y = 0$. Then

$$\lambda_1 + \lambda_2 = 0,$$
$$xz - 2\lambda_1 = 0,$$
$$x - z = 3,$$
$$x - z = 8,$$

which has no solutions.

Example 4.22. We will find the local constrained extreme points of the function

$$f(x_1, \ldots, x_n) = \sum_{j=1}^{n} x_j, \quad (x_1, \ldots, x_n) \in \mathbb{R}^n,$$

under the constraint

$$g(x_1, \ldots, x_n) = \sum_{j=1}^{n} x_j^2 - 1, \quad (x_1, \ldots, x_n) \in \mathbb{R}^n.$$

Take $r_1 = 1$. Then

$$g_{x_1}(x_1, \ldots, x_n) = 2x_1$$
$$\neq 0, \quad x_1 \neq 0.$$

Therefore we can apply the method of Lagrange multipliers. Consider the function

$$F(x_1, \ldots, x_n) = \sum_{j=1}^{n} x_j - \lambda \left(\sum_{j=1}^{n} x_j^2 - 1 \right), \quad (x_1, \ldots, x_n) \in \mathbb{R}^n,$$

where $\lambda \in \mathbb{R}$ is a constant. We have

$$F_{x_l}(x_1, \ldots, x_n) = 1 - 2\lambda x_l,$$
$$F_{x_l x_m}(x_1, \ldots, x_n) = 0, \quad l \neq m, \quad l, m \in \{1, \ldots, n\},$$
$$F_{x_l x_l}(x_1, \ldots, x_n) = -2\lambda, \quad l \in \{1, \ldots, n\}.$$

Hence

$$d^2 F(x_1, \ldots, x_n) = -2\lambda (dx_1^2 + \cdots + dx_n^2), \quad (x_1, \ldots, x_n) \in \mathbb{R}^n.$$

Consider the system

$$F_{x_l}(x_1, \ldots, x_n) = 0, \quad l \in \{1, \ldots, n\},$$
$$g(x_1, \ldots, x_n) = 0,$$

or

$$1 - 2\lambda x_l = 0, \quad l \in \{1, \ldots, n\},$$
$$\sum_{j=1}^{n} x_j^2 = 1,$$

whereupon

$$x_l = \frac{1}{2\lambda}, \quad l \in \{1, \ldots, n\},$$
$$\sum_{j=1}^{n} \frac{1}{4\lambda^2} = 1,$$

and

$$\lambda^2 = \frac{n}{4},$$

or

$$\lambda_{1,2} = \pm\frac{\sqrt{n}}{2},$$

and thus

$$x_l = \pm\frac{1}{\sqrt{n}}, \quad l \in \{1, \ldots, n\}.$$

1. Let $\lambda = \frac{\sqrt{n}}{2}$. Then

$$(x, \ldots, x_n) = \left(\frac{1}{\sqrt{n}}, \ldots, \frac{1}{\sqrt{n}}\right),$$
$$d^2F(x, \ldots, x_n) = -\sqrt{n}(dx^2 + \cdots + dx_n^2)$$
$$\leq 0.$$

Thus $(\frac{1}{\sqrt{n}}, \ldots, \frac{1}{\sqrt{n}})$ is a local constrained maximum point, and

$$f\left(\frac{1}{\sqrt{n}}, \ldots, \frac{1}{\sqrt{n}}\right) = \sum_{j=1}^{n} \frac{1}{\sqrt{n}}$$
$$= \sqrt{n}.$$

2. Let $\lambda = -\frac{\sqrt{n}}{2}$. Then

$$(x, \ldots, x_n) = \left(-\frac{1}{\sqrt{n}}, \ldots, -\frac{1}{\sqrt{n}}\right),$$
$$d^2F(x, \ldots, x_n) = \sqrt{n}(dx^2 + \cdots + dx_n^2)$$
$$\geq 0.$$

So $(-\frac{1}{\sqrt{n}}, \ldots, -\frac{1}{\sqrt{n}})$ is a local constrained maximum point, and

$$f\left(-\frac{1}{\sqrt{n}}, \ldots, -\frac{1}{\sqrt{n}}\right) = -\sum_{j=1}^{n} \frac{1}{\sqrt{n}}$$
$$= -\sqrt{n}.$$

Exercise 4.11. Find the local constrained extreme points of the function f under constraints (4.50), where

1.

$$f(x, y, z) = 2x^2 + 3y^2 + 4z^2,$$
$$g(x, y, z) = x + y + z - 13, \quad (x, y, z) \in \mathbb{R}^3.$$

2.

$$f(x, y, z) = xy^2z^3,$$
$$g(x, y, z) = x + y + z - 12, \quad (x, y, z) \in \mathbb{R}^3.$$

3.
$$f(x,y,z) = x^2 y^3 z^4,$$
$$g(x,y,z) = 2x + 3y + 4z - 18, \quad (x,y,z) \in \mathbb{R}^3.$$

4.
$$f(x,y,z) = \sin x \sin y \sin z,$$
$$g(x,y,z) = x + y + z - \frac{\pi}{2}, \quad (x,y,z) \in \mathbb{R}^3, \quad x > 0, \quad y > 0, \quad z > 0.$$

5.
$$f(x,y,z) = x - 2y + 2z,$$
$$g(x,y,z) = x^2 + y^2 + z^2 - 9, \quad (x,y,z) \in \mathbb{R}^3.$$

6.
$$f(x,y,z) = xyz,$$
$$g_1(x,y,z) = xy + xz + yz,$$
$$g_2(x,y,z) = x + y + z - 5, \quad (x,y,z) \in \mathbb{R}^3.$$

7.
$$f(x,y,z) = xy + yz,$$
$$g_1(x,y,z) = x^2 + y^2 - 2,$$
$$g_2(x,y,z) = y + z - 2, \quad (x,y,z) \in \mathbb{R}^3, \quad y > 0.$$

8.
$$f(x,y,z) = x^2 + y^2 + z^2,$$
$$g_1(x,y,z) = \frac{x^2}{4} + y^2 + z^2 - 1,$$
$$g_2(x,y,z) = x + y + z, \quad (x,y,z) \in \mathbb{R}^3.$$

9.
$$f(x_1,\ldots,x_n) = \sum_{j=1}^{n} a_j x_j^2,$$
$$g(x_1,\ldots,x_n) = \sum_{j=1}^{n} x_j - 1, \quad (x,\ldots,x_n) \in \mathbb{R}^n, \quad a_j > 0, \quad j \in \{1,\ldots,n\}.$$

10.
$$f(x_1,\ldots,x_n) = \sum_{j=1}^{n} x_j^2,$$
$$g(x_1,\ldots,x_n) = \sum_{j=1}^{n} \frac{x_j}{a_j} - 1, \quad (x,\ldots,x_n) \in \mathbb{R}^n, \quad a_j > 0, \quad j \in \{1,\ldots,n\}.$$

Now consider the system

$$a_{j1}x_1 + \cdots + a_{jn}x_n = 0, \quad a_{jl} \in \mathbb{R}, \quad j \in \{1,\ldots,m\}, \quad l \in \{1,\ldots,n\}, \tag{4.58}$$

and the linear equation

$$b_1x_1 + \cdots + b_nx_n = 0, \quad b_j \in \mathbb{R}, \quad j \in \{1, \ldots, n\}. \tag{4.59}$$

We have the following result.

Theorem 4.21. *System (4.58) and equation (4.59) are equivalent if and only if equation (4.59) is a linear combination of the equations of system (4.58).*

Proof. Let $A = (a_{ij})$. Suppose that rank $A = p$. Then $p \le m$. If $p < m$, then $m - p$ equations of system (4.58) are linear combinations of the rest. We abstract these $m - p$ equations that are linear combination of the rest. Thus we get a system of p linearly independent equations, and the obtained system is equivalent to system (4.58). Hence equation (4.59) is equivalent to system (4.58) if and only if it is a linear combination of these p equations. Therefore, without loss of generality, we can suppose that $m = p$. Firstly, suppose that (4.58) and (4.58)–(4.59) are equivalent. Then the spaces of their solutions coincide and have the same dimension s. Let

$$a_j = \begin{pmatrix} a_{j1} \\ \vdots \\ a_{jn} \end{pmatrix},$$

$$b = \begin{pmatrix} b_1 \\ \vdots \\ b_n \end{pmatrix}.$$

Note that the rank of system (4.58)–(4.59) is m. Then $s = n - m$, and the vectors

$$b, \quad a_1, \quad \ldots, \quad a_m$$

are linearly dependent. Consequently, there are constants $\mu_j, j \in \{0, 1, \ldots, n\}, \mu_0 \ne 0$, such that

$$\mu_0 b + \mu_1 a_1 + \cdots + \mu_m a_m = 0.$$

Conversely, the vector b is a linear combination of the vectors $a_1, \ldots, a_n$, i. e.,

$$b = \lambda_1 a_1 + \cdots + \lambda_m a_m, \quad \lambda_j \in \mathbb{R}, \quad j \in \{1, \ldots, m\},$$

if and only if the rank of the matrix A and the rank of system (4.58)–(4.59) are equal if and only if systems (4.58) and (4.58)–(4.59) are equivalent. This completes the proof. $\square$

Corollary 4.5. *The vector b is a linear combination of the vectors $a_1, \ldots, a_m$ in the proof of Theorem 4.21 if and only if any solution of (4.58) is a solution to equation (4.59).*

4.10 Advanced practical problems

Problem 4.1. Find $f_{xx}(a, b), f_{xy}(a, b), f_{yy}(a, b)$ where

1.

$$f(x, y) = \frac{x}{x + y}, \quad (x, y) \in X,$$
$$X = \{(x, y) \in \mathbb{R}^2 : x \neq -y\},$$
$$(a, b) = (1, 0).$$

2.

$$f(x, y) = y^2(1 - e^x), \quad (x, y) \in X,$$
$$X = \mathbb{R}^2,$$
$$(a, b) = (0, 1).$$

3.

$$f(x, y) = \log(x^2 + y), \quad (x, y) \in X,$$
$$X = \{(x, y) \in \mathbb{R}^2, \, x^2 + y > 0\},$$
$$(a, b) = (0, 1).$$

4.

$$f(x, y) = y \sin\left(\frac{y}{x}\right), \quad (x, y) \in X,$$
$$X = \{(x, y) \in \mathbb{R}^2 : x \neq 0\},$$
$$(a, b) = (2, \pi).$$

5.

$$f(x, y) = \cos(xy - \cos y), \quad (x, y) \in X,$$
$$X = \mathbb{R}^2,$$
$$(a, b) = \left(0, \frac{\pi}{2}\right).$$

6.

$$f(x, y) = \arctan\left(\frac{x}{y}\right), \quad (x, y) \in X,$$
$$X = \left\{(x, y) \in \mathbb{R}^2 : \frac{x}{y} \in \left(0, \frac{\pi}{2}\right), y \neq 0\right\},$$
$$(a, b) = (1, 1).$$

7.

$$f(x, y) = \arcsin\left(\frac{x}{\sqrt{x^2 + y^2}}\right), \quad (x, y) \in X,$$
$$X = \left\{(x, y) \in \mathbb{R}^2 : (x, y) \neq (0, 0), \frac{x}{\sqrt{x^2 + y^2}} \in \left(-\frac{\pi}{2}, \frac{\pi}{2}\right)\right\},$$
$$(a, b) = (1, -1).$$

8.
$$f(x,y) = (xy)^{x+y}, \quad (x,y) \in X,$$
$$X = \{(x,y) \in \mathbb{R}^2 : x > 0, \, y > 0\},$$
$$(a,b) = (1,1).$$

Problem 4.2. Find the second-order derivatives of the following functions:
1. $f(x,y,z) = x(1+y^2z^3), (x,y,z) \in \mathbb{R}^3$.
2. $f(x,y,z) = \sin(x+y+z), (x,y,z) \in \mathbb{R}^3$.

Problem 4.3. Find $f_{xz}(a,b,c)$, $(a,b,c) \in X$, where
1.
$$f(x,y,z) = \log(x^2 + y^2 + z^2), \quad (x,y,z) \in X,$$
$$X = \mathbb{R}^3,$$
$$(a,b,c) = (1,1,1).$$

2.
$$f(x,y,z) = x^{y^z}, \quad (x,y,z) \in X,$$
$$X = \{(x,y,z) \in \mathbb{R}^3 : x > 0, \, y > 0, \, z > 0\},$$
$$(a,b,c) = (e,1,1).$$

Problem 4.4. Find $\frac{\partial^3 f}{\partial y \partial x^2}(x,y)$, $(x,y) \in X$, where
1.
$$f(x,y) = \frac{x^4 + 8xy^3}{x+2y}, \quad (x,y) \in X,$$
$$X = \{(x,y) : x \neq -2y \in \mathbb{R}^2\}.$$

2.
$$f(x,y) = \log(x+y), \quad (x,y) \in X,$$
$$X = \{(x,y) \in \mathbb{R}^2 : x+y > 0\}.$$

3.
$$f(x,y) = \sin(x + \cos y), \quad (x,y) \in X,$$
$$X = \mathbb{R}^2.$$

4.
$$f(x,y) = \cos(e^{2y} - 2x), \quad (x,y) \in X,$$
$$X = \mathbb{R}^2.$$

Problem 4.5. Find $f_{xyz}(x,y,z)$, $(x,y,z) \in X$, where
1.
$$f(x,y,z) = e^{xyz}, \quad (x,y,z) \in X,$$
$$X = \mathbb{R}^3.$$

2.

$$f(x, y, z) = \arctan \frac{x + y + z - xyz}{1 - xy - yz - xz}, \quad (x, y, z) \in X,$$

$$X = \left\{ (x, y, z) \in \mathbb{R}^3 : xy + xz + yz \neq 1, \ \frac{x + y + z - xyz}{1 - xy - xz - yz} \in \left(-\frac{\pi}{2}, \frac{\pi}{2} \right) \right\}.$$

Problem 4.6. Find $\frac{\partial^{m+k} f}{\partial x^m \partial y^k}(x, y)$, $(x, y) \in X$, where

1.

$$f(x, y) = \frac{x + y}{x - y}, \quad (x, y) \in X,$$

$$X = \{(x, y) \in \mathbb{R}^2 : x \neq y\}.$$

2.

$$f(x, y) = (x^2 + y^2) e^{x+y}, \quad (x, y) \in X,$$

$$X = \mathbb{R}^2.$$

3.

$$f(x, y) = \log(x^x y^y), \quad (x, y) \in X,$$

$$X = \{(x, y) \in \mathbb{R}^2 : x > 0, \ y > 0\}.$$

Problem 4.7. Find

$$\frac{\partial^{m+k+l} f}{\partial x^m \partial y^k \partial z^l}(x, y, z), \quad (x, y, z) \in X,$$

where

$$f(x, y, z) = xyz e^{x+y+z}, \quad (x, y, z) \in \mathbb{R}^3.$$

Problem 4.8. Find

$$f_{x_1 x_2 x_3 x_4}(x_1, x_2, x_3, x_4), \quad (x_1, x_2, x_3, x_4) \in X,$$

where

$$f(x_1, x_2, x_3, x_4) = \log \frac{1}{\sqrt{(x_1 - x_3)^2 + (x_2 - x_4)^2}}, \quad (x_1, x_2, x_3, x_4) \in X,$$

$$X = \{(x_1, x_2, x_3, x_4) \in \mathbb{R}^4 : x_1 \neq x_3, \ x_2 \neq x_4\}.$$

Problem 4.9. Find $d^2 f(a, b)$, where

1.

$$f(x, y) = 4y^2 + (\sin(x - y))^2, \quad (x, y) \in \mathbb{R}^2, \quad (a, b) = (0, 0).$$

2.

$$f(x, y) = (2x + y) \log\left(\frac{x}{y}\right), \quad (x, y) \in \mathbb{R}^2, \quad \frac{x}{y} > 0, \quad y \neq 0, \quad (a, b) = (1, 1).$$

3.
$$f(x,y) = e^{y \log x}, \quad (x,y) \in \mathbb{R}^2, \quad x > 0, \quad (a,b) = (2,1).$$

4.
$$f(x,y) = e^{xy - \pi \sin y}, \quad (x,y) \in \mathbb{R}^2, \quad (a,b) = (0,0).$$

5.
$$f(x,y) = \log \frac{2 - y - x^2}{2 + y + x^2}, \quad (x,y) \in \mathbb{R}^2, \quad \frac{2 - y - x^2}{2 + y + x^2} > 0,$$

$2 + y + x^2 \ne 0, (a,b) = (1,0).$

6.
$$f(x,y) = \arctan(x^2 - 2y), \quad (x,y) \in \mathbb{R}^2, \quad (a,b) = (1,0).$$

7.
$$f(x,y) = y \arctan \frac{x}{1 + 2y}, \quad (x,y) \in \mathbb{R}^2, \quad \frac{x}{1 + 2y} \in \left(-0\frac{\pi}{2}, \frac{\pi}{2}\right), \quad 1 + 2y \ne 0,$$

$(a,b) = (0,0).$

8.
$$f(x,y) = (x + y)^{xy}, \quad (x,y) \in \mathbb{R}^2, \quad x + y > 0, \quad (a,b) = (1,0).$$

9.
$$f(x,y) = (\sin x)^{\cos y}, \quad (x,y) \in \mathbb{R}^2, \quad x \in (0,\pi), \quad (a,b) = \left(\frac{\pi}{6}, \frac{\pi}{2}\right).$$

Problem 4.10. Find $d^2 f(x,y,z), (x,y,z) \in \mathbb{R}^3$, where

1.
$$f(x,y,z) = xy + xz + yz, \quad (x,y,z) \in \mathbb{R}^3.$$

2.
$$f(x,y,z) = \log(x + y + z), \quad (x,y,z) \in \mathbb{R}^3, \quad x + y + z > 0.$$

Problem 4.11. Find $d^2 f(a,b,c)$, where

1.
$$f(x,y,z) = x^4 + 2y^3 + 3z^2 - 2xy + 4xz + 2yz,$$

$(x,y,z) \in \mathbb{R}^3, (a,b,c) = (0,0,0).$

2.
$$f(x,y,z) = (1 + x)^{\alpha}(1 + y)^{\beta}(1 + z)^{\gamma}, \quad (x,y,z) \in \mathbb{R}^3, \quad x,y,z > -1,$$

$\alpha, \beta, \gamma \in \mathbb{R}, (a,b,c) = (0,0,0).$

3.
$$f(x,y,z) = \frac{z}{x^2 + y^2}, \quad (x,y,z) \in X,$$
$$X = \{(x,y,z) \in \mathbb{R}^3 : x^2 + y^2 \ne 0\},$$
$$(a,b,c) = (1,1,1).$$

4.

$$f(x,y,z) = \left(\frac{x}{y}\right)^{\frac{1}{z}}, \quad (x,y,z) \in X,$$

$$X = \{(x,y,z) \in \mathbb{R}^3 : x,y,z > 0\},$$

$$(a,b,c) = (1,1,1).$$

Problem 4.12. Find $d^2f(1,\ldots,1)$, where

$$f(x_1,\ldots,x_n) = \log\left(\sum_{j=1}^{n} x_j^2\right), \quad (x,\ldots,x_n) \in \mathbb{R}^n, \quad x^2 + \cdots + x_n^2 > 0.$$

Problem 4.13. Find $d^3f(a,b)$, where

1.

$$f(x,y) = e^{x^2y}, \quad (x,y) \in \mathbb{R}^2, \quad (a,b) = (0,1).$$

2.

$$f(x,y) = \sin(2x+y), \quad (x,y) \in \mathbb{R}^2, \quad (a,b) = (0,\pi).$$

3.

$$f(x,y) = x\cos y + y\sin x, \quad (x,y) \in \mathbb{R}^2, \quad (a,b) = (0,0).$$

Problem 4.14. Find $d^6f(\pi,0)$, where

$$f(x,y) = \cos x \cosh y, \quad (x,y) \in \mathbb{R}^2.$$

Problem 4.15. Find $d^4f(x,y,z)$, $(x,y,z) \in \mathbb{R}^3$, where

1.

$$f(x,y,z) = x^4 + 4x^3y + 2xy^2z - 3xyz^2, \quad (x,y,z) \in \mathbb{R}^3.$$

2.

$$f(x,y,z) = x^4 + 5x^3y - x^2y + z^3 + xz^3, \quad (x,y,z) \in \mathbb{R}^3.$$

Problem 4.16. Find the Taylor formula for the function $f = f(x,\ldots,x_n)$, $(x,\ldots,x_n) \in \mathbb{R}^n$, in a neighborhood of the point $(a_1,\ldots,a_n)$, where

1.

$$f(x,y) = x^3 - 2y^3 + 3xy, \quad (x,y) \in \mathbb{R}^2, \quad (a_1,a_2) = (1,2).$$

2.

$$f(x,y) = x^3 - 5x^2 - xy + y^2 + 10x + 5y, \quad (x,y) \in \mathbb{R}^2, \quad (a_1,a_2) = (2,-1).$$

3.

$$f(x,y) = \frac{1}{x-y}, \quad (x,y) \in \mathbb{R}^2, \quad x \neq y, \quad (a_1,a_2) = (2,1).$$

4.

$$f(x,y) = \sqrt{x+y}, \quad (x,y) \in \mathbb{R}^2, \quad x+y \geq 0, \quad (a_1,a_2) = (2,2).$$

5.
$$f(x,y) = \arctan\left(\frac{x}{y}\right), \quad (x,y) \in \mathbb{R}^2, \quad y \neq 0, \quad \frac{x}{y} \in (1,1),$$

$(a_1, a_2) = (1, 1).$

6.
$$f(x,y,z) = xyz, \quad (x,y,z) \in \mathbb{R}^3, \quad (a_1, a_2, a_3) = (1, 2, 3).$$

7.
$$f(x,y,z) = x^3 + y^3 + z^3 - 3xyz, \quad (x,y,z) \in \mathbb{R}^3, \quad (a_1, a_2, a_3) = (1, 0, 1).$$

Problem 4.17. Find the Taylor formula to $o(d^2)$ for the function $f = f(x,\ldots,x_n)$, $(x,\ldots,x_n) \in \mathbb{R}^n$, in a neighborhood of the point $(a_1,\ldots,a_n)$, where

1.
$$f(x,y) = \frac{\cos x}{\cos y}, \quad (x,y) \in \mathbb{R}^2, \quad (a_1, a_2) = (0, 0).$$

2.
$$f(x,y) = \arctan\frac{1+x}{1+y}, \quad (x,y) \in \mathbb{R}^2, \quad (a_1, a_2) = (0, 0).$$

3.
$$f(x,y) = \arctan(x^2 y - 2e^{x-1}), \quad (x,y) \in \mathbb{R}^2, \quad (a_1, a_2) = (1, 3).$$

4.
$$f(x,y) = \arcsin\left(2x - \frac{3}{2}xy\right), \quad (x,y) \in \mathbb{R}^2, \quad (a_1, a_2) = (-1, 1).$$

5.
$$f(x,y) = \cos(3\arcsin x + y^2 - 2xy), \quad (x,y) \in \mathbb{R}^2, \quad (a_1, a_2) = \left(\frac{1}{2}, 1\right).$$

6.
$$f(x,y) = \log\left(\pi - 4\arctan x + \frac{x^2}{y}\right), \quad (x,y) \in \mathbb{R}^2, \quad (a_1, a_2) = (1, 1).$$

7.
$$f(x,y,z) = \cos x \cos y \cos z - \cos(x + y + z), \quad (x,y,z) \in \mathbb{R}^3, \quad (a_1, a_2, a_3) = (0, 0, 0).$$

8.
$$f(x,y,z) = \log(xy + z^2), \quad (x,y,z) \in \mathbb{R}^3, \quad (a_1, a_2, a_3) = (0, 0, 1).$$

Problem 4.18. Find the Taylor formula to $o(d^4)$ for the function $f = f(x,\ldots,x_n)$, $(x,\ldots,x_n) \in \mathbb{R}^n$, in a neighborhood of the point $(a_1,\ldots,a_n)$, where

1.
$$f(x,y) = \frac{1}{(1-x)(1-y)}, \quad (x,y) \in \mathbb{R}^2, \quad (a_1, a_2) = (0, 0).$$

2.
$$f(x,y) = \sqrt{1 - x^2 - y^2}, \quad (x,y) \in \mathbb{R}^2, \quad (a_1, a_2) = (0, 0).$$

3.
$$f(x,y) = \cos x \cos y, \quad (x,y) \in \mathbb{R}^2, \quad (a_1, a_2) = (0, 0).$$

4.
$$f(x,y) = \frac{\sin x}{\cos y}, \quad (x,y) \in \mathbb{R}^2, \quad (a_1,a_2) = (0,0).$$

5.
$$f(x,y) = e^x \sin y, \quad (x,y) \in \mathbb{R}^2, \quad (a_1,a_2) = (0,0).$$

6.
$$f(x,y) = e^{2x} \log(1+y), \quad (x,y) \in \mathbb{R}^2, \quad (a_1,a_2) = (0,0).$$

7.
$$f(x,y) = \frac{x}{y}, \quad (x,y) \in \mathbb{R}^2, \quad (a_1,a_2) = (0,2).$$

8.
$$f(x,y) = \sin x \log y, \quad (x,y) \in \mathbb{R}^2, \quad (a_1,a_2) = (0,2).$$

Problem 4.19. Find the local extreme points of the following functions:

1.
$$f(x,y) = 3(x^2+y^2) - x^3 + 4y, \quad (x,y) \in \mathbb{R}^2.$$

2.
$$f(x,y) = 3x^2y + y^3 - 12x - 15y + 3, \quad (x,y) \in \mathbb{R}^2.$$

3.
$$f(x,y) = 2x^3 + xy^2 + 5x^2 + y^2, \quad (x,y) \in \mathbb{R}^2.$$

4.
$$f(x,y) = 3x^3 + y^3 - 3y^2 - x - 1, \quad (x,y) \in \mathbb{R}^2.$$

5.
$$f(x,y) = x^2y^2 - 2xy^2 - 6x^2y + 12xy, \quad (x,y) \in \mathbb{R}^2.$$

6.
$$f(x,y) = x^4 + y^4 - 2x^2, \quad (x,y) \in \mathbb{R}^2.$$

7.
$$f(x,y) = x^4 + y^4 - 2(x-y)^2, \quad (x,y) \in \mathbb{R}^2.$$

8.
$$f(x,y) = 2x^4 + y^4 - x^2 - 2y^2, \quad (x,y) \in \mathbb{R}^2.$$

9.
$$f(x,y) = xy^2(12 - x - y), \quad (x,y) \in \mathbb{R}^2.$$

10.
$$f(x,y) = \sin x + \cos y + \cos(x-y), \quad x,y \in \left(0, \frac{\pi}{2}\right).$$

11.
$$f(x,y,z) = \frac{256}{x} + \frac{x^2}{y} + \frac{y^2}{z} + z^2, \quad (x,y,z) \in \mathbb{R}^3.$$

12.
$$f(x,y,z) = x + \frac{y^2}{4x} + \frac{z^2}{y} + \frac{2}{z}, \quad (x,y,z) \in \mathbb{R}^3.$$

13.
$$f(x, y, z) = \sin x + \sin y + \sin z - \sin(x + y + z), \quad (x, y, z) \in \mathbb{R}^3.$$

14.
$$f(x, y, z) = (x + 7z)e^{-(x^2 + y^2 + z^2)}, \quad (x, y, z) \in \mathbb{R}^3.$$

15.
$$f(x, y, z) = 2\log x + 3\log y + 5\log z + \log(22 - x - y - z), \quad (x, y, z) \in \mathbb{R}^3.$$

Problem 4.20. Find $u_x(a, b)$ and $u_y(a, b)$, where

1.
$$u - x = y \cot(u - x), \quad (a, b) = \left(\frac{\pi}{4}, \frac{\pi}{4}\right), \quad y\left(\frac{\pi}{4}, \frac{\pi}{4}\right) = \frac{\pi}{2}.$$

2.
$$u^3 - xu + y = 0, \quad (a, b) = (3, -2), \quad y(3, -2) = 2.$$

3.
$$x^3 + 2y^3 + u^3 - 3xyu + 2y - 3 = 0, \quad (a, b) = (1, 1), \quad y(1, 1) = 1.$$

Problem 4.21. Find the local extreme points of the function f under the constraints (4.50), where

1.
$$f(x, y, z) = x - y + 2z,$$
$$g(x, y, z) = x^2 + y^2 + 2z^2 - 16, \quad (x, y, z) \in \mathbb{R}^3.$$

2.
$$f(x, y, z) = xyz,$$
$$g(x, y, z) = x^2 + y^2 + z^2 - 3, \quad (x, y, z) \in \mathbb{R}^3.$$

3.
$$f(x, y, z) = xy + 2xz + 2yz,$$
$$g(x, y, z) = xyz - 108, \quad (x, y, z) \in \mathbb{R}^3.$$

4.
$$f(x, y, z) = x^2 + y^2 + z^2,$$
$$g(x, y, z) = \frac{x^2}{a^2} + \frac{y^2}{b^2} + \frac{z^2}{c^2} - 1, \quad (x, y, z) \in \mathbb{R}^3, \quad a > 0, \quad b > 0, \quad c > 0.$$

5.
$$f(x, y, z) = x + y + z,$$
$$g(x, y, z) = \frac{a}{x} + \frac{b}{y} + \frac{c}{z} - 1, \quad (x, y, z) \in \mathbb{R}^3, \quad a > 0, \quad b > 0, \quad c > 0.$$

6.
$$f(x, y, z) = (x - 1)^2 + (y - 2)^2 + (z - 3)^2,$$
$$g_1(x, y, z) = x^2 + y^2 + z^2 - 21,$$
$$g_2(x, y, z) = 3x + 2y + z, \quad (x, y, z) \in \mathbb{R}^3.$$

7.
$$f(x, y, z) = \frac{x^2}{4} + y^2 + z^2,$$
$$g_1(x, y, z) = x^2 + y^2 + z^2 - 1,$$
$$g_2(x, y, z) = x + 2y + 3z, \quad (x, y, z) \in \mathbb{R}^3.$$

8.
$$f(x_1, \dots, x_n) = \sum_{j=1}^{n} x_j^a,$$
$$g(x_1, \dots, x_n) = \sum_{j=1}^{n} x_j - a, \quad (x, \dots, x_n) \in \mathbb{R}^n, \quad a > 0, \quad a > 0.$$

9.
$$f(x_1, \dots, x_n) = \sum_{j=1}^{n} \frac{a_j}{x_j},$$
$$g(x_1, \dots, x_n) = \sum_{j=1}^{n} b_j x_j - 1, \quad (x, \dots, x_n) \in \mathbb{R}^n, \quad a_j > 0, \quad b_j > 0, \quad x_j > 0,$$

$j \in \{1, \dots, n\}.$

10.
$$f(x_1, \dots, x_n) = \prod_{j=1}^{n} x_j^{a_j},$$
$$g(x_1, \dots, x_n) = \sum_{j=1}^{n} x_j - a, \quad (x, \dots, x_n) \in \mathbb{R}^n, \quad a > 0, \quad x_j > 0,$$

$j \in \{1, \dots, n\}.$

5 Solutions, hints, and answers to the exercises

Chapter 1

Exercise 1.1. Answer: 4.

Exercise 1.2. Answer: 1.

Exercise 1.3. Answer:
1. $(\frac{1}{4}, 3, 0)$.
2. $(0, \frac{1}{3}, e^3, e^{-2})$.
3. $(1, 1)$.
4. $(0, -\frac{1}{4}, 2, 3, 4)$.
5. $(0, \frac{1}{4\sqrt{2}}, 1, 3)$.

Exercise 1.5. *Solution.* Let $y \in B$ be arbitrarily chosen. Then

$$d(y, x^0) > \epsilon.$$

Set

$$\delta_y = d(y, x^0) - \epsilon.$$

Consider $U(y, \frac{\delta_y}{2})$. Take arbitrary $z \in U(y, \frac{\delta_y}{2})$. Then

$$d(z, y) < \frac{\delta_y}{2}$$

and

$$d(y, x^0) \le d(z, y) + d(z, x^0),$$

whereupon

$$d(z, x^0) \ge d(y, x^0) - d(z, y)$$
$$> \epsilon + \delta_y - \frac{\delta_y}{2}$$
$$= \epsilon + \frac{\delta_y}{2}$$
$$> \epsilon.$$

Therefore $z \in B$. Because $z \in U(y, \frac{\delta_y}{2})$ was arbitrarily chosen and it is an element of B, we get

$$U\left(y, \frac{\delta_y}{2}\right) \subset B.$$

https://doi.org/10.1515/9783112218082-005

Since $y \in B$ was arbitrarily chosen and there is a neighborhood of it contained in B, we conclude that B is an open set.

Exercise 1.6. *Solution.* Let $x \in B$. Then

$$d(x,0) = \sqrt{x_1^2 + x_2^2}$$
$$< \epsilon.$$

Take arbitrary $y \in B$. Set

$$\delta_y = \epsilon - d(y,0).$$

Define

$$\tilde{U}(y,\delta_y) = \{x \in \mathbb{R}^n : (x_1 - y_1)^2 + (x_2 - y_2)^2 < \delta_y^2, \quad x_j = 0, \quad j \in \{3,\ldots,n\}\}.$$

Let $z \in \tilde{U}(y,\delta_y)$ be arbitrarily chosen. Then $z_j = 0, j \in \{3,\ldots,n\}$,

$$d(z,y) = \sqrt{(z_1 - y_1)^2 + (z_2 - y_2)^2}$$
$$< \delta_y,$$

and

$$d(z,0) \le d(z,y) + d(y,0)$$
$$< \epsilon - \delta(y,0) + \delta(y,0)$$
$$= \epsilon.$$

Therefore $z \in B$. Because $z \in \tilde{U}(y,\delta_y)$ was arbitrarily chosen and it is an element of B, we obtain

$$\tilde{U}(y,\delta_y) \subset B.$$

Since $y \in B$ was arbitrarily chosen and its δ_y-neighborhood is contained in B, we conclude that B is an open set in $\mathbb{R}^n$.

Exercise 1.7. *Solution.* Let $x \in B$ be arbitrarily chosen. Set

$$\epsilon = f(x) - x^0.$$

Since f is a continuous function on $\mathbb{R}$, there is $\delta = \delta(\epsilon) > 0$ such that the inequality $|x - y| < \delta$ implies the inequality

$$|f(x) - f(y)| < \epsilon.$$

Hence, for $z \in U(x, \delta)$, we have $|x - z| < \delta$ and

$$f(z) > f(x) - \epsilon$$
$$= f(x) - f(x) + x^0$$
$$= x^0,$$

and thus $z \in B$. Because $z \in U(x, \delta)$ was arbitrarily chosen and it is an element of B, we obtain

$$U(x, \delta) \subset B.$$

Since $x \in B$ was arbitrarily chosen and its δ-neighborhood is contained in B, we conclude that B is an open set in $\mathbb{R}$.

Exercise 1.8. *Solution.* Note that

$$B \backslash C = C^c \cap B. \tag{5.1}$$

Since C is a closed set, C^c is an open set by Corollary 1.4. Now by Theorem 1.14 we get that $C^c \cap B$ is an open set. Hence by (5.1) we conclude that $B \backslash C$ is an open set.

Exercise 1.9. Answer:

$$d(A, B) = \sqrt{\frac{3}{10}},$$
$$x = \left(\frac{9}{10}, -\frac{1}{10}, \frac{19}{10}, \frac{1}{10} \right),$$
$$y = \left(1, \frac{3}{10}, \frac{8}{5}, \frac{3}{10} \right).$$

Exercise 1.10. Answer:
1. yes, yes, yes, yes.
2. yes, yes, not, not.
3. not, not, yes, not.

Exercise 1.12. Answer:
1. $(5, 2, 2, 1)$.
2. $(1, -4, 0, -1)$.
3. $(-2, -14, -2, -4)$.

Exercise 1.13. Answer:
1. 7.
2. 1.
3. -5.

Exercise 1.14. Answer: $a = 1$.

Exercise 1.15. Answer: $a = 0, b = 1$.

Exercise 1.16. Answer: $-\frac{3}{2}$.

Exercise 1.17. Answer: 0.

Chapter 2

Exercise 2.1. Answer:
1.
$$\{(x,y) \in \mathbb{R}^2 : x > 4\sqrt{y^2 + 1}\}.$$
2.
$$\{(x,y) \in \mathbb{R}^2 : x^2 + y^2 \le 1\}.$$
3.
$$\{(x,y) \in \mathbb{R}^2 : x^2 + y^2 < 1\}.$$
4.
$$\{(x,y) \in \mathbb{R}^2 : x^2 + y^2 > 1\}.$$
5.
$$\left\{(x,y) \in \mathbb{R}^2 : x < \frac{1}{4}y^2 + 2\right\}.$$

Exercise 2.2. Answer:
1.
$$\{(x,y,z) \in \mathbb{R}^3 : x + y + z < 1\}.$$
2.
$$\{(x,y,z) \in \mathbb{R}^3 : |x| + |y| + |z| \le 1\}.$$
3.
$$\{(x,y,z) \in \mathbb{R}^3 : x > 0, \quad y > 0\}.$$
4.
$$\{(x,y,z) \in \mathbb{R}^3 : x^2 + y^2 + z^2 < 1\}.$$
5.
$$\{(x,y,z) \in \mathbb{R}^3 : x^2 + z^2 \le 16\}.$$

Exercise 2.3. Answer:
1.
$$\{(x_1,\ldots,x_4) : x_1^2 + x_2^2 + x_3^2 + x_4^2 \ne 1\}.$$
2.
$$\{(x_1,\ldots,x_4) : -1 \le x_1 \le 1, \quad -2 \le x_2 \le 2, \quad -3 \le x_3 \le 3, \quad -4 \le x_4 \le 4\}.$$

3.
$$\left\{(x_1,\ldots,x_n) : \sum_{j=1}^{b} x_j^2 \le 1\right\}.$$

4. $\mathbb{R}^n$.

5.
$$\{(x_1,\ldots,x_n) : x_1 \le 1\}.$$

Exercise 2.4. Answer:
1. $[0, \frac{3}{2}]$.
2. $(-\infty, -2] \cup [2, \infty)$.
3. $[\frac{1}{4}, 1]$.

Exercise 2.5. Answer:
$$f(x,y) = \frac{x+y}{4}\sqrt{4z^2 - (x-y)^2},$$

where
$$x > 0, \quad x > 0, \quad 2z > |x-y|.$$

Exercise 2.6. Answer:
1.
$$\frac{x}{y} - xy, \quad (x,y) \in \mathbb{R}^2.$$

2.
$$x + \sqrt{y} - 1, \quad (x,y) \in \mathbb{R}^2.$$

3.
$$(\text{sign } x)\sqrt{x^2 + y^2}, \quad (x,y) \in \mathbb{R}^2.$$

4.
$$\left(1 - \frac{z}{x}\right)(1 - ze^{-y}), \quad (x,y,z) \in \mathbb{R}^3, \quad x > 0, \quad z > 0.$$

Exercise 2.7. Answer:
1.
$$y - x = c, \quad c \in \mathbb{R}, \quad (x,y) \in \mathbb{R}^2.$$

2.
$$x^2 + y^2 = \frac{1}{c}, \quad c > 0, \quad (x,y) \in \mathbb{R}^2.$$

3.
$$x^2 + y^2 = 1 - e^c, \quad c < 0, \quad (x,y) \in \mathbb{R}^2, \quad (x,y) \ne (0,0).$$

4.
$$4x^2 + 9y^2 = 36 - c^2, \quad c \in [0,36], \quad (x,y) \in \mathbb{R}^2, \quad 4x^2 + 9y^2 \le 36.$$

5.
$$x^2 - y^2 = \frac{1}{c^2}, \quad c \in \mathbb{R}, \quad c \neq 0, \quad (x,y) \in \mathbb{R}^2, \quad x \neq \pm y.$$

Exercise 2.8. Answer:

1.
$$x + 2y + 3z = \log c, \quad c > 0, \quad (x,y,z) \in \mathbb{R}^3.$$

2.
$$(x+1)^2 + y^2 + z^2 = 1 + \frac{1}{c}, \quad c > 0, \quad (x,y,z) \in \mathbb{R}^3, \quad (x+1)^2 + y^2 + z^2 \neq 1.$$

3.
$$x^2 + y^2 + \left(z - \frac{1}{c}\right)^2 = \frac{1}{c^2}, \quad c \in \mathbb{R}, \quad c \neq 0, \quad (x,y,z) \in \mathbb{R}^3, \quad (x,y,z) \neq (0,0,0).$$

4.
$$x^2 + y^2 + z^2 = \left(\frac{e^c - 1}{e^c + 1}\right)^2, \quad c > 0, \quad x^2 + y^2 + z^2 \neq 1.$$

5.
$$x^2 + y^2 = z^2 (\sin c)^2, \quad c \in \mathbb{R}, \quad (x,y,z) \in \mathbb{R}^3, \quad z \neq 0.$$

Exercise 2.9. Answer:
1. Degree of positive homogeneity 1.
2. Degree of homogeneity $\sqrt{2}$.
3. Degree of homogeneity 10.

Exercise 2.11. *Solution.* Let $(x_1,\ldots,x_n) \in X$ and $\lambda \in \mathbb{R}$ be such that $(\lambda x_1,\ldots,\lambda x_n) \in X$. Then
$$f(\lambda x_1,\ldots,\lambda x_n) = \lambda^k f(x_1,\ldots,x_n),$$
$$g(x_1,\ldots,x_n) = \lambda^k g(x_1,\ldots,x_n),$$

and
$$(f \pm g)(\lambda x_1,\ldots,\lambda x_n) = f(\lambda x_1,\ldots,\lambda x_n) \pm g(\lambda x_1,\ldots,\lambda x_n)$$
$$= \lambda^k f(x_1,\ldots,x_n) \pm \lambda^k g(x_1,\ldots,x_n)$$
$$= \lambda^k (f(x_1,\ldots,x_n) \pm g(x_1,\ldots,x_n))$$
$$= \lambda^k (f \pm g)(x_1,\ldots,x_n).$$

Thus $f \pm g : X \to Y$ is a homogeneous function of degree k.

Exercise 2.15. *Solution.* Take arbitrary $(x_1,\ldots,x_n) \in X$. Then $(\frac{x_1}{x_n},\ldots,\frac{x_{n-1}}{x_n},1) \in X$, and $x_n(\frac{x_1}{x_n},\ldots,\frac{x_{n-1}}{x_n},1) \in X$. Set
$$F\left(\frac{x_1}{x_n},\ldots,\frac{x_{n-1}}{x_n}\right) = f\left(\frac{x_1}{x_n},\ldots,\frac{x_{n-1}}{x_n},1\right).$$

Then

$$x_n^k F\left(\frac{x_1}{x_n}, \ldots, \frac{x_{n-1}}{x_n}\right) = x_n^k f\left(\frac{x_1}{x_n}, \ldots, \frac{x_{n-1}}{x_n}, 1\right)$$
$$= f(x_1, \ldots, x_n).$$

Exercise 2.16. Answer:
1. 6.
2. 1.
3. 2.
4. e.
5. -2.

Exercise 2.17. Answer:
1. $0, 0$, does not exist.
2. $-1, 1$, does not exist.
3. $0, 0, 0$.
4. $0, \infty$, does not exist.
5. $\frac{\sqrt{3}}{2}, 0$, does not exist.

Exercise 2.18. Answer: 0.

Exercise 2.19. Answer:
1. -1.
2. Such a does not exist.

Exercise 2.21. Answer:
1. No.
2. Yes.

Exercise 2.22. Answer:
1. $\omega(\delta, f, X) = \infty, \delta > 0$.
2. $\omega(\delta, f, X) = 0, \delta > 0$.
3. $\omega(\delta, f, X) = \infty, \delta > 0$.
4. $\omega(\delta, f, X) = 4\delta, \delta > 0$.

Chapter 3

Exercise 3.1. Answer:
1.

$$f_x(x,y) = \frac{2x - y}{y^2},$$

$$f_y(x,y) = \frac{xy - 2x^2}{y^3}, \quad (x,y) \in \mathbb{R}^2, \quad y \neq 0.$$

2.
$$f_x(x,y,z) = y + z,$$
$$f_y(x,y,z) = x + z,$$
$$f_z(x,y,z) = x + y, \quad (x,y,z) \in \mathbb{R}^3.$$

3.
$$f_{x_j}(x_1,\ldots,x_n) = 2x_j f(x_1,\ldots,x_n), \quad (x_1,\ldots,x_n) \in \mathbb{R}^n, \quad j \in \{1,\ldots,n\}.$$

Exercise 3.2. Answer:
$$f_x(1,1) = 1,$$
$$f_y(1,1) = -2.$$

Exercise 3.3. Answer:
1. 0.
2. 2.

Exercise 3.6. Answer: $(1,4)$.

Exercise 3.7. Answer:
1.
$$df(x,y) = (8x^3 - 6xy^2 + 3x^2y)dx + (x^3 - 6x^2y)dy, \quad (x,y) \in X.$$

2.
$$df(x,y,z) = \frac{1}{\sqrt{x^2 + y^2 + z^2}}(xdx + ydy + zdz), \quad (x,y,z) \in X.$$

3.
$$df(x_1,\ldots,x_n) = 3\sum_{j=1}^{n} x_j^2 dx_j, \quad (x_1,\ldots,x_n) \in X.$$

Exercise 3.10. Answer:
1. $(1, 2(1 + \log 2))$.
2. $(\frac{1}{4}, 0, 0)$.

Exercise 3.11. Answer:
1. -1.
2. $\frac{5}{9}$.
3. 2.

Chapter 4

Exercise 4.1. Answer:
$$f_{xx}(x,y) = y^2 e^{xy},$$

$$f_{xy}(x,y) = (1 + xy)e^{xy},$$
$$f_{yy}(x,y) = x^2 e^{xy}, \quad (x,y) \in X.$$

Exercise 4.2. Answer: $m!k!$.

Exercise 4.3. *Solution.* For the function f_{x_j}, apply Theorem 4.1, we get

$$f_{x_j x_l x_k} = f_{x_j x_k x_l}. \tag{5.2}$$

For the function f_{x_l}, applying Theorem 4.1, we get

$$f_{x_l x_j x_k} = f_{x_l x_k x_j}. \tag{5.3}$$

For the function f_{x_k}, applying Theorem 4.1, we get

$$f_{x_k x_l x_j} = f_{x_k x_j x_l}. \tag{5.4}$$

For the function f, applying Theorem 4.1, we get

$$f_{x_j x_l} = f_{x_l x_j}, \tag{5.5}$$
$$f_{x_j x_k} = f_{x_k x_j}, \tag{5.6}$$

and

$$f_{x_l x_k} = f_{x_k x_l}. \tag{5.7}$$

Now differentiating equations (5.5)–(5.7) with respect to x_k, x_l, and x_j, respectively, we get

$$f_{x_j x_l x_k} = f_{x_l x_j x_k},$$
$$f_{x_j x_k x_l} = f_{x_k x_j x_l},$$
$$f_{x_l x_k x_j} = f_{x_k x_l x_j}.$$

By the last equations and equations (5.2)–(5.4) we get equation (5.2).

Exercise 4.5. Answer:

1.
$$e^{-1}(dx^2 + dy^2).$$

2.
$$e(6dx^2 - 8dxdy + 3dy^2).$$

3.
$$-2dxdy.$$

4.
$$2(dx^2 - dy^2).$$

5.
$$-2(dx^2 - \pi dxdy).$$

Exercise 4.6. Answer:
1.
$$f(x,y) = 2(x-1)^2 - (x-1)(y+2) - (y+2)^2, \quad (x,y) \in \mathbb{R}^2.$$

2.
$$f(x,y,z) = 2 - 4(y-1) + 7(z-2) + x^2 + 3(z-2)^2 - 2(y-1)(z-2),$$

$(x,y,z) \in \mathbb{R}^3.$

Exercise 4.7. Answer:
1. Local minimum at $(7,-2)$,
$$f(7,-2) = -39.$$

2. Local maximum at $(1,0)$,
$$f(1,0) = 4.$$

3. No local extremum.
4. Local minimum at $(x, 2x+1), x \in \mathbb{R}$,
$$f(x, 2x+1) = 0.$$

5. Local minimum at $(-\frac{2}{3}, -\frac{1}{3}, -1)$,
$$f\left(-\frac{2}{3}, -\frac{1}{3}, -1\right) = -\frac{1}{3}.$$

6. Local maximum at $(-3, 2, -1)$,
$$f(-3, 2, 1) = 22.$$

7. No local extremum.
8. Local minimum at $(6, -18, 2)$,
$$f(6, -18, 2) = -112.$$

9. Local maximum at $(4, 4, 2)$,
$$f(4, 4, 2) = 128.$$

10. Local minimum at $(1,1,1)$, local maximum at $(-1,1,-1)$,

$$f(1,1,1) = 5,$$
$$f(-1,1,-1) = -3.$$

Exercise 4.8. Answer:

1.
$$u_x(1,-1) = -\frac{1}{1+u_0},$$
$$u_y(1,-1) = -\frac{1}{1+u_0},$$

where u_0 satisfies the equation

$$u_0 + \log u_0 = 0.$$

2.
$$u_x(5,4) = \frac{5u_0}{9},$$
$$u_y(5,4) = -\frac{4u_0}{9},$$

where u_0 satisfies the equation

$$\frac{u_0}{3} = 1 + \arctan \frac{u_0}{3}.$$

3.
$$u_x(0,1) = -\cos 1,$$
$$u_y(0,1) = -1.$$

Exercise 4.11. Answer:
1. Local constrained minimum at $(6,4,3)$,

$$f(6,4,3) = 156.$$

2. Local constrained maximum at $(2,4,6)$,

$$f(2,4,6) = 6912.$$

3. Local constrained maximum at $(2,2,2)$,

$$f(2,2,2) = 512.$$

4. Local constrained maximum at $(\frac{\pi}{6}, \frac{\pi}{6}, \frac{\pi}{6})$,

$$f\left(\frac{\pi}{6}, \frac{\pi}{6}, \frac{\pi}{6}\right) = \frac{1}{8}.$$

5. Local constrained minimum at $(-1, 2, -2)$ and local constrained maximum at $(1, -2, 2)$,

$$f(-1, 2, -2) = -9,$$
$$f(1, -2, 2) = 9.$$

6. Local constrained minimum at $(2, 2, 1)$, $(2, 1, 2)$, $(1, 2, 2)$ and local constrained maximum at $(\frac{4}{3}, \frac{4}{3}, \frac{7}{3})$, $(\frac{4}{3}, \frac{7}{3}, \frac{4}{3})$, $(\frac{7}{3}, \frac{4}{3}, \frac{4}{3})$,

$$f(2, 2, 1) = f(2, 1, 2)$$
$$= f(1, 2, 2)$$
$$= 4,$$
$$f\left(\frac{4}{3}, \frac{4}{3}, \frac{7}{3}\right) = f\left(\frac{4}{3}, \frac{7}{3}, \frac{4}{3}\right)$$
$$= f\left(\frac{7}{3}, \frac{4}{3}, \frac{4}{3}\right)$$
$$= \frac{112}{27}.$$

7. Local constrained maximum at $(1, 1, 1)$ and local constrained minimum at $(-1, 1, 1)$,

$$f(1, 1, 1) = 2,$$
$$f(-1, 1, 1) = 0.$$

8. Local constrained minimum at $(0, \pm\frac{1}{\sqrt{2}}, \mp\frac{1}{\sqrt{2}})$ and local constrained maximum at $(\pm\frac{2}{\sqrt{3}}, \mp\frac{1}{\sqrt{3}}, \mp\frac{1}{\sqrt{3}})$,

$$f\left(0, \pm\frac{1}{\sqrt{2}}, \mp\frac{1}{\sqrt{2}}\right) = 1,$$
$$f\left(\pm\frac{2}{\sqrt{3}}, \mp\frac{1}{\sqrt{3}}, \mp\frac{1}{\sqrt{3}}\right) = 2.$$

9. Local constrained minimum at $(\frac{A}{a_1}, \ldots, \frac{A}{a_n})$, $A = (\sum_{j=1}^{n} \frac{1}{a_j})^{-1}$,

$$f\left(\frac{A}{a_1}, \ldots, \frac{A}{a_n}\right) = A.$$

10. Local constrained minimum at $(\frac{A}{a_1}, \ldots, \frac{A}{a_n})$, $A = \sum_{j=1}^{n} \frac{1}{a_j^2}$,

$$f\left(\frac{A}{a_1}, \ldots, \frac{A}{a_n}\right) = A.$$

6 Solutions, hints, and answers to the problems

Chapter 1

Problem 1.1. Answer:
  1. $\sqrt{19}$.
  2. $\sqrt{5}$.
  3. $\sqrt{22}$.

Problem 1.2. Answer:
  1. $\pm\sqrt{\frac{5}{2}}$.
  2. $\pm\sqrt{\frac{21}{2}}$.
  3. $\pm\sqrt{\frac{77}{2}}$.

Problem 1.3. Answer:
  1. $(\frac{2}{3}, 1, 1)$.
  2. $(\frac{2}{3}, -4)$.

Problem 1.5. Answer:

$$\{(0, x_2) : x_2 \in [-1, 1]\}.$$

Problem 1.6. *Solution.* Let $n = 1$ and $B = \{\frac{1}{2^k}\}_{k\in\mathbb{N}}$. Fix arbitrary $k \in \mathbb{N}$. Set $\epsilon_k = \frac{1}{2^{k+2}}$. Consider $U(\frac{1}{2^k}, \epsilon_k)$. For any $x \in U(\frac{1}{2^k}, \epsilon_k)$, we have

$$\left| x - \frac{1}{2^k} \right| < \frac{1}{2^{k+2}},$$

or

$$\frac{1}{2^k} - \frac{1}{2^{k+2}} < x < \frac{1}{2^k} + \frac{1}{2^{k+1}},$$

or

$$\frac{3}{2^{k+2}} < x < \frac{3}{2^{k+1}}.$$

Note that

$$\frac{3}{2^{k+2}} > \frac{1}{2^{k+1}}$$

and

$$\frac{3}{2^{k+1}} < \frac{1}{2^{k-1}}.$$

https://doi.org/10.1515/9783112218082-006

Therefore $U(\frac{1}{2^k}, \epsilon_k)$ contains only $\frac{1}{2^k}$ and does not contain other points of B. Hence $\frac{1}{2^k}$ is an isolated point of B. Because $k \in \mathbb{N}$ was arbitrarily chosen, we conclude that all points of B are isolated. Moreover, we have

$$\lim_{k\to\infty} \frac{1}{2^k} = 0.$$

Thus the set of all limit points of B is not empty.

Problem 1.7. *Solution.* Let B be the set of all isolated points of A. Then, for any $x \in B$, there are $\alpha_x^j, \beta_x^j \in \mathbb{Q}, j \in \{1, \ldots, n\}$, such that

$$((\alpha_x^1, \beta_x^1) \times \ldots (\alpha_x^n, \beta_x^n)) \cap A = \{x\}.$$

Define the map $\phi : B \to \mathbb{Q}^{2n}$ as follows:

$$\phi(x) = (\alpha_x^1, \beta_x^1, \ldots, \alpha_x^n, \beta_x^n).$$

Take arbitrary $x, y \in B$ such that $x \neq y$. Assume that

$$(\alpha_x^1, \beta_x^1, \ldots, \alpha_x^n, \beta_x^n) = (\alpha_y^1, \beta_y^1, \ldots, \alpha_y^n, \beta_y^n).$$

Then

$$\alpha_x^j = \alpha_y^j,$$
$$\beta_x^j = \beta_y^j, \quad j \in \{1, \ldots, n\}.$$

Hence

$$x, y \in ((\alpha_x^1, \beta_x^1) \times \cdots \times (\alpha_x^n, \beta_x^n)) \cap A,$$

which is a contradiction. Therefore $\phi : B \to \mathbb{Q}^{2n}$ is an injection, and B is at most countable.

Problem 1.8. *Solution.* Let $n = 1$ and

$$x_m = \sum_{j=1}^m \frac{1}{j}.$$

Define $A = \{x_m\}_{m\in\mathbb{N}}$. Since the series $\sum_{j=1}^\infty \frac{1}{j}$ is divergent, the set A has no any limit points. Note that

$$\inf_{x,y\in A} = \inf_{m,n\in\mathbb{N}} \left| \sum_{j=1}^m \frac{1}{j} - \sum_{j=1}^k \frac{1}{j} \right|$$
$$= 0.$$

Now fix arbitrary $m \in \mathbb{N}$ and set

$$\epsilon_m = \frac{1}{2(m+1)}.$$

Consider $U(x_m, \epsilon_m)$. For any $x \in U(x_m, \epsilon_m)$, we have

$$|x - x_m| < \epsilon_m,$$

or

$$\left| x - \sum_{j=1}^{m} \frac{1}{j} \right| < \frac{1}{2(m+1)},$$

or

$$\sum_{j=1}^{m} \frac{1}{j} - \frac{1}{2(m+1)} < x < \sum_{j=1}^{m} \frac{1}{j} + \frac{1}{2(m+1)}.$$

Note that

$$\frac{1}{2(m+1)} < \frac{1}{m+1} < \frac{1}{m}.$$

Therefore $U(x_m, \epsilon_m)$ contains only x_m, and there are no other points of A. Thus x_m is an isolated point of A. Since $m \in \mathbb{N}$ was arbitrarily chosen, we conclude that all points of A are isolated.

Problem 1.11. *Solution.* We have that

$$B^c = \{x \in \mathbb{R} : f(x) > y_0\}.$$

By Exercise 1.10 we have that B^c is an open set. Hence by Corollary 1.4 we conclude that B is a closed set. Next,

$$C^c = \{x \in \mathbb{R} : f(x) < y_0\}.$$

By Problem 1.10 we have that C^c is an open set. Applying Corollary 1.4, we get that C is a closed set.

Problem 1.12. *Solution.* We have

$$B^c = \{x \in \mathbb{R} : f(x) < z_0\} \cup \{x \in \mathbb{R} : f(x) > y_0\}$$
$$= B_1 \cup B_2.$$

By Problem 1.10 we have that B_1 is an open set. By Exercise 1.10 we get that B_2 is an open set. By Theorem 1.13 we get that $B^c = B_1 \cup B_2$ is an open set. Hence by Corollary 1.4 we conclude that B is a closed set.

Problem 1.14. Answer:
1. Yes.
2. No if $\{\epsilon_k\}_{k\in\mathbb{N}}$ is a bounded set and yes if $\{\epsilon_k\}_{k\in\mathbb{N}}$ is an unbounded set.

Problem 1.15. Answer: No if $\{\epsilon_k\}_{k\in\mathbb{N}}$ is a bounded set and yes if $\{\epsilon_k\}_{k\in\mathbb{N}}$ is an unbounded set.

Problem 1.17. Answer: $A = [1,2]\bigcup\{3\}$.

Problem 1.18. Answer: Any open interval (a,b).

Problem 1.19. Answer:
1. $\sqrt{n-1}$.
2. $\frac{7\sqrt{2}}{8}$.
3. $\frac{4}{\sqrt{13}}$.

Problem 1.20. Answer:
1. yes, yes, not, not.
2. not, not, yes, not.
3. yes, yes, not, not.

Problem 1.21. Answer:
1. not.
2. yes.
3. yes.

Problem 1.22. Answer:
1. 6.
2. ∞.

Problem 1.24. Answer:
1. 6.
2. 6.
3. 6.

Problem 1.25. Answer: 5.

Problem 1.26. Answer: ± 3.

Problem 1.27. Answer:
1. $(-4,6,2)$.
2. $(-2,2,0)$.
3. $(-8,10,2)$.

Problem 1.28. Answer:
1. -1.
2. 5.
3. $\frac{19}{18}$.

Problem 1.29. Answer: $a = \pm 3$.

Problem 1.30. Answer: $a = 2, b = \pm 2\sqrt{2}$.

Problem 1.32. Answer: $a_{1,2} = 1 \pm \sqrt{13}$.

Problem 1.33. Answer: $\frac{2}{\sqrt{114}}$.

Chapter 2

Problem 2.1. Answer:

$$f(x,y) = \frac{y^2}{24\pi^2}\sqrt{4\pi^2 x^2 - y^2},$$

where

$$x > 0, \quad y > 0, \quad y < 2\pi x.$$

Problem 2.2. Answer:

1.
$$\left\{(x,y) \in \mathbb{R}^2 : 3 > x > 0, \ 3(1-x) < y < \frac{3}{2}(x+2)\right\}.$$

2.
$$\{(x,y) \in \mathbb{R}^2 : (x-1)^2 + 4y^2 > 4\}.$$

3.
$$\{(x,y) \in \mathbb{R}^2 : 2x < x^2 + y^2 < x, \ x < 0\} \cup \{(x,y) \in \mathbb{R}^2 : x < x^2 + y^2 < 2x, \ x > 0\}.$$

4.
$$\{(x,y) \in \mathbb{R}^2 : |x| + |y| < 1\}.$$

5.
$$\{(x,y) \in \mathbb{R}^2 : x > 0, \quad y > 0, \quad x + y < 1\}.$$

6.
$$\{(x,y) \in \mathbb{R}^2 : x^2 + y^2 < 1\}.$$

7.
$$\{(x,y) \in \mathbb{R}^2 : 1 < x^2 + y^2 < 2\}.$$

8.
$$\left\{(x,y) \in \mathbb{R}^2 : x > 3, \ 3 - 3x < y < \frac{3}{2}x + 3\right\}.$$

Problem 2.3. Answer:

1.
$$\{(x,y,z) \in \mathbb{R}^3 : z > x^2 + y^2\}.$$

2.
$$\left\{(x,y,z) \in \mathbb{R}^3 : x^2 + \frac{y^2}{2} - \frac{z^2}{3} < -1\right\}.$$

3. $\mathbb{R}^3$.

4.
$$\{(x,y,z) \in \mathbb{R}^3 : 4 < x^2 + y^2 + z^2 \le 16\}.$$

5.
$$\{(x,y,z) \in \mathbb{R}^3 : x^2 + y^2 + z^2 \ge 1\}.$$

6.
$$\{(x,y,z) \in \mathbb{R}^3 : x > 0,\ y > 1,\ z > 0,\quad x + y + z < 5\}.$$

7.
$$\left\{(x,y,z) \in \mathbb{R}^3 : x \ge 0,\ y \ge 0,\ z \ge 0,\ x + y \le 1,\ x + \tfrac{1}{3}y \ge z\right\}.$$

8.
$$\left\{(x,y,z) \in \mathbb{R}^3 : x^2 + y^2 < \tfrac{1}{2},\ x^2 + y^2 < z^2 < 1 - x^2 - y^2\right\}.$$

9.
$$\emptyset.$$

10.
$$\{(x,y,z) \in \mathbb{R}^3 : xy \le z \le 1 - \sqrt{x^2 + y^2}\}.$$

Problem 2.4. Answer:

1.
$$\{(x_1,\ldots,x_n) : |x_j| \le 1,\ j \in \{1,\ldots,n\}\}.$$

2.
$$\left\{(x_1,\ldots,x_n) : x_j \ge 0,\ j \in \{1,\ldots,n\},\ \sum_{j=1}^{n} x_j \le 1\right\}.$$

3.
$$\left\{(x_1,\ldots,x_n) : \sum_{j=1}^{n}(x_j - j)^2 \le 1\right\}.$$

4.
$$\left\{(x_1,\ldots,x_n) : \sum_{j=1}^{n} x_j^2 + \sum_{i<j, j\ne 1} x_i x_j > 0\right\}.$$

5.
$$\{(x_1,\ldots,x_n) : j - 1 \le x_j \le j + 1,\ j \in \{1,\ldots,n\}\}.$$

6.
$$\left\{(x_1,\ldots,x_n) : -\tfrac{1}{j} \le x_j \le \tfrac{1}{j},\ j \in \{1,\ldots,n\}\right\}.$$

7.
$$\{(x_1,\ldots,x_n) : x_j \ge -j,\ j \in \{1,\ldots,n\}\}.$$

8. $\mathbb{R}^n$.
9. $\mathbb{R}^n$.
10.
$$\{(x_1,\ldots,x_n) : x_j \ne 3,\ j \in \{1,\ldots,n\}\}.$$

Problem 2.5. Answer:
1. $[\log 3, \infty)$.
2. $[\frac{7}{4}, \infty)$.
3. $[-1, 9]$.
4. $[-15, 15]$.
5. $\{0, \pi\}$.
6. $[-50, 150]$.
7. $[\log(\frac{24}{5}), \log(12)]$.

Problem 2.6. Answer:
1.
$$x, \quad (x, y) \in \mathbb{R}^2.$$

2.
$$\sin\left(\frac{\pi x}{2y}\right), \quad (x, y) \in \mathbb{R}^2.$$

3.
$$x^2 + (e^x + y - 1)^2, \quad (x, y) \in \mathbb{R}^2.$$

4.
$$x + \frac{x - y}{z}, \quad (x, y, z) \in \mathbb{R}^3, \quad x > 0, \quad z > 0.$$

5.
$$\frac{2x^2 - 2y^2 + z^2}{x^2}, \quad (x, y, z) \in \mathbb{R}^3, \quad x > 0, \quad z > 0.$$

Problem 2.7. Answer:
1.
$$y = \sin x + c^2, \quad c \geq 0.$$

2.
$$y = e^c \sin x, \quad c \in \mathbb{R}, \quad (x, y) \in \mathbb{R}^2, \quad x > 0, \quad y \in [2k\pi, (2k + 1)\pi], \quad k \in \mathbb{Z}.$$

3.
$$y = x \sin c \backslash \{(0, 0)\}, \quad c \in \mathbb{R}, \quad |c| \leq \frac{\pi}{2}.$$

Problem 2.8. Answer:
1.
$$|x| + |y| + |z| = 1 - c, \quad c \in [0, 1], \quad (x, y, z) \in \mathbb{R}^3.$$

2.
$$cx + cy + (c - 1)z = c, \quad (x, y, z) \in \mathbb{R}^3, \quad c \in \mathbb{R}.$$

3.
$$x^2 + y^2 + e^c = z^2, \quad c \in \mathbb{R}, \quad (x, y, z) \in \mathbb{R}^3.$$

Problem 2.9. Answer:
1. Degree of positive homogeneity 1.

2. Degree of positive homogeneity 1.
3. Degree of homogeneity −2.

Problem 2.11. Answer:

1. 0.
2. $\frac{\sqrt{2}}{8}$.
3. 0.
4. 0.
5. 1.
6. 1.
7. Does not exist.
8. 1.
9. $\sqrt{e}$.
10. 2.
11. $\frac{13}{4}$.

Problem 2.12. Answer:

1. 0, 0, does not exist.
2. 1, 1, 1.
3. 0, does not exist, 0.
4. does not exist, does not exist, 0.
5. 0, 1, does not exist.
6. 1, ∞, does not exist.
7. 0, 0, 0.

Problem 2.14. Answer:

1. a. 0 if $\phi \in (\frac{\pi}{2}, \frac{3\pi}{2})$.
   b. 1 if $\phi = 0, \frac{\pi}{2}, \pi, \frac{3\pi}{2}$.
   c. ∞ if $\phi \in (0, \frac{\pi}{2}) \cup (\frac{3\pi}{2}, 2\pi)$.
2. a. 0 if $\phi \in (\frac{\pi}{4}, \frac{3\pi}{4}) \cup (\frac{3\pi}{4}, \frac{5\pi}{4}) \cup (\frac{5\pi}{4}, \frac{7\pi}{4})$, $\phi = 0, \phi = \pi$.
   b. does not exist in the other cases.
3. a. $\cos \phi + \sin \phi$ if $\phi \neq 0, \frac{\pi}{2}, \pi, \frac{3\pi}{2}$.
   b. does not exist if $\phi = 0, \frac{\pi}{2}, \pi, \frac{3\pi}{2}$.

Problem 2.15. Answer: 0.

Problem 2.16. Answer:

1. 0.
2. $\frac{a}{a^2+1}$.
3. Such a does not exist.

Problem 2.17. Answer: $a \in \mathbb{R}, b = 0$.

Problem 2.18. Answer: No.

Problem 2.19. Answer: $a = -1, b = 0$.

Problem 2.23. *Solution.* Since (2.14) holds, there is $\epsilon > 0$ such that

$$f(x^0) > c + \epsilon.$$

Assume the contrary, i. e., in any neighborhood $U(x^0)$ of the point x^0, there is $x \in U(x^0)$ such that

$$f(x) \le c.$$

Since f is continuous at x^0, there is $\delta > 0$ such that $x^1 \in U(x^0, \delta)$ implies that

$$|f(x^1) - f(x^0)| < \epsilon,$$

whereupon

$$f(x^0) - f(x^1) < \epsilon, \quad x^1 \in U(x^0, \delta).$$

By our assumption it follows that there is $x^2 \in U(x^0, \delta)$ such that

$$f(x^2) \le c.$$

Hence

$$\epsilon > f(x^0) - f(x^2)$$
$$> c + \epsilon - c$$
$$= \epsilon,$$

which is a contradiction.

Problem 2.25. *Solution.* Let $X \subset \mathbb{R}^n, c \in \mathbb{R}, f : X \to \mathbb{R}$, and $f \in C(X)$. Consider

$$Y = \{x \in X : f(x) = c\}.$$

Take $\{x^k\}_{k \in \mathbb{N}} \subset X$ such that

$$\lim_{k \to \infty} x^k = x^0$$

for some $x^0 \in X$. Then

$$f(x^k) = c, \quad k \in \mathbb{N}.$$

Since $f \in C(X)$, we have that

$$c = \lim_{k \to \infty} f(x^k)$$
$$= f(x^0),$$

i.e., $x^0 \in Y$. Thus Y is a closed set.

Problem 2.26. *Solution.* 1. Let $x^0 \in X$ be arbitrarily chosen. Then

$$f(x^0) < c.$$

By Problem 2.24 it follows that there is a neighborhood $U(x^0)$ of x^0 such that

$$f(x) < c, \quad x \in U(x^0).$$

Thus $U(x^0) \subset X$. Hence we conclude that x^0 is an interior point of X. Because $x^0 \in X$ was arbitrarily chosen, we conclude that X is an open set.

2. Let $\{y^k\}_{k \in \mathbb{N}} \subset Y$ be such that

$$\lim_{k \to \infty} y^k = y^0$$

for some $y^0 \in \mathbb{R}^n$. We have

$$f(y^k) \le c, \quad k \in \mathbb{N}.$$

Then using that $f \in C(\mathbb{R}^n)$, we obtain

$$c \ge \lim_{k \to \infty} f(y^k)$$
$$= f(y^0),$$

i.e., $y^0 \in Y$, and Y is a closed set.

Problem 2.27. *Solution.* Let $\{x^k\}_{k \in \mathbb{N}} \subset Y$ be such that

$$\lim_{k \to \infty} x^k = x^0$$

for some $x^0 \in \mathbb{R}^n$. Since X is a closed set in $\mathbb{R}^n$, $Y \subset X$, and $\{x^k\}_{k \in \mathbb{N}} \subset Y$, we conclude that $x^0 \in X$. Note that

$$f(x^k) \ge c, \quad k \in \mathbb{N}.$$

Now using that $f \in C(X)$, we get

$$c \le \lim_{k \to \infty} f(x^k)$$
$$= f(x^0),$$

whereupon $x^0 \in Y$, and Y is a closed set.

Problem 2.28. *Solution.* Let $(x^0, y^0) \in X$ be arbitrarily chosen. Take arbitrary $\epsilon > 0$. Since $f(x, y)$ is continuous in x, there is $\delta_1 > 0$ such that

$$\left| f(x, y^0) - f(x^0, y^0) \right| < \frac{\epsilon}{2}. \tag{6.1}$$

Since X is a domain, we have that $(x, y^0) \in X$. Now using that $f(x, y)$ is continuous in y uniformly with respect to x, we find $\delta_2 > 0$ such that the inequality

$$\left| y - y^0 \right| < \delta_2$$

implies the inequality

$$\left| f(x, y) - f(x, y^0) \right| < \frac{\epsilon}{2} \tag{6.2}$$

for all x that satisfy (6.1). Let

$$\delta = \min\{\delta_1, \delta_2\}.$$

Then, for $(x, y) \in X$ for which

$$d((x, y), (x^0, y^0)) < \delta,$$

we have

$$\left| x - x^0 \right| < \delta$$
$$\leq \delta_1,$$
$$\left| y - y^0 \right| < \delta$$
$$\leq \delta_2,$$

and hence, using (6.1) and (6.2), we arrive at

$$\begin{aligned} \left| f(x, y) - f(x^0, y^0) \right| &= \left| f(x, y) - f(x, y^0) + f(x, y^0) - f(x^0, y^0) \right| \\ &\leq \left| f(x, y) - f(x, y^0) \right| + \left| f(x, y^0) - f(x^0, y^0) \right| \\ &< \frac{\epsilon}{2} + \frac{\epsilon}{2} \\ &= \epsilon. \end{aligned}$$

Thus f is continuous at (x^0, y^0). Since (x^0, y^0) was arbitrarily chosen, we conclude that $f \in C(X)$.

Problem 2.29. *Solution.* Take $(x^0, y^0) \in X$ and arbitrary $\epsilon > 0$. Since $f(x, y)$ is continuous in x, there is $\delta_1 > 0$ such that the inequality

$$\left| x - x^0 \right| < \delta_1 \tag{6.3}$$

implies the inequality

$$|f(x,y^0) - f(x^0,y^0)| < \frac{\epsilon}{2}. \tag{6.4}$$

Because X is a domain, we have that $(x,y^0) \in X$. Next, for any y such that

$$|y - y^0| < \frac{\epsilon}{2L},$$

we have

$$|f(x,y) - f(x,y^0)| \le L|y - y^0|$$
$$< L\frac{\epsilon}{2L} \tag{6.5}$$
$$= \frac{\epsilon}{2}$$

for all x that satisfy (6.3). Since X is a domain, we have that $(x,y) \in X$. Let

$$\delta = \min\left\{\delta_1, \frac{\epsilon}{2L}\right\}.$$

Then, for $(x,y) \in X$ for which

$$d((x,y), (x^0,y^0)) < \delta,$$

we have

$$|x - x^0| < \delta$$
$$\le \delta_1,$$
$$|y - y^0| < \delta$$
$$\le \frac{\epsilon}{2L},$$

and hence, using (6.4) and (6.5), we arrive at

$$|f(x,y) - f(x^0,y^0)| = |f(x,y) - f(x,y^0) + f(x,y^0) - f(x^0,y^0)|$$
$$\le |f(x,y) - f(x,y^0)| + |f(x,y^0) - f(x^0,y^0)|$$
$$< \frac{\epsilon}{2} + \frac{\epsilon}{2}$$
$$= \epsilon.$$

Thus f is continuous at (x^0,y^0). Since (x^0,y^0) was arbitrarily chosen, we conclude that $f \in C(X)$.

Chapter 3

Problem 3.1. Answer:

1.

$$f_x(x,y) = \cos x - 2xy,$$
$$f_y(x,y) = -x^2, \quad (x,y) \in \mathbb{R}^2.$$

2.

$$f_x(x,y) = \frac{1}{y}\cos\frac{x}{y}\cos\frac{y}{x} + \frac{y}{x^2}\sin\frac{x}{y}\sin\frac{y}{x},$$
$$f_y(x,y) = -\frac{x}{y^2}\cos\frac{x}{y}\cos\frac{y}{x} - \frac{1}{x}\sin\frac{x}{y}\sin\frac{y}{x},$$

$(x,y) \in \mathbb{R}^2, \ x \neq 0, y \neq 0.$

3.

$$f_x(x,y) = e^x(x\sin y + \sin y + \cos x),$$
$$f_y(x,y) = e^x(x\cos y - \sin y), \quad (x,y) \in \mathbb{R}^2.$$

4.

$$f_x(x,y) = -\frac{2}{\sqrt{x^2+y^2}},$$

$$f_y(x,y) = \frac{2x}{y\sqrt{x^2+y^2}}, \quad (x,y) \in \mathbb{R}^2, \quad x \neq 0, \quad y \neq 0.$$

5.

$$f_x(x,y) = \frac{xy^2\sqrt{2x^2-2y^2}}{|x|(x^4-y^4)},$$

$$f_y(x,y) = \frac{yx^2\sqrt{2x^2-2y^2}}{|y|(y^4-x^4)}, \quad (x,y) \in \mathbb{R}^2, \quad x \neq 0, \quad y \neq 0, \quad x^2 \neq y^2.$$

6.

$$f_x(x,y) = \sin(2x)\log y(1+(\sin x)^2)^{\log y - 1},$$

$$f_y(x,y) = \frac{1}{y}(1+(\sin x)^2)^{\log y}\log(1+(\sin x)^2),$$

$(x,y) \in \mathbb{R}^2, y > 0.$

7.

$$f_x(x,y,z) = -\frac{x}{(\sqrt{x^2+y^2+z^2})^3},$$

$$f_y(x,y,z) = -\frac{y}{(\sqrt{x^2+y^2+z^2})^3},$$

$$f_z(x,y,z) = -\frac{z}{(\sqrt{x^2+y^2+z^2})^3}, \quad (x,y,z) \in \mathbb{R}^3, \quad (x,y,z) \neq (0,0,0).$$

8.
$$f_x(x,y,z) = \frac{1}{z} - \frac{z}{x^2},$$
$$f_y(x,y,z) = 0,$$
$$f_z(x,y,z) = \frac{1}{x} - \frac{x}{z^2}, \quad (x,y,z) \in \mathbb{R}^3, \quad x \neq 0, \quad z \neq 0.$$

9.
$$f_x(x,y,z) = 0,$$
$$f_y(x,y,z) = \frac{1}{z},$$
$$f_z(x,y,z) = -\frac{y}{z^2}, \quad (x,y,z) \in \mathbb{R}^3, \quad z \neq 0.$$

10.
$$f_x(x,y,z) = yz^{xy} \log z,$$
$$f_y(x,y,z) = xz^{xy} \log z,$$
$$f_z(x,y,z) = xyz^{xy-1}, \quad (x,y,z) \in \mathbb{R}^3, \quad z > 0.$$

11.
$$f_x(x,y,z) = \frac{z}{x}\left(\frac{x}{y}\right)^z,$$
$$f_y(x,y,z) = -\frac{z}{y}\left(\frac{x}{y}\right)^z,$$
$$f_z(x,y,z) = \log\frac{x}{y}\left(\frac{x}{y}\right)^z, \quad (x,y,z) \in \mathbb{R}^3, \quad x > 0, \quad y > 0.$$

12.
$$f_{x_j}(x_1,\ldots,x_n) = -\sin(2x_j), \quad j \in \{1,\ldots,n\}, \quad (x_1,\ldots,x_n) \in \mathbb{R}^n.$$

Problem 3.2. Answer:
$$f_x(1,2) = \frac{1}{3},$$
$$f_y(1,2) = -\frac{1}{6}.$$

Problem 3.3. Answer:
$$f_x(1,1) = 1 - \pi,$$
$$f_y(1,1) = 1 - \pi.$$

Problem 3.4. Answer:
1. 0.
2. 1.

Problem 3.5. Answer: $\frac{3}{2}$.

Problem 3.6. Answer:
1. Yes.
2. Yes.
3. Yes.
4. Yes.
5. No.
6. No.
7. No.
8. No.
9. No.
10. No.
11. Yes.
12. Yes.
13. Yes.
14. Yes.
15. Yes.
16. No.
17. No.
18. No.
19. No.
20. No.
21. Yes.
22. Yes.
23. Yes.
24. No.
25. Yes.

Problem 3.7. Answer:
1. $[0, \frac{5}{2}]$.
2. $[\frac{1}{2}, \infty)$.

Problem 3.8. Answer:
1.
$$df(x,y) = \frac{x^2 - y^2}{xy}\left(\frac{1}{x}dx - \frac{1}{y}dy\right), \quad (x,y) \in X.$$

2.
$$df(x,y) = y(x^2 + y^2)^{-\frac{3}{2}}(ydx - xdy), \quad (x,y) \in X.$$

3.
$$df(x,y) = 2^{-\frac{y}{x}}\frac{\log 2}{x^2}(ydx - xdy), \quad (x,y) \in X.$$

4.
$$df(x,y) = \frac{1}{\sqrt{x^2 + y^2}}\left(dx + \frac{y}{x + \sqrt{x^2 + y^2}}dy\right), \quad (x,y) \in X.$$

5.
$$df(x,y) = \frac{1}{\sqrt{y}} \cot\left(\frac{x+1}{\sqrt{y}}\right)\left(dx - \frac{x+1}{2y}dy\right), \quad (x,y) \in X.$$

6.
$$df(x,y) = 0, \quad (x,y) \in X.$$

7.
$$df(x,y) = \frac{1}{x^2 + y^2}(xdx - ydy), \quad (x,y) \in X.$$

8.
$$df(x,y,z) = e^{xy \sin z}(y \sin z dx + x \sin z dy + xy \cos z dz), \quad (x,y,z) \in X.$$

9.
$$df(x,y,z) = (xy)^{z-1}(yzdx + xzdy + xy \log(xy)dz), \quad (x,y,z) \in X.$$

10.
$$df(x_1,\dots,x_n) = -2\sin\left(\sum_{j=1}^{n} x_j^2\right)\left(\sum_{j=1}^{n} x_j dx_j\right), \quad (x_1,\dots,x_n) \in X.$$

Problem 3.10. Answer:
1. $(2,0,0)$.
2. $(-\frac{4}{3},0,0)$.

Problem 3.11. Answer:
1. -18.
2. $\frac{52}{5}$.
3. $\frac{1}{5}$.
4. 0.

Problem 3.12. Answer:
1. $\sqrt{290}$.
2. $\frac{\sqrt{29}}{2}$.
3. $\frac{7}{6}$.
4. $\sqrt{\frac{137}{8}}$.

Problem 3.13. Answer:
1. $(-\frac{4}{\sqrt{41}}, \frac{5}{\sqrt{41}})$.
2. $(\frac{1}{\sqrt{2}}, -\frac{1}{\sqrt{2}})$.
3. $(\frac{2}{\sqrt{23}}, \frac{4}{\sqrt{23}}, -\frac{\sqrt{3}}{\sqrt{23}})$.
4. $(\frac{1}{\sqrt{37}}, 0, -\frac{6}{\sqrt{37}})$.

Chapter 4

Problem 4.1. Answer:

1.

$$f_{xx}(a, b) = 0,$$
$$f_{xy}(a, b) = 1,$$
$$f_{yy}(a, b) = 2.$$

2.

$$f_{xx}(a, b) = -1,$$
$$f_{xy}(a, b) = -2,$$
$$f_{yy}(a, b) = 0.$$

3.

$$f_{xx}(a, b) = 2,$$
$$f_{xy}(a, b) = 0,$$
$$f_{yy}(a, b) = -1.$$

4.

$$f_{xx}(a, b) = -\frac{\pi^3}{16},$$
$$f_{xy}(a, b) = \frac{\pi^2}{8},$$
$$f_{yy}(a, b) = -\frac{\pi}{4}.$$

5.

$$f_{xx}(a, b) = -\frac{\pi^2}{4},$$
$$f_{xy}(a, b) = -\frac{\pi}{2},$$
$$f_{yy}(a, b) = -1.$$

6.

$$f_{xx}(a, b) = -\frac{1}{2},$$
$$f_{xy}(a, b) = 0,$$
$$f_{yy}(a, b) = \frac{1}{2}.$$

7.

$$f_{xx}(a, b) = \frac{1}{2},$$
$$f_{xy}(a, b) = 0,$$
$$f_{yy}(a, b) = \frac{1}{2}.$$

8.

$$f_{xx}(a, b) = 4,$$
$$f_{xy}(a, b) = 6,$$
$$f_{yy}(a, b) = 4.$$

Problem 4.2. Answer:

1.

$$f_{xx}(x, y, z) = 0,$$
$$f_{xy}(x, y, z) = 2yz^3,$$
$$f_{xz}(x, y, z) = 3y^2z^2,$$
$$f_{yy}(x, y, z) = 2xz^3,$$
$$f_{yz}(x, y, z) = 6xyz^2,$$
$$f_{zz}(x, y, z) = 6xy^2z, \quad (x, y, z) \in \mathbb{R}^3.$$

2. $-\sin(x + y + z), (x, y, z) \in \mathbb{R}^3.$

Problem 4.3. Answer:
1. $-\frac{4}{9}.$
2. $0.$

Problem 4.4. Answer:
1. $-4.$
2. $2(x + y)^{-3}, (x, y) \in X.$
3. $\sin y \cos(x + \cos y), (x, y) \in X.$
4. $8e^{2y} \sin(e^{2y} - 2x), (x, y) \in X.$

Problem 4.5. Answer:
1. $(1 + 3xyz + x^2y^2z^2)e^{xyz}, (x, y, z) \in X.$
2. $0.$

Problem 4.6. Answer:
1.

$$\frac{2(-1)^m(m + k - 1)!(kx + my)}{(x - y)^{m+k-1}}, \quad (x, y) \in X.$$

2.

$$(x^2 + y^2 + 2mx + 2ky + m^2 - m + k^2 - k)e^{x+y}, \quad (x, y) \in X.$$

3. $0.$

Problem 4.7. Answer:

$$(x + m)(y + k)(z + l)e^{x+y+z}, \quad (x, y, z) \in \mathbb{R}^3.$$

Problem 4.8. Answer:

$$\frac{48(x_1 - x_3)^2(x_2 - x_4)^2}{((x_1 - x_3)^2 + (x_2 - x_4)^2)^{\frac{3}{2}}} - \frac{6}{((x_1 - x_3)^2 + (x_2 - x_4)^2)^2}, \quad (x_1, x_2, x_3, x_4) \in X.$$

Problem 4.9. Answer:

1.
$$2(dx^2 - 2dxdy + 5dy^2).$$

2.
$$(dx - dy)^2.$$

3.
$$2(1 + \log 2)dxdy + 2(\log 2)^2 dy^2.$$

4.
$$2dxdy + \pi^2 dy^2.$$

5.
$$-\frac{8}{9}(7dx^2 + 4dxdy + dy^2).$$

6.
$$-dx^2 + 4dxdy - 2dy^2.$$

7.
$$2dxdy.$$

8.
$$2(dxdy + dy^2).$$

9.
$$-2\sqrt{3}dxdy + (\log 2)^2 dy^2.$$

Problem 4.10. Answer:

1.
$$2(dxdy + dxdz + dydz), \quad (x, y, z) \in \mathbb{R}^3.$$

2.
$$-\left(\frac{dx + dy + dz}{x + y + z}\right)^2, \quad (x, y, z) \in \mathbb{R}^3, \quad x + y + z > 0.$$

Problem 4.11. Answer:

1.
$$6dz^2 - 4dxdy + 8dxdz + 4dydz.$$

2.
$$\alpha(\alpha - 1)dx^2 + \beta(\beta - 1)dy^2 + \gamma(\gamma - 1)dz^2 + 2\alpha\beta dxdy$$
$$+ 2\beta\gamma dydz + 2\alpha\gamma dxdz.$$

3.
$$\frac{1}{2}dx^2 + \frac{1}{2}dy^2 + 2dxdy - dxdz - dydz.$$

4.
$$2(dy^2 - dxdy + dydz - dxdz).$$

Problem 4.12. Answer:
$$\frac{2}{n^2}\left((n-2)\sum_{j=1}^{n} dx_j^2 - 4 \sum_{i,j=1,i<j}^{n} dx_i dx_j\right).$$

Problem 4.13. Answer:
1.
$$6dx^2 dy.$$

2.
$$(2dx + dy)^3.$$

3.
$$-3dxdy^2.$$

Problem 4.14. Answer:
$$dx^6 - 15dx^4 dy^2 + 15dx^2 dy^4 - dy^6.$$

Problem 4.15. Answer:
1.
$$24(dx^4 + 4dx^3 dy + 2dxdy^2 dz - 3dxdydz^2), \quad (x,y,z) \in \mathbb{R}^3.$$

2.
$$24(dx^4 + 5dx^3 dy + dxdy^3), \quad (x,y,z) \in \mathbb{R}^3.$$

Problem 4.16. Answer:
1.
$$f(x,y) = -9 + 9(x-1) - 21(y-2) + 3(x-1)^2 + 3(x-1)(y-2)$$
$$- 12(y-2)^2 + (x-1)^3 - 2(y-2)^3, \quad (x,y) \in \mathbb{R}^3.$$

2.
$$f(x,y) = 6 + 3(x-2) + (y+1) + (x-2)^2 - (x-2)(y+1)$$
$$+ (y+1)^2 + (x-2)^3, \quad (x,y) \in \mathbb{R}^2.$$

3.
$$f(x,y) = 1 - (x-2) + (y-1) + (x-2)^2 - 2(x-2)(y-1)$$
$$+ (y-1)^2, \quad (x,y) \in \mathbb{R}^2.$$

4.
$$f(x,y) = 2 + \frac{1}{4}(x-2) + \frac{1}{4}(y-2) - \frac{1}{64}(x-2)^2$$
$$- \frac{1}{32}(x-2)(y-2) - \frac{1}{64}(y-2)^2, \quad (x,y) \in \mathbb{R}^2.$$

5.
$$f(x,y) = \frac{\pi}{4} + \frac{1}{2}(x-1) - \frac{1}{2}(y-1) - \frac{1}{4}(x-1)^2 + \frac{1}{4}(y-1)^2, \quad (x,y) \in \mathbb{R}^2.$$

6.

$$f(x,y,z) = 6 + 6(x-1) + 3(y-2) + 2(z-3) + 3(x-1)(y-2)$$
$$+ 2(x-1)(z-3) + (y-2)(z-3) + (x-1)(y-2)(z-3), \quad (x,y,z) \in \mathbb{R}^3.$$

7.

$$f(x,y,z) = 2 + 3(x-1) - 3y + 3(z-1) + 3(x-1)^2 + 3(z-1)^2$$
$$- 3(x-1)y - 3y(z-1) + (x-1)^3 + y^3$$
$$+ (z-1)^3 - 3(x-1)y(z-1), \quad (x,y,z) \in \mathbb{R}^3.$$

Problem 4.17. Answer:

1.
$$f(x,y) = 1 - \frac{1}{2}(x^2 - y^2) + o(d^2), \quad (x,y) \in \mathbb{R}^2.$$

2.
$$f(x,y) = \frac{\pi}{4} + \frac{1}{2}(x-y) - \frac{1}{4}(x^2 - y^2) + o(d^2), \quad (x,y) \in \mathbb{R}^2.$$

3.
$$f(x,y) = \frac{\pi}{4} + 2(x-1) + \frac{1}{3}(y-3) - 3(x-1)^2 - \frac{1}{4}(y-3)^2$$
$$- (x-1)(y-3) + o(d^2), \quad (x,y) \in \mathbb{R}^2.$$

4.
$$f(x,y) = -\frac{\pi}{6} + \frac{x+1}{\sqrt{3}} + \sqrt{3}(y-1) - \frac{(x+1)^2}{6\sqrt{3}}$$
$$- \frac{\sqrt{3}}{2}(y-1)^2 - \frac{4}{\sqrt{3}}(x+1)(y-1) + o(d^2), \quad (x,y) \in \mathbb{R}^2.$$

5.
$$f(x,y) = (2 - 2\sqrt{3})\left(x - \frac{1}{2}\right) - (y-1) - \frac{2}{\sqrt{3}}(x-1)^2 - (y-1)^2$$
$$+ 2\left(x - \frac{1}{2}\right)(y-1) + o(d^2), \quad (x,y) \in \mathbb{R}^2.$$

6.
$$f(x,y) = -(y-1) + 2(x-1)^2 + \frac{1}{2}(y-1)^2 - 2(x-1)(y-1) + o(d^2), \quad (x,y) \in \mathbb{R}^2.$$

7.
$$f(x,y,z) = xy + xz + yz + o(d^2), \quad (x,y,z) \in \mathbb{R}^3.$$

8.
$$f(x,y,z) = 2(z-1) - (z-1)^2 + xy + o(d^2), \quad (x,y,z) \in \mathbb{R}^3.$$

Problem 4.18. Answer:

1.
$$f(x,y) = 1 + x + y + x^2 + xy + y^2 + x^3 + x^2y + xy^2 + y^3$$
$$+ x^4 + x^3y + x^2y^2 + xy^3 + y^4 + o(d^4), \quad (x,y) \in \mathbb{R}^2.$$

2.
$$f(x,y) = 1 - \frac{1}{2}(x^2 + y^2) - \frac{1}{8}(x^4 + 2x^2y^2 + y^4) + o(d^4), \quad (x,y) \in \mathbb{R}^2.$$

3.
$$f(x,y) = 1 - \frac{1}{2}(x^2 + y^2) + \frac{1}{24}(x^4 + 6x^2y^2 + y^4) = o(d^4), \quad (x,y) \in \mathbb{R}^2.$$

4.
$$f(x,y) = x - \frac{x^3}{6} + \frac{xy^2}{2} + o(d^4), \quad (x,y) \in \mathbb{R}^2.$$

5.
$$f(x,y) = y + xy + \frac{x^2y}{2} - \frac{y^3}{6} - \frac{xy^3}{6} + \frac{yx^3}{6} + o(d^4), \quad (x,y) \in \mathbb{R}^2.$$

6.
$$f(x,y) = y + 2xy - \frac{y^2}{2} - xy^2 + 2x^2y + \frac{y^3}{3} + \frac{4}{3}x^3y$$
$$- x^2y^2 + \frac{2}{3}xy^3 - \frac{y^4}{4} + o(d^4), \quad (x,y) \in \mathbb{R}^2.$$

7.
$$f(x,y) = \frac{x}{2} - \frac{x(y-2)}{4} + \frac{x(y-2)^2}{8} - \frac{x(y-2)^3}{16} + o(d^4), \quad (x,y) \in \mathbb{R}^2.$$

8.
$$f(x,y) = x\log 2 + \frac{x(y-2)}{2} - \frac{\log 2}{6}x^3 - \frac{x(y-2)^2}{8}$$
$$- \frac{x^3(y-2)}{12} + \frac{x(y-2)^3}{24} + o(d^4), \quad (x,y) \in \mathbb{R}^2.$$

Problem 4.19. Answer:
1. Local minimum at $(0, -\frac{2}{3})$,
$$f\left(0, -\frac{2}{3}\right) = -\frac{4}{3}.$$

2. Local minimum at $(1,2)$ and local maximum at $(-1,-2)$,
$$f(1,2) = -25,$$
$$f(-1,-2) = 31.$$

3. Local minimum at $(0,0)$ and local maximum at $(-\frac{5}{3},0)$,
$$f(0,0) = 0,$$
$$f\left(-\frac{5}{3},0\right) = \frac{125}{27}.$$

4. Local minimum at $(\frac{1}{3},2)$ and local maximum at $(-\frac{1}{3},0)$,
$$f\left(\frac{1}{3},2\right) = -\frac{47}{9},$$
$$f\left(-\frac{1}{3},0\right) = -\frac{7}{9}.$$

5. Local maximum at $(1, 3)$,

$$f(1, 3) = 9.$$

6. Local minimum at $(\pm 1, 0)$,

$$f(\pm 1, 0) = -1.$$

7. Local minimum at $(\sqrt{2}, -\sqrt{2})$,

$$f(\sqrt{2}, -\sqrt{2}) = f(-\sqrt{2}, \sqrt{2}) = -8.$$

8. Local maximum at $(0, 0)$ and local minimum at $(\pm\frac{1}{2}, \pm 1)$,

$$f(0, 0) = 0,$$
$$f\left(\pm\frac{1}{2}, \pm 1\right) = -\frac{9}{8}.$$

9. Local maximum at $(3, 6)$,

$$f(3, 6) = 324.$$

10. Local maximum at $(\frac{\pi}{3}, \frac{\pi}{6})$,

$$f\left(\frac{\pi}{3}, \frac{\pi}{6}\right) = \frac{3\sqrt{3}}{2}.$$

11. Local minimum at $(8, 4, 2)$,

$$f(8, 4, 2) = 60.$$

12. Local maximum at $(-\frac{1}{2}, -1, -1)$ and local minimum at $(\frac{1}{2}, 1, 1)$,

$$f\left(\frac{1}{2}, 1, 1\right) = 4,$$
$$f\left(-\frac{1}{2}, -1, -1\right) = -4.$$

13. Local maximum at $(\frac{\pi}{2}, \frac{\pi}{2}, \frac{\pi}{2})$,

$$f\left(\frac{\pi}{2}, \frac{\pi}{2}, \frac{\pi}{2}\right) = 4.$$

14. Local maximum at $(\frac{1}{10}, 0, \frac{7}{10})$ and local minimum at $(-\frac{1}{10}, 0, -\frac{7}{10})$,

$$f\left(\frac{1}{10}, 0, \frac{7}{10}\right) = \frac{5}{\sqrt{e}},$$

$$f\left(-\frac{1}{10}, 0, -\frac{7}{10}\right) = -\frac{5}{\sqrt{e}}.$$

15. Local maximum at $(4, 6, 10)$,

$$f(4, 6, 10) = 13\log 2 + 3\log 3 + 5\log 5.$$

Problem 4.20. Answer:

1.

$$u_x\left(\frac{\pi}{4}, \frac{\pi}{4}\right) = 1,$$

$$u_y\left(\frac{\pi}{4}, \frac{\pi}{4}\right) = \frac{2}{2 + \pi}.$$

2.

$$u_x(3, -2) = \frac{2}{9},$$

$$u_y(3, -2) = -\frac{1}{9}.$$

3. $u_x(1, 1)$ and $u_y(1, 1)$ do not exist.

Problem 4.21. Answer:

1. Local constrained minimum at $(-2, 2, -2)$ and local constrained maximum at $(2, -2, 2)$,

$$f(-2, 2, -2) = -8,$$
$$f(2, -2, 2) = 8.$$

2. Local constrained minimum at $(1, 1, -1)$, $(1, -1, 1)$, $(-1, 1, 1)$, $(-1, -1, -1)$ and local constrained maximum at $(1, 1, 1)$, $(1, -1, -1)$, $(-1, -1, 1)$, $(-1, 1, -1)$,

$$
\begin{aligned}
f(1, 1, -1) &= f(1, -1, 1) \\
&= f(-1, 1, 1) \\
&= f(-1, -1, -1) \\
&= -1, \\
f(1, 1, 1) &= f(1, -1, -1) \\
&= f(-1, -1, 1) \\
&= f(-1, 1, -1) \\
&= 1.
\end{aligned}
$$

3. Local constrained minimum at $(6, 6, 3)$,

$$f(6, 6, 3) = 108.$$

4. Local constrained minimum at $(0,0,\pm c)$ and local constrained maximum at $(\pm a,0,0)$,

$$f(0,0,\pm c) = c^2,$$
$$f(\pm a,0,0) = a^2.$$

5. Local constrained minimum at $(d\sqrt{a}, d\sqrt{b}, d\sqrt{c})$, $d = \sqrt{a} + \sqrt{b} + \sqrt{c}$,

$$f(d\sqrt{a}, d\sqrt{b}, d\sqrt{c}) = d^2.$$

6. Local constrained minimum at $(-2,1,4)$ and local constrained maximum at $(2,-1,-4)$,

$$f(-2,1,4) = 11,$$
$$f(2,-1,-4) = 59.$$

7. Local constrained minimum at $(\pm\frac{13}{\sqrt{182}}, \mp\frac{2}{\sqrt{182}}, \mp\frac{3}{\sqrt{182}})$ and local constrained maximum at $(0, \pm\frac{3}{\sqrt{13}}, \mp\frac{2}{\sqrt{13}})$,

$$f\left(\pm\frac{13}{\sqrt{182}}, \mp\frac{2}{\sqrt{182}}, \mp\frac{3}{\sqrt{182}}\right) = \frac{17}{56},$$
$$f\left(0, \pm\frac{3}{\sqrt{13}}, \mp\frac{2}{\sqrt{13}}\right) = 1.$$

8. Local constrained minimum at $(\frac{a}{n}, \ldots, \frac{a}{n})$,

$$f\left(\frac{a}{n}, \ldots, \frac{a}{n}\right) = \frac{a^a}{n^{a-1}}.$$

9. Local constrained minimum at $(\frac{1}{A}\sqrt{\frac{a_1}{b_1}}, \ldots, \frac{1}{A}\sqrt{\frac{a_n}{b_n}})$,

$$f\left(\frac{1}{A}\sqrt{\frac{a_1}{b_1}}, \ldots, \frac{1}{A}\sqrt{\frac{a_n}{b_n}}\right) = A^2, \quad A = \sum_{j=1}^{n}\sqrt{a_j b_j}.$$

10. Local constrained maximum at $(a\frac{a_1}{A}, \ldots, a\frac{a_n}{A})$,

$$f\left(a\frac{a_1}{A}, \ldots, a\frac{a_n}{A}\right) = \left(\frac{a}{A}\right)^A \prod_{j=1}^{n} a_j^{a_j}, \quad A = \sum_{j=1}^{n} a_j.$$

A The gamma function

A.1 Definition of the gamma function

Definition A.1. The gamma function Γ is defined by the integral

$$\Gamma(z) = \int_0^\infty e^{-t} t^{z-1} dt. \qquad (A.1)$$

Note that $\Gamma(z)$ converges in the right-half of the complex plane $\mathrm{Re}(z) > 0$. Indeed, for $z = x + iy$, $x, y \in \mathbb{R}$, we have

$$
\begin{aligned}
\Gamma(x + iy) &= \int_0^\infty e^{-t} t^{x+iy-1} dt \\
&= \int_0^\infty e^{-t} t^{x-1} t^{iy} dt \\
&= \int_0^\infty e^{-t} t^{x-1} e^{iy \log t} dt \\
&= \int_0^\infty e^{-t} t^{x-1} (\cos(y \log t) + i \sin(y \log t)) dt.
\end{aligned}
\qquad (A.2)
$$

Since

$$\cos(y \log t) + i \sin(y \log t)$$

is bounded for all t, the convergence of (A.2) at infinity is provided by e^{-t}, and for the convergence at $t = 0$, we must have $x = \mathrm{Re}(z) > 0$.

A.2 Some properties of the gamma function

Theorem A.1. *Let* $\mathrm{Re}(z) > 0$. *Then*

$$\Gamma(z + 1) = z\Gamma(z).$$

Proof. We have

$$\Gamma(z + 1) = \int_0^\infty e^{-t} t^z dt$$

https://doi.org/10.1515/9783112218082-007

$$= -\int_0^\infty t^z \, de^{-t}$$

$$= -t^z e^{-t}\big|_{t=0}^{t=\infty} + z \int_0^\infty e^{-t} t^{z-1} \, dt$$

$$= z\Gamma(z).$$

This completes the proof. □

We have

$$\Gamma(1) = \int_0^\infty e^{-t} \, dt$$

$$= -e^{-t}\big|_{t=0}^{t=\infty}$$

$$= 1,$$

$$\Gamma(2) = \Gamma(1+1)$$

$$= 1\Gamma(1)$$

$$= 1$$

$$= 1!,$$

$$\Gamma(3) = \Gamma(2+1)$$

$$= 2\Gamma(1)$$

$$= 2$$

$$= 2!.$$

Suppose that

$$\Gamma(n) = (n-1)! \tag{A.3}$$

for some $n \in \mathbb{N}$. We will prove that

$$\Gamma(n+1) = n!.$$

Indeed, by Theorem A.1 and (A.3) we get

$$\Gamma(n+1) = n\Gamma(n)$$
$$= n(n-1)!$$
$$= n!.$$

Consequently, (A.3) holds for all $n \in \mathbb{N}$.

We will recall some useful definitions from complex analysis.

Definition A.2. Given a complex-valued function f of a single complex variable, the derivative of f at a point z_0 in its domain is defined by the limit

$$f'(z_0) = \lim_{z \to z_0} \frac{f(z) - f(z_0)}{z - z_0}.$$

If f is complex differentiable at every point z_0 in an open set U, then we say that f is holomorphic on U. We say that f is holomorphic at the point z_0 if it is holomorphic on some neighborhood of the point z_0. We say that f is holomorphic on some nonopen set A if it is holomorphic in an open set containing A.

Definition A.3. An entire function, also called an integral function, is a complex-valued function that is holomorphic at all finite points over the whole complex plane.

Definition A.4. Suppose U is an open subset of the complex plane $\mathbb{C}$, p is an element of U, and $f : U\backslash\{p\} \to \mathbb{C}$ is a function holomorphic over its domain. If there exist a holomorphic function $g : U \to \mathbb{C}$ such that $g(p)$ is nonzero and a positive integer n such that for all $z \in U\backslash\{p\}$,

$$f(z) = \frac{g(z)}{(z - p)^n},$$

then p is called a pole of f. The smallest such n is called the order of the pole. A pole of order 1 is called a simple pole.

Theorem A.2. *The gamma function has simple poles at the points* $z = -n, n \in \mathbb{N}_0$.

Proof. We rewrite (A.1) in the following way:

$$\Gamma(z) = \int_0^1 e^{-t} t^{z-1} dt + \int_1^\infty e^{-t} t^{z-1} dt, \quad \mathrm{Re}(z) > 0. \tag{A.4}$$

Let $\mathrm{Re}(z) = x > 0$. Then

$$\mathrm{Re}(z + n) = x + n > 0,$$

and

$$t^{z+n} = t^{x+iy+n}|_{t=0}$$
$$= 0, \quad n \in \mathbb{N}_0.$$

Therefore

$$\int_0^1 e^{-t} t^{z-1} dt = \int_0^1 \left(\sum_{k=}^\infty \frac{(t)^k}{k!} \right) t^{z-1} dt$$

$$= \sum_{k=0}^{\infty} \frac{(-1)^k}{k!} \int_0^1 t^{k+z-1} dt$$

$$= \sum_{k=0}^{\infty} \frac{(-1)^k}{k!} \frac{1}{k+z} t^{k+z} \Big|_{t=0}^{t=1} \tag{A.5}$$

$$= \sum_{k=0}^{\infty} \frac{(1)^k}{k!(k+z)}.$$

We set

$$\phi(z) = \int_1^{\infty} e^{-t} t^{z-1} dt.$$

Then

$$\phi(z) = \int_1^{\infty} e^{-t+(z-1)\log t} dt.$$

Note that $e^{-t+(z-1)\log t}$ is a continuous function of z and t for arbitrary z and $t \geq 1$. Also, for $t \geq 1$, it is an entire function of z. Let D be an arbitrary closed domain in the complex plane and denote $x_0 = \max_{z \in D} \operatorname{Re}(z)$. Then

$$\left| e^{-t} t^{z-1} \right| = \left| e^{(z-1)\log t - t} \right|$$
$$= \left| e^{(x-1)\log t - t + iy \log t} \right|$$
$$= \left| e^{(x-1)\log t - t} \right| \left| e^{iy \log t} \right|$$
$$= e^{(x-1)\log t - t}$$
$$\leq e^{(x_0-1)\log t - t}$$
$$= e^{-t} t^{x_0-1}.$$

Therefore $\phi(z)$ converges uniformly in D, and the differentiation under the integral sign is allowed. Because D was arbitrarily chosen, we conclude that the function $\phi(z)$ has the above properties in the whole complex plane. Consequently, $\phi(z)$ is an entire function allowing differentiation under the integral sign. Hence by (A.4) and (A.5) we conclude that

$$\Gamma(z) = \sum_{k=0}^{\infty} \frac{(-1)^k}{k!(k+z)} + \text{entire function.}$$

Thus $\Gamma(z)$ has only simple poles at the points $z = -n$, $n \in \mathbb{N}_0$. This completes the proof. $\square$

A.3 Limit representation of the gamma function

Theorem A.3. *The gamma function can be represented as*

$$\Gamma(z) = \lim_{n \to \infty} \frac{n! n^z}{z(z+1)\dots(z+n)}, \quad \text{Re}(z) > 0. \tag{A.6}$$

Proof. Suppose that $\text{Re}(z) > 0$, $z = x + iy$, x and y are real variables. Let

$$f_n(z) = \int_0^n \left(1 - \frac{t}{n}\right)^n t^{z-1} dt.$$

Substituting $\tau = \frac{t}{n}$, we get

$$t = n\tau,$$
$$dt = n d\tau,$$

$$f_n(z) = \int_0^1 (1-\tau)^n n^{z-1} \tau^{z-1} n d\tau$$

$$= n^z \int_0^1 (1-\tau)^n \tau^{z-1} d\tau$$

$$= \frac{n^z}{z}(1-\tau)^n \tau^z \Big|_{\tau=0}^{\tau=1} + \frac{n^z}{z} n \int_0^1 (1-\tau)^{n-1} \tau^z d\tau$$

$$= n^z \frac{n}{z} \int_0^1 (1-\tau)^{n-1} \tau^z d\tau \tag{A.7}$$

$$= n^z \frac{n}{z} \frac{1}{z+1} \left((1-\tau)^{n-1} \tau^{z+1} \Big|_{\tau=0}^{\tau=1} + (n-1) \int_0^1 (1-\tau)^{n-2} \tau^{z+1} d\tau \right)$$

$$= n^z \frac{n}{z} \frac{n-1}{z+1} \int_0^1 (1-\tau)^{n-2} \tau^{z+1} d\tau$$

$$\vdots$$

$$= \frac{n^z n!}{z(z+1)\dots(z+n-1)(z+n)}.$$

Note that

$$\lim_{n \to \infty} \left(1 - \frac{t}{n}\right)^n = e^{-t}.$$

We set

$$\psi(z) = \int_0^\infty e^{-t} t^{z-1} dt - f_n(z).$$

Then

$$\psi(z) = \int_0^\infty e^{-t} t^{z-1} dt - \int_0^n \left(1 - \frac{t}{n}\right) t^{z-1} dt$$

$$= n \int_0^n \left(e^{-t} - \left(1 - \frac{t}{n}\right)^n\right) t^{z-1} dt + \int_n^\infty e^{-t} t^{z-1} dt.$$

(A.8)

Let $\epsilon > 0$ be arbitrarily chosen. Then there exists $N = N(\epsilon) \in \mathbb{N}$ such that

$$\int_n^\infty e^{-t} t^{x-1} dt < \frac{\epsilon}{3}, \quad \int_n^\infty \left| e^{-t} - \left(1 - \frac{t}{n}\right)^n \right| t^{x-1} dt < \frac{\epsilon}{3}, \quad \frac{1}{2n(x+2)} N^{x+1} < \frac{\epsilon}{3}$$

for all $n > N$. We fix N and take $n > N$. Then using (A.8), we obtain

$$\psi(z) = \int_0^N \left(e^{-t} - \left(1 - \frac{t}{n}\right)^n\right) t^{z-1} dt$$

$$+ \int_N^n \left(e^{-t} - \left(1 - \frac{t}{n}\right)^n\right) t^{z-1} dt$$

(A.9)

$$+ \int_n^\infty e^{-t} t^{z-1} dt.$$

We have

$$\left| \int_N^n \left(e^{-t} - \left(1 - \frac{t}{n}\right)^n\right) t^{z-1} dt \right| \leq \int_N^n \left| \left(e^{-t} - \left(1 - \frac{t}{n}\right)^n\right) t^z \right| dt$$

$$= \int_N^n \left| e^{-t} - \left(1 - \frac{t}{n}\right)^n \right| t^{x-1} dt$$

(A.10)

$$\leq \int_N^\infty \left| e^{-t} - \left(1 - \frac{t}{n}\right)^n \right| t^{x-1} dt$$

$$< \frac{\epsilon}{3}$$

and

$$\left|\int_n^\infty e^{-t}t^{z-1}dt\right| \le \int_n^\infty e^{-t}|t^{z-1}|dt$$

$$\le \int_n^\infty e^{-t}t^{x-1}dt \tag{A.11}$$

$$< \frac{e}{3}.$$

Note that

$$\int_0^t e^\tau\left(1-\frac{\tau}{n}\right)^{n-1}\frac{\tau}{n}d\tau = -\int_0^t e^\tau\left(1-\frac{\tau}{n}\right)^n d\tau + \int_0^t e^\tau\left(1-\frac{\tau}{n}\right)^{n-1}d\tau$$

$$= -e^\tau\left(1-\frac{\tau}{n}\right)^n\Big|_{\tau=0}^{\tau=t} - \int_0^t e^\tau\left(1-\frac{\tau}{n}\right)^{n-1}d\tau + \int_0^t e^\tau\left(1-\frac{\tau}{n}\right)^{n-1}d\tau$$

$$= 1 - e^t\left(1-\frac{t}{n}\right)^n, \quad n \in \mathbb{N}.$$

Hence

$$0 < 1 - e^t\left(1-\frac{t}{n}\right)^n$$

$$= \int_0^t e^\tau\left(1-\frac{\tau}{n}\right)^{n-1}\frac{\tau}{n}d\tau$$

$$\le \int_0^t e^\tau\frac{\tau}{n}d\tau$$

$$< e^t\int_0^t \frac{\tau}{n}d\tau$$

$$= e^t\frac{t^2}{2n}, \quad 0 < t < n,$$

and

$$\left|\int_0^N\left(e^{-t}-\left(1-\frac{t}{n}\right)^n\right)t^{z-1}dt\right| \le \int_0^N\left|e^{-t}-\left(1-\frac{t}{n}\right)^n\right|t^{x-1}dt$$

$$\le \frac{1}{2n}\int_0^N t^{x+1}dt \tag{A.12}$$

$$= \frac{1}{2n(x+2)} N^{x+1}$$

$$< \epsilon.$$

Therefore, using (A.9), (A.10), (A.11), and (A.12), we get

$$|\psi(z)| < \frac{\epsilon}{3} + \frac{\epsilon}{3} + \frac{\epsilon}{3}$$

$$= \epsilon$$

and

$$\lim_{n\to\infty} f_n(z) = \lim_{n\to\infty} \int_0^n \left(1 - \frac{t}{n}\right)^n t^{z-1} dt$$

$$= \int_0^\infty e^{-t} t^{z-1} dt$$

$$= \Gamma(z),$$

from which by (A.8) we obtain (A.7). This completes the proof. □

Remark A.1. The condition $\mathrm{Re}(z) > 0$ in (A.7) can be weakened for $z \neq 0, -1, -2, \ldots$, in the following manner. If $-m < \mathrm{Re}(z) < -m+1$ for some $m \in \mathbb{N}$, then

$$\Gamma(z) = \frac{\Gamma(z+m)}{z(z+1)\ldots(z+m-1)}$$

$$= \frac{1}{z(z+1)\ldots(z+m-1)} \lim_{n\to\infty} \frac{n^{z+m} n!}{(z+m)\ldots(z+m+n)}$$

$$= \frac{1}{z(z+1)\ldots(z+m-1)} \lim_{n\to\infty} \frac{(n-m)^{z+m}(n-m)!}{(z+m)(z+m+1)\ldots(z+n)}$$

$$= \lim_{n\to\infty} \frac{n^z n!}{z(z+1)\ldots(z+n)}.$$

B The beta function

B.1 Definition of the beta function

Definition B.1. Let $z, w \in \mathbb{C}$, $\mathrm{Re}(z) > 0$, $\mathrm{Re}(w) > 0$. The beta function $B(z, w)$ is defined as follows:

$$B(z, w) = \int_0^1 t^{z-1}(1 - t)^{w-1} dt.$$

B.2 Properties of the beta function

Suppose that $z, w \in \mathbb{C}$, $\mathrm{Re}(z) > 0$, $\mathrm{Re}(w) > 0$. In this section, we will derive some of the properties of the beta function.

1. Let

$$\tau = 1 - t.$$

Then

$$d\tau = -dt, \quad t = 1 - \tau.$$

Therefore

$$B(z, w) = \int_1^0 (1 - \tau)^{z-1} \tau^{w-1}(-d\tau)$$

$$= \int_0^1 \tau^{w-1}(1 - \tau)^{z-1} d\tau$$

$$= B(w, z),$$

i. e.,

$$B(z, w) = B(w, z). \tag{B.1}$$

2. Let

$$y = (1 - t)^{w-1}, \quad x = \frac{t^z}{z}.$$

Then

$$dy = -(w - 1)(1 - t)^{w-2} dt,$$

https://doi.org/10.1515/9783112218082-008

$$dx = t^{z-1}dt,$$

and

$$t^z = t^{z-1} - t^{z-1}(1-t).$$

Then

$$B(z,w) = \int_0^1 t^{z-1}(1-t)^{w-1}dt$$

$$= \frac{1}{z}\int_0^1 (1-t)^{w-1}d(t^z)$$

$$= \frac{1}{z}(1-t)^{w-1}t^z\Big|_{t=0}^{t=1} + \frac{w-1}{z}\int_0^1 t^z(1-t)^{w-2}dt$$

$$= \frac{w-1}{z}\int_0^1 t^z(1-t)^{w-2}dt$$

$$= \frac{w-1}{z}\int_0^1 (t^{z-1} - t^{z-1}(1-t))(1-t)^{w-2}dt$$

$$= \frac{w-1}{z}\int_0^1 t^{z-1}(1-t)^{w-2}dt - \frac{w-1}{z}\int_0^1 t^{z-1}(1-t)^{w-1}dt$$

$$= \frac{w-1}{z}B(z,w-1) - \frac{w-1}{z}B(z,w).$$

Hence

$$B(z,w) + \frac{w-1}{z}B(z,w) = \frac{w-1}{z}B(z,w-1),$$

or

$$\frac{w+z-1}{z}B(z,w) = \frac{w-1}{z}B(z,w-1),$$

or

$$B(z,w) = \frac{w-1}{w+z-1}B(z,w-1).$$

As before, using (B.1), we obtain

$$B(z,w) = B(w,z)$$

$$= \frac{z-1}{w+z-1} B(w, z-1)$$

$$= \frac{z-1}{w+z-1} B(z-1, w).$$

Therefore

$$\frac{z-1}{w+z-1} B(z-2, w) = \frac{w-1}{w+z-1} B(z, w-1),$$

or

$$B(z-1, w) = \frac{w-1}{z-1} B(z, w-1).$$

Setting

$$p = z - 1, \quad q = w - 1,$$

we get

$$B(p, q+1) = \frac{q}{p} B(p+1, q).$$

Let $w = n, n \in \mathbb{N}$. Then

$$B(z, n) = \frac{n-1}{z+n-1} B(z, n-1)$$

$$= \frac{n-1}{z+n-1} \frac{n-2}{z+n-2} B(z, n-2)$$

$$\vdots$$

$$= \frac{n-1}{z+n-1} \frac{n-2}{z+n-2} \cdots \frac{1}{z+1} B(z, 1).$$

On the other hand,

$$B(z, 1) = \int_0^1 t^{z-1} dt$$

$$= \frac{t^z}{z} \Big|_{t=0}^{t=1}$$

$$= \frac{1}{z}.$$

Consequently,

$$B(z, n) = \frac{(n-1)!}{z(z+1)\ldots(z+n-1)}.$$

For $m \in \mathbb{N}$, we have

$$B(m,n) = \frac{(n-1)!}{m(m+1)\dots(m+n-1)}$$
$$= \frac{(n-1)!(m-1)!}{(m+n-1)!}.$$

3. Now consider $B(z,z)$. We have

$$B(z,z) = \int_0^1 t^{z-1}(1-t)^{z-1}dt$$

$$= \int_0^1 \left(\frac{1}{4} - \left(\frac{1}{2}-t\right)^2\right)^{z-1}dt$$

$$= \int_0^{\frac{1}{2}} \left(\frac{1}{4} - \left(\frac{1}{2}-t\right)^2\right)^{z-1}dt + \int_{\frac{1}{2}}^1 \left(\frac{1}{4} - \left(\frac{1}{2}-t\right)^2\right)^{z-1}dt$$

$$= \int_0^{\frac{1}{2}} \left(\frac{1}{4} - \left(\frac{1}{2}-t\right)^2\right)^{z-1}dt + \int_0^{\frac{1}{2}} \left(\frac{1}{4} - \left(\frac{1}{2}-t\right)^2\right)^{z-1}dt$$

$$= 2\int_0^{\frac{1}{2}} \left(\frac{1}{4} - \left(\frac{1}{2}-t\right)^2\right)^{z-1}dt.$$

For $t \in [0,\frac{1}{2}]$, we set

$$\frac{\sqrt{y}}{2} = \frac{1}{2} - t.$$

Then

$$t = \frac{1-\sqrt{y}}{2},$$
$$dt = -\frac{1}{4\sqrt{y}}dy, \quad t \in \left[0,\frac{1}{2}\right].$$

Hence

$$B(z,z) = 2\int_1^0 \left(\frac{1}{4} - \frac{y}{4}\right)^{z-1}\left(-\frac{1}{4\sqrt{y}}\right)dy$$

$$= \frac{2}{2^{2z}}\int_0^1 y^{-\frac{1}{2}}(1-y)^{z-1}dy$$

$$= \frac{1}{2^{2z-1}} \int_0^1 y^{\frac{1}{2}-1}(1-y)^{z-1}dy$$

$$= \frac{1}{2^{2z-1}} B\left(\frac{1}{2}, z\right),$$

i. e.,

$$B(z,z) = \frac{1}{2^{2z-1}} B\left(\frac{1}{2}, z\right).$$

4. Let

$$t = \frac{y}{y+1}, \quad t \in [0,1].$$

Then

$$1 - t = 1 - \frac{y}{1+y}$$

$$= \frac{1+y-y}{1+y}$$

$$= \frac{1}{1+y},$$

$$dt = \frac{1}{1+y}dy - \frac{1}{(1+y)^2}dy$$

$$= \frac{1+y-y}{(1+y)^2}dy$$

$$= \frac{1}{(1+y)^2}dy.$$

Hence

$$B(z,w) = \int_0^1 t^{z-1}(1-t)^{w-1}dt$$

$$= \int_0^\infty \frac{y^{z-1}}{(1+y)^{z-1}} \frac{1}{(1+y)^{w-1}} \frac{1}{(1+y)^2}dy$$

$$= \int_0^\infty \frac{y^{z-1}}{(1+y)^{z+w}}dy.$$

5. By (A.1) we have

$$\Gamma(z) = \int_0^\infty t^{z-1}e^{-t}dt.$$

Let

$$t = xy, \quad x > 0.$$

Then

$$dt = x\,dy$$

and

$$\Gamma(z) = \int_0^\infty x^{z-1} y^{z-1} e^{-xy} x\,dy$$

$$= \int_0^\infty x^z y^{z-1} e^{-xy}\,dy,$$

whereupon

$$\frac{\Gamma(z)}{x^2} = \int_0^\infty y^{z-1} e^{-xy}\,dy.$$

Hence

$$\frac{\Gamma(z+w)}{(1+x)^{z+w}} = \int_0^\infty y^{z+w-1} e^{-(1+x)y}\,dy.$$

6. We have

$$\Gamma(z+w)B(z,w) = \Gamma(z+w)\int_0^\infty \frac{t^{z-1}}{(1+t)^{z+w}}\,dt$$

$$= \int_0^\infty \frac{\Gamma(z+w)}{(1+t)^{z+w}} t^{z-1}\,dt$$

$$= \int_0^\infty \left(\int_0^\infty y^{z+w-1} e^{-(1+t)y}\,dy \right) t^{z-1}\,dt$$

$$= \int_0^\infty \left(\int_0^\infty y^{z+w-1} e^{-y} e^{-ty}\,dy \right) t^{z-1}\,dt$$

$$= \int_0^\infty \left(\int_0^\infty t^{z-1} e^{-ty}\,dt \right) y^{z+w-1} e^{-y}\,dy$$

$$= \int_0^\infty \left(\int_0^\infty (ty)^{z-1} e^{-ty} dt \right) y^w e^{-y} dy$$

$$= \int_0^\infty \left(\int_0^\infty p^{z-1} e^{-p} dp \right) y^{w-1} e^{-y} dy$$

$$= \Gamma(z) \int_0^\infty y^{w-1} e^{-y} dy$$

$$= \Gamma(z)\Gamma(w).$$

Therefore

$$B(z, w) = \frac{\Gamma(z)\Gamma(w)}{\Gamma(z + w)}.$$

7. Let

$$e^{-x} = y.$$

Then

$$x = \log \frac{1}{y}$$

$$= \lim_{n \to \infty} \left(n(1 - y^{\frac{1}{n}}) \right)$$

$$= \lim_{n \to \infty} \frac{1 - y^{\frac{1}{n}}}{\frac{1}{n}}$$

$$= \lim_{n \to \infty} \frac{-y^{\frac{1}{n}} \log y (-\frac{1}{n^2})}{-\frac{1}{n^2}}$$

$$= -\log y$$

$$= \log \frac{1}{y}.$$

Therefore

$$\Gamma(z) = \int_0^\infty x^{z-1} e^{-x} dx$$

$$= -\int_1^0 \lim_{n \to \infty} \left(n(1 - y^{\frac{1}{n}}) \right)^{z-1} y \frac{dy}{y}$$

$$= \lim_{n \to \infty} \left(n^{z-1} \int_0^1 (1 - y^{\frac{1}{n}})^{z-1} dy \right).$$

Let now

$$t^n = y.$$

Then

$$dy = nt^{n-1}dt,$$

and

$$\Gamma(z) = \lim_{n\to\infty}\left(n^{z-1}\int_0^1 (1-t)^{z-1}nt^{n-1}dt\right)$$

$$= \lim_{n\to\infty}\left(n^z\int_0^1 t^{n-1}(1-t)^{z-1}dt\right)$$

$$= \lim_{n\to\infty}\left(n^z B(n,z)\right)$$

$$= \lim_{n\to\infty} n^z \frac{(n-1)!}{z(z+1)\dots(z+n-1)}$$

$$= \lim_{n\to\infty} n^z \frac{n!}{z(z+1)\dots(z+n)}$$

$$= \lim_{n\to\infty} n^z \frac{1.2\dots n}{z(z+1)\dots(z+n)}$$

$$= \lim_{n\to\infty} n^z \frac{1.2\dots n}{z(1+\frac{z}{1})2(1+\frac{z}{2})\dots n(1+\frac{z}{n})}$$

$$= \frac{1}{z} \lim_{n\to\infty} \frac{n^z}{(1+\frac{z}{1})(1+\frac{z}{2})\dots(1+\frac{z}{n})}$$

$$= \frac{1}{z} \lim_{n\to\infty} \frac{(1+\frac{1}{1})^z(1+\frac{1}{2})^z\dots(1+\frac{1}{n-1})^z}{(1+\frac{z}{1})(1+\frac{z}{2})\dots(1+\frac{z}{n})}$$

$$= \frac{1}{z} \lim_{n\to\infty} \frac{(1+\frac{1}{1})^z(1+\frac{1}{2})^z\dots(1+\frac{1}{n})^z}{(1+\frac{z}{1})(1+\frac{z}{2})\dots(1+\frac{z}{n})}$$

$$= \frac{1}{z} \prod_{n=1}^{\infty} \frac{(1+\frac{1}{n})^z}{1+\frac{z}{n}},$$

i. e.,

$$z\Gamma(z) = \prod_{n=1}^{\infty} \frac{(1+\frac{1}{n})^z}{1+\frac{z}{n}},$$

whereupon

$$\Gamma(z+1) = \prod_{n=1}^{\infty} \frac{(1+\frac{1}{n})^z}{1+\frac{z}{n}}.$$ (B.2)

8. By (B.2) we get

$$\Gamma(1-z) = \prod_{n=1}^{\infty} \frac{(1+\frac{1}{n})^{-z}}{1-\frac{z}{n}}.$$

Hence

$$\Gamma(z)\Gamma(1-z) = \left(\frac{1}{z}\prod_{n=1}^{\infty} \frac{(1+\frac{1}{n})^z}{1+\frac{z}{n}}\right)\left(\prod_{n=-1}^{\infty} \frac{(1+\frac{1}{n})^{-z}}{1+\frac{z}{n}}\right)$$

$$= \frac{1}{z\prod_{n=1}^{\infty}(1-\frac{z^2}{n^2})}.$$

Now using that

$$\sin(\pi z) = \pi z \prod_{n=1}^{\infty}\left(1 - \frac{z^2}{n^2}\right),$$

we get

$$\Gamma(z)\Gamma(1-z) = \frac{\pi}{\sin(\pi z)}.$$ (B.3)

9. Putting $z = \frac{1}{2}$ in (B.3), we get

$$\left(\Gamma\left(\frac{1}{2}\right)\right)^2 = \frac{\pi}{\sin\frac{\pi}{2}}$$

$$= \pi.$$

Therefore

$$\Gamma\left(\frac{1}{2}\right) = \sqrt{\pi}.$$

10. We have

$$\Gamma\left(z + \frac{1}{2}\right) = \lim_{n\to\infty} n^{z+\frac{1}{2}} \frac{1.2\ldots n}{(z+\frac{1}{2})(z+\frac{1}{2}+1)\ldots(z+\frac{1}{2}+n)},$$

$$\Gamma(2z) = \lim_{n\to\infty} (2n)^{2z} \frac{1.2\ldots(2n)}{2z(2z+1)(2z+2)\ldots(2z+2n)}.$$

Hence

$$2^{2z-1}\frac{\Gamma(z)\Gamma(z+\frac{1}{2})}{\Gamma(2z)} = 2^{2z-1}\lim_{n\to\infty}\frac{n^{2z+\frac{1}{2}}(n!)^2}{(2n)!(2n)^{2z}}$$

$$\times \frac{2z(2z+1)\dots(2z+2n)}{z(z+1)\dots(z+n)(z+\frac{1}{2})(z+\frac{1}{2}+1)\dots(z+\frac{1}{2}+n)}$$

$$= \frac{1}{2}\lim_{n\to\infty}\frac{(n!)^2 n^{\frac{1}{2}}}{(2n)!}$$

$$\times \frac{2^{2n+2}z(2z+1)(z+1)(2z+3)\dots(z+n)}{z(z+1)\dots(z+n)(2z+1)(2z+3)\dots(2z+2n+1)}$$

$$= \lim_{n\to\infty}\frac{(n!)^2 n^{\frac{1}{2}} 2^{2n+1}}{(2n)!(2z+2n+1)}$$

$$= \lim_{n\to\infty}\frac{(n!)^2 (2n)2^{2n}}{(2n)!(2z+2n+1)n^{\frac{1}{2}}}$$

$$= \lim_{n\to\infty}\frac{2^{2n}(n!)^2}{(2n)!n^{\frac{1}{2}}},$$

i. e.,

$$2^{2z-1}\frac{\Gamma(z)\Gamma(z+\frac{1}{2})}{\Gamma(2z)} = \lim_{n\to\infty}\frac{2^{2n}(n!)^2}{(2n)!n^{\frac{1}{2}}}. \tag{B.4}$$

11. Putting $z = \frac{1}{2}$ in (B.4), we get

$$\lim_{n\to\infty}\frac{2^{2n}(n!)^2}{(2n)!n^{\frac{1}{2}}} = \sqrt{\pi}.$$

B.3 An application

Consider the integral

$$I = \int_0^{\frac{\pi}{2}}(\sin x)^n(\cos x)^m dx.$$

We set

$$y = (\sin x)^2, \quad x \in \left[0, \frac{\pi}{2}\right].$$

Then

$$(\sin x)^n = ((\sin x)^2)^{\frac{n}{2}}$$

$$= y^{\frac{n}{2}},$$

$$(\cos x)^m = ((\cos x)^2)^{\frac{m}{2}}$$

$$= (1 - (\sin x)^2)^{\frac{m}{2}}$$

$$= (1 - y)^{\frac{m}{2}},$$

$$dy = 2 \sin x \cos x \, dx$$

$$= 2((\sin x)^2)^{\frac{1}{2}}((\cos x)^2)\frac{1}{2} dx$$

$$= 2y^{\frac{1}{2}}(1 - y)^{\frac{1}{2}} dx,$$

$$dx = \frac{1}{2} y^{-\frac{1}{2}}(1 - y)^{-\frac{1}{2}} dy.$$

Hence

$$I = \frac{1}{2} \int_0^1 y^{\frac{n}{2}}(1 - y)^{\frac{m}{2}} y^{-\frac{1}{2}}(1 - y)^{-\frac{1}{2}} dy$$

$$= \frac{1}{2} \int_0^1 y^{\frac{n+1}{2}-1}(1 - y)^{\frac{m+1}{2}-1} dy$$

$$= \frac{1}{2} B\left(\frac{n+1}{2}, \frac{m+1}{2} \right)$$

$$= \frac{1}{2} \frac{\Gamma(\frac{n+1}{2})\Gamma(\frac{m+1}{2})}{\Gamma(\frac{m+n}{2} + 1)}.$$

Index

https://doi.org/10.1515/9783112218082-010

www.ingramcontent.com/pod-product-compliance
Lightning Source LLC
Chambersburg PA
CBHW080916220326
41598CB00034B/5586